MINNESOTA STUDIES IN THE PHILOSOPHY OF SCIENCE

Minnesota Studies in the
PHILOSOPHY OF SCIENCE

HERBERT FEIGL AND GROVER MAXWELL, GENERAL EDITORS

VOLUME VII

Language, Mind, and Knowledge

EDITED BY

KEITH GUNDERSON

FOR THE MINNESOTA CENTER FOR PHILOSOPHY OF SCIENCE

UNIVERSITY OF MINNESOTA PRESS, MINNEAPOLIS

Library of Congress Catalog Card Number: 74-22836
ISBN 0-8166-0742-7

Preface

To publish a collection of papers in any field as susceptible to change as the philosophy of language currently is, incurs the editorial risk of endorsing some odd package of intellectual curios flattened into irredeemable obsolescence by the whirring past of Time's wingèd chariot. (This calls to mind an analogous impasse in appreciating contemporary art: by the time something seems worthy of display, it's *not* indicative of what's going on.) Fortunately for the readers of this volume, the editor, by biding his time, resisting all pressures to rush into print — some felt as early as 1969 — has been able to select only papers with a high density of timeless truths. This latter quality was not, unfortunately, always perceived by the authors themselves. Their resultant impatience — typified by such banter as "When *is* that volume with *my* article going to appear?" — I prefer, however, not to parade in public, any more than I would seek to puff up my own importance as editor by making conspicuous whatever number of compliments I have received in private for my judicious "holding back" until the contents were as they should be. (Note: what was to be my own contribution to this volume — a criticism of Professor Dennett's piece — was, after years of scrutiny, judged unworthy of inclusion and, hence, self-rejected.)

Some of the papers, in one form or another, were presented at or grew out of a conference in the philosophy of language sponsored by the Minnesota Center for the Philosophy of Science in the summer of 1968 under the directorship of Professor Herbert Feigl, and funded by a grant from the Carnegie Foundation. Other pieces were written more or less from scratch by participants in the conference a year or so afterward. Some were generously donated to the volume by invitees to the conference who had been unable to attend (the papers by Noam Chom-

v

Preface

sky and Jerrold Katz), and some were simply solicited from the unwary by the editor along the way.

The aim of this volume, like that of the conference, was to assemble a wide variety of approaches to issues in the philosophy of language, with an eye to scope and to the ways in which the salient issues involved have bearing on other matters such as linguistic theory, cognitive psychology, the philosophy of mind, and epistemology. Although it has proved impossible to cover such a wide and variegated waterfront within the confines of one book, I hope the collection displays some of the more fascinating wharves to be found.

For help in preparing this volume, special thanks are due Caroline Cohen, the secretary for the Minnesota Center, and, as always, to Herbert Feigl, Regents' Professor of Philosophy Emeritus and former director, and to Professor Grover Maxwell, current director of the Center. I also wish to extend my gratitude to Christopher Swoyer and Ernesto Pasquale LePore for preparing the index.

Keith Gunderson, *Research Associate*
MINNESOTA CENTER FOR PHILOSOPHY OF SCIENCE

February 1975

Contents

vii

MINNESOTA STUDIES IN THE PHILOSOPHY OF SCIENCE

Languages and Language

Thesis

What is a language? Something which assigns meanings to certain strings of types of sounds or of marks. It could therefore be a function, a set of ordered pairs of strings and meanings. The entities in the domain of the function are certain finite sequences of types of vocal sounds, or of types of inscribable marks; if σ is in the domain of a language £, let us call σ a *sentence* of £. The entities in the range of the function are meanings; if σ is a sentence of £, let us call £(σ) the *meaning of* σ *in* £. What could a meaning of a sentence be? Something which, when combined with factual information about the world — or factual information about *any* possible world — yields a truth-value. It could therefore be a function from worlds to truth-values — or more simply, a set of worlds. We can say that a sentence σ is *true in* a language £ at a world w if and only if w belongs to the set of worlds £(σ). We can say that σ is *true in* £ (without mentioning a world) if and only if our actual world belongs to £(σ). We can say that σ is *analytic in* £ if and only if every possible world belongs to £(σ). And so on, in the obvious way.

Antithesis

What is language? A social phenomenon which is part of the natural history of human beings; a sphere of human action, wherein people utter strings of vocal sounds, or inscribe strings of marks, and wherein people respond by thought or action to the sounds or marks which they observe to have been so produced.

This verbal activity is, for the most part, rational. He who produces

AUTHOR'S NOTE: This paper was originally prepared in 1968 and was revised in 1972. The 1968 draft appears in Italian translation as "Lingue e lingua," *Versus*, 4(1973): 2–21.

certain sounds or marks does so for a reason. He knows that someone else, upon hearing his sounds or seeing his marks, is apt to form a certain belief or act in a certain way. He wants, for some reason, to bring about that belief or action. Thus his beliefs and desires give him a reason to produce the sounds or marks, and he does. He who responds to the sounds or marks in a certain way also does so for a reason. He knows how the production of sounds or marks depends upon the producer's state of mind. When he observes the sounds or marks, he is therefore in a position to infer something about the producer's state of mind. He can probably also infer something about the conditions which caused that state of mind. He may merely come to believe these conclusions, or he may act upon them in accordance with his other beliefs and his desires.

Not only do both have reasons for thinking and acting as they do; they know something about each other, so each is in a position to replicate the other's reasons. Each one's replication of the other's reasons forms part of his own reason for thinking and acting as he does; and each is in a position to replicate the other's replication of his own reasons. Therefore the Gricean mechanism[1] operates: X intends to bring about a response on the part of Y by getting Y to recognize that X intends to bring about that response; Y does recognize X's intention, and is thereby given some sort of reason to respond just as X intended him to.

Within any suitable population, various regularities can be found in this rational verbal activity. There are regularities whereby the production of sounds or marks depends upon various aspects of the state of mind of the producer. There are regularities whereby various aspects of responses to sounds or marks depend upon the sounds or marks to which one is responding. Some of these regularities are accidental. Others can be explained, and different ones can be explained in very different ways.

Some of them can be explained as conventions of the population in which they prevail. Conventions are regularities in action, or in action and belief, which are arbitrary but perpetuate themselves because they serve some sort of common interest. Past conformity breeds future conformity because it gives one a reason to go on conforming; but there is some alternative regularity which could have served instead, and

---

[1] H. P. Grice, "Meaning," *Philosophical Review*, 66(1957):377–388.

4

would have perpetuated itself in the same way if only it had got started.

More precisely: a regularity R, in action or in action and belief, is a *convention* in a population P if and only if, within P, the following six conditions hold. (Or at least they almost hold. A few exceptions to the "everyone"s can be tolerated.)

(1) Everyone conforms to R.

(2) Everyone believes that the others conform to R.

(3) This belief that the others conform to R gives everyone a good and decisive reason to conform to R himself. His reason may be that, in particular, those of the others he is now dealing with conform to R; or his reason may be that there is general or widespread conformity, or that there has been, or that there will be. His reason may be a practical reason, if conforming to R is a matter of acting in a certain way; or it may be an epistemic reason, if conforming to R is a matter of believing in a certain way. First case: according to his beliefs, some desired end may be reached by means of some sort of action in conformity to R, provided that the others (all or some of them) also conform to R; therefore he wants to conform to R if they do. Second case: his beliefs, together with the premise that others conform to R, deductively imply or inductively support some conclusion; and in believing this conclusion, he would thereby conform to R. Thus reasons for conforming to a convention by believing something—like reasons for belief in general—are believed premises tending to confirm the truth of the belief in question. Note that I am *not* speaking here of practical reasons for acting so as to somehow produce in oneself a certain desired belief.

(4) There is a general preference for general conformity to R rather than slightly-less-than-general conformity — in particular, rather than conformity by all but any one. (This is not to deny that some state of *widespread* nonconformity to R might be even more preferred.) Thus everyone who believes that at least almost everyone conforms to R will want the others, as well as himself, to conform. This condition serves to distinguish cases of convention, in which there is a predominant coincidence of interest, from cases of deadlocked conflict. In the latter cases, it may be that each is doing the best he can by conforming to R, given that the others do so; but each wishes the others did not conform to R, since he could then gain at their expense.

(5) R is not the only possible regularity meeting the last two condi-

tions. There is at least one alternative R' such that the belief that the others conformed to R' would give everyone a good and decisive practical or epistemic reason to conform to R' likewise; such that there is a general preference for general conformity to R' rather than slightly-less-than-general conformity to R'; and such that there is normally no way of conforming to R and R' both. Thus the alternative R' could have perpetuated itself as a convention instead of R; this condition provides for the characteristic arbitrariness of conventions.

(6) Finally, the various facts listed in conditions (1) to (5) are matters of common (or mutual) knowledge: they are known to everyone, it is known to everyone that they are known to everyone, and so on. The knowledge mentioned here may be merely potential: knowledge that would be available if one bothered to think hard enough. Everyone must potentially know that (1) to (5) hold; potentially know that the others potentially know it; and so on. This condition ensures stability. If anyone tries to replicate another's reasoning, perhaps including the other's replication of his own reasoning, . . . , the result will reinforce rather than subvert his expectation of conformity to R. Perhaps a negative version of (6) would do the job: no one disbelieves that (1) to (5) hold, no one believes that others disbelieve this, and so on.

This definition can be tried out on all manner of regularities which we would be inclined to call conventions. It is a convention to drive on the right. It is a convention to mark poisons with skull and crossbones. It is a convention to dress as we do. It is a convention to train beasts to turn right on "gee" and left on "haw." It is a convention to give goods and services in return for certain pieces of paper or metal. And so on.

The common interests which sustain conventions are as varied as the conventions themselves. Our convention to drive on the right is sustained by our interest in not colliding. Our convention for marking poisons is sustained by our interest in making it easy for everyone to recognize poisons. Our conventions of dress might be sustained by a common aesthetic preference for somewhat uniform dress, or by the low cost of mass-produced clothes, or by a fear on everyone's part that peculiar dress might be thought to manifest a peculiar character, or by a desire on everyone's part not to be too conspicuous, or — most likely — by a mixture of these and many other interests.

It is a platitude — something only a philosopher would dream of denying — that there are conventions of language, although we do not find it easy to say what those conventions are. If we look for the fundamental difference in verbal behavior between members of two linguistic communities, we can be sure of finding something which is arbitrary but perpetuates itself because of a common interest in coordination. In the case of conventions of language, that common interest derives from our common interest in taking advantage of, and in preserving, our ability to control others' beliefs and actions to some extent by means of sounds and marks. That interest in turn derives from many miscellaneous desires we have; to list them, list the ways you would be worse off in Babel.

Synthesis

What have languages to do with language? What is the connection between what I have called *languages*, functions from strings of sounds or of marks to sets of possible worlds, semantic systems discussed in complete abstraction from human affairs, and what I have called *language*, a form of rational, convention-governed human social activity? We know what to *call* this connection we are after: we can say that a given language £ is *used* by, or is a (or the) language *of*, a given population P. We know also that this connection holds by virtue of the conventions of language prevailing in P. Under suitably different conventions, a different language would be used by P. There is some sort of convention whereby P uses £ — but what is it? It is worthless to call it a convention to use £, even if it can correctly be so described, for we want to know what it is to use £.

My proposal[2] is that the convention whereby a population P uses a language £ is a convention of *truthfulness* and *trust* in £. To be truthful in £ is to act in a certain way: to try never to utter any sentences of £ that are not true in £. Thus it is to avoid uttering any sentence of £ unless one believes it to be true in £. To be trusting in £ is to form beliefs in a certain way: to impute truthfulness in £ to others, and thus to tend to respond to another's utterance of any sentence of £ by coming to believe that the uttered sentence is true in £.

[2] This proposal is adapted from the theory given in Erik Stenius, "Mood and Language-Game," *Synthese*, 17(1967):254–274.

David Lewis

Suppose that a certain language £ is used by a certain population P. Let this be a perfect case of normal language use. Imagine what would go on; and review the definition of a convention to verify that there does prevail in P a convention of truthfulness and trust in £.

(1) There prevails in P at least a regularity of truthfulness and trust in £. The members of P frequently speak (or write) sentences of £ to one another. When they do, ordinarily the speaker (or writer) utters one of the sentences he believes to be true in £; and the hearer (or reader) responds by coming to share that belief of the speaker's (unless he already had it), and adjusting his other beliefs accordingly.

(2) The members of P believe that this regularity of truthfulness and trust in £ prevails among them. Each believes this because of his experience of others' past truthfulness and trust in £.

(3) The expectation of conformity ordinarily gives everyone a good reason why he himself should conform. If he is a speaker, he expects his hearer to be trusting in £; wherefore he has reason to expect that by uttering certain sentences that are true in £ according to his beliefs — by being truthful in £ in a certain way — he can impart certain beliefs that he takes to be correct. Commonly, a speaker has some reason or other for wanting to impart some or other correct beliefs. Therefore his beliefs and desires constitute a practical reason for acting in the way he does: for uttering some sentence truthfully in £.

As for the hearer: he expects the speaker to be truthful in £, wherefore he has good reason to infer that the speaker's sentence is true in £ according to the speaker's beliefs. Commonly, a hearer also has some or other reason to believe that the speaker's beliefs are correct (by and large, and perhaps with exceptions for certain topics); so it is reasonable for him to infer that the sentence he has heard is probably true in £. Thus his beliefs about the speaker give him an epistemic reason to respond trustingly in £.

We have coordination between truthful speaker and trusting hearer. Each conforms as he does to the prevailing regularity of truthfulness and trust in £ because he expects complementary conformity on the part of the other.

But there is also a more diffuse and indirect sort of coordination. In coordinating with his present partner, a speaker or hearer also is coordinating with all those whose past truthfulness and trust in £ have contributed to his partner's present expectations. This indirect coordination is a four-way affair: between present speakers and past speakers, present speakers and past hearers, present hearers and past speakers, and present hearers and past hearers. And whereas the direct coordination between a speaker and his hearer is a coordination of truthfulness with trust for a single sentence of £, the indirect coordination with one's partner's previous partners (and with *their* previous partners, etc.) may involve various sentences of £. It may happen that a hearer, say, has never before encountered the sentence now addressed to him; but he forms the appropriate belief on hearing it — one such that he has responded trustingly in £ — because his past experience with truthfulness in £ has involved many sentences grammatically related to this one.

(4) There is in P a general preference for general conformity to the regularity of truthfulness and trust in £. Given that most conform, the members of P want all to conform. They desire truthfulness and trust in £ from each other, as well as from themselves. This general preference is sustained by a common interest in communication. Everyone wants occasionally to impart correct beliefs and bring about appropriate actions in others by means of sounds and marks. Everyone wants to preserve his ability to do so at will. Everyone wants to be able to learn about the parts of the world that he cannot observe for himself by observing instead the sounds and marks of his fellows who have been there.

(5) The regularity of truthfulness and trust in £ has alternatives. Let £' be any language that does not overlap £ in such a way that it is possible to be truthful and trusting simultaneously in £ and in £', and that is rich and convenient enough to meet the needs of P for communication. Then the regularity of truthfulness and trust in £' is an alternative to the prevailing regularity of truthfulness and trust in £. For the alternative regularity, as for the actual one, general conformity by the others would give one a reason to conform; and general conformity would be generally preferred over slightly-less-than-general conformity.

(6) Finally, all these facts are common knowledge in P. Everyone knows them, everyone knows that everyone knows them, and so on. Or at any rate none believes that another doubts them, none believes that another believes that another doubts them, and so on.

In any case in which a language £ clearly is used by a population P, then, it seems that there prevails in P a convention of truthfulness and trust in £, sustained by an interest in communication. The converse is supported by an unsuccessful search for counterexamples: I have not been able to think of any case in which there is such a convention and yet the language £ is clearly not used in the population P. Therefore I adopt this definition, claiming that it agrees with ordinary usage in the cases in which ordinary usage is fully determinate:

> a language £ is *used by* a population P if and only if there prevails in P a convention of truthfulness and trust in £, sustained by an interest in communication.

Such conventions, I claim, provide the desired connection between languages and language-using populations.

Once we understand how languages are connected to populations, whether by conventions of truthfulness and trust for the sake of communication or in some other way, we can proceed to redefine relative to a population all those semantic concepts that we previously defined relative to a language. A string of sounds or of marks is a *sentence of P* if and only if it is a sentence of some language £ which is used in P. It has a certain *meaning in P* if and only if it has that meaning in some language £ which is used in P. It is *true in P* at a world w if and only if it is true at w in some language £ which is used in P. It is *true in P* if and only if it is true in some language £ which is used in P.

The account just given of conventions in general, and of conventions of language in particular, differs in one important respect from the account given in my book *Convention*.[3]

Formerly, the crucial clause in the definition of convention was stated in terms of a conditional preference for conformity: each prefers to conform if the others do, and it would be the same for the alternatives to the actual convention. (In some versions of the definition, this condition was subsumed under a broader requirement of general preference for general conformity.) The point of this was to explain why the belief that others conform would give everyone a reason for conforming likewise, and so to explain the rational self-perpetuation of conventions. But a reason involving preference in this way must be a practical reason for acting, not an epistemic reason for believing. Therefore I said that conventions were regularities in action alone. It made no sense to speak of believing something in conformity to convention. (Except in the peculiar case that others' conformity to the convention gives one a prac-

---

[3] Cambridge, Mass.: Harvard University Press, 1969. A similar account was given in the original version of this paper, written in 1968.

tical reason to conform by acting to somehow produce a belief in oneself; but I knew that this case was irrelevant to ordinary language use.) Thus I was cut off from what I now take to be the primary sort of conventional coordination in language use: that between the action of the truthful speaker and the responsive believing of his trusting hearer. I resorted to two different substitutes.

Sometimes it is common knowledge how the hearer will want to act if he forms various beliefs, and we can think of the speaker not only as trying to impart beliefs but also as trying thereby to get the hearer to act in a way that speaker and hearer alike deem appropriate under the circumstances that the speaker believes to obtain. Then we have speaker-hearer coordination of action. Both conform to a convention of truthfulness for the speaker plus appropriate responsive action by the hearer. The hearer's trustful believing need not be part of the content of the convention, though it must be mentioned to explain why the hearer acts in conformity. In this way we reach the account of "signaling" in *Convention*, chapter IV.

But signaling was all-too-obviously a special case. There may be no appropriate responsive action for the hearer to perform when the speaker imparts a belief to him. Or the speaker and hearer may disagree about how the hearer ought to act under the supposed circumstances. Or the speaker may not know how the hearer will decide to act; or the hearer may not know that he knows; and so on. The proper hearer's response to consider is *believing*, but that is not ordinarily an action. So in considering language use in general, in *Convention*, chapter V, I was forced to give up on speaker-hearer coordination. I took instead the diffuse coordination between the present speaker and the past speakers who trained the present hearer. Accordingly, I proposed that the convention whereby a population P used a language £ was simply a convention of truthfulness in £. Speakers conform; hearers do not, until they become speakers in their turn, if they ever do.

I think now that I went wrong when I went beyond the special case of signaling. I should have kept my original emphasis on speaker-hearer coordination, broadening the definition of convention to fit. It was Jonathan Bennett[4] who showed me how that could be done: by restat-

---

[4] Personal communication, 1971. Bennett himself uses the broadened concept of convention differently, wishing to exhibit conventional meaning as a special case of

11

ing the crucial defining clause not in terms of preference for conformity but rather in terms of reasons for conformity — practical or epistemic reasons. The original conditional preference requirement gives way now to clause (3): the belief that others conform gives everyone a reason to conform likewise, and it would be the same for the alternatives to the actual convention. Once this change is made, there is no longer any obstacle to including the hearer's trust as part of the content of a convention.

(The old conditional preference requirement is retained, however, in consequence of the less important clause (4). Clause (3) as applied to practical reasons, but not as applied to epistemic reasons, may be subsumed under (4).)

Bennett pointed out one advantage of the change: suppose there is only one speaker of an idiolect, but several hearers who can understand him. Shouldn't he and his hearers comprise a population that uses his idiolect? More generally, what is the difference between (a) someone who does not utter sentences of a language because he does not belong to any population that uses it, and (b) someone who does not utter sentences of the language although he does belong to such a population because at present — or always, perhaps — he has nothing to say? Both are alike, so far as action in conformity to a convention of truthfulness goes. Both are vacuously truthful. In *Convention* I made it a condition of truthfulness in £ that one sometimes does utter sentences of £, though not that one speaks up on any particular occasion. But that is unsatisfactory: what degree of truthful talkativeness does it take to keep up one's active membership in a language-using population? What if someone just never thought of anything worth saying?

(There is a less important difference between my former account and the present one. Then and now, I wanted to insist that cases of convention are cases of predominant coincidence of interest. I formerly provided for this by a defining clause that seems now unduly restrictive: in any instance of the situation to which the convention applies, everyone has approximately the same preferences regarding all possible combinations of actions. Why *all*? It may be enough that they agree in preferences to the extent specified in my present clause (4). Thus I have left out the further agreement-in-preferences clause.)

Gricean meaning. See his "The Meaning-Nominalist Strategy," *Foundations of Language*, 10(1973):141–168.

Objections and Replies

Objection: Many things which meet the definition of a language given in the thesis — many functions from strings of sounds or of marks to sets of possible worlds — are not really possible languages. They could not possibly be adopted by any human population. There may be too few sentences, or too few meanings, to make as many discriminations as language-users need to communicate. The meanings may not be anything language-users would wish to communicate about. The sentences may be very long, impossible to pronounce, or otherwise clumsy. The language may be humanly unlearnable because it has no grammar, or a grammar of the wrong kind.

Reply: Granted. The so-called languages of the thesis are merely an easily specified superset of the languages we are really interested in. A language in a narrower and more natural sense is any one of these entities that could possibly — possibly in some appropriately strict sense — be used by a human population.

Objection: The so-called languages discussed in the thesis are excessively simplified. There is no provision for indexical sentences, dependent on features of the context of their utterance: for instance, tensed sentences, sentences with personal pronouns or demonstratives, or anaphoric sentences. There is no provision for ambiguous sentences. There is no provision for non-indicative sentences: imperatives, questions, promises and threats, permissions, and so on.

Reply: Granted. I have this excuse: the phenomenon of language would be not too different if these complications did not exist, so we cannot go too far wrong by ignoring them. Nevertheless, let us sketch what could be done to provide for indexicality, ambiguity, or non-indicatives. In order not to pile complication on complication we shall take only one at a time.

We may define an *indexical language* $£$ as a function that assigns sets of possible worlds not to its sentences themselves, but rather to sentences paired with possible occasions of their utterance. We can say that σ is true in $£$ at a world w on a possible occasion o of the utterance of σ if and only if w belongs to $£(\sigma, o)$. We can say that σ is true in $£$ on o (without mentioning a world) if and only if the world in which o is located — our actual world if o is an actual occasion of utterance of σ, or some other world if not — belongs to $£(\sigma, o)$. We can say that a

13

speaker is truthful in £ if he tries not to utter any sentence σ of £ unless σ would be true in £ on the occasion of his utterance of σ. We can say that a hearer is trusting in £ if he believes an uttered sentence of £ to be true in £ on its occasion of utterance.

We may define an *ambiguous language* £ as a function that assigns to its sentences not single meanings, but finite sets of alternative meanings. (We might or might not want to stipulate that these sets are non-empty.) We can say that a sentence σ is true in £ at w under some meaning if and only if w belongs to some member of £(σ). We can say that σ is true in £ under some meaning if and only if our actual world belongs to some member of £(σ). We can say that someone is (minimally) truthful in £ if he tries not to utter any sentence σ of £ unless σ is true in £ under some meaning. He is trusting if he believes an uttered sentence of £ to be true in £ under some meaning.

We may define a *polymodal language* £ as a function which assigns to its sentences meanings containing two components: a set of worlds, as before; and something we can call a *mood*: indicative, imperative, etc. (It makes no difference what things these are — they might, for instance, be taken as code numbers.) We can say that a sentence σ is indicative, imperative, etc., in £ according as the mood-component of the meaning £(σ) is indicative, imperative, etc. We can say that a sentence σ is true in £, regardless of its mood in £, if and only if our actual world belongs to the set-of-worlds-component of the meaning £(σ). We can say that someone is truthful in £ with respect to indicatives if he tries not to utter any indicative sentence of £ which is not true in £; truthful in £ with respect to imperatives if he tries to act in such a way as to make true in £ any imperative sentence of £ that is addressed to him by some-one in a relation of authority to him; and so on for other moods. He is trusting in £ with respect to indicatives if he believes uttered indicative sentences of £ to be true in £; trusting in £ with respect to imperatives if he expects his utterance of an imperative sentence of £ to result in the addressee's acting in such a way as to make that sentence true in £, provided he is in a relation of authority to the addressee; and so on. We can say simply that he is truthful and trusting in £ if he is so with respect to all moods that occur in £. It is by virtue of the various ways in which the various moods enter into the definition of truthfulness and of trust that they deserve the familiar names we have given them. (I am deliberating stretching the ordinary usage of "true," "truthfulness,"

14

and "trust" in extending them to non-indicatives. For instance, truthfulness with respect to imperatives is roughly what we might call *obedience* in £.)

Any natural language is simultaneously indexical, ambiguous, and polymodal; I leave the combination of complications as an exercise. Henceforth, for the most part, I shall lapse into ignoring indexicality, ambiguity, and non-indicatives.

Objection: We cannot always discover the meaning of a sentence in a population just by looking into the minds of the members of the population, no matter what we look for there. We may also need some information about the causal origin of what we find in their minds. So, in particular, we cannot always discover the meaning of a sentence in a population just by looking at the conventions prevailing therein. Consider an example: What is the meaning of the sentence "Mik Karthee was wise" in the language of our 137th-century descendants, if all we can find in any of their minds is the inadequate dictionary entry: "Mik Karthee: controversial American politician of the early atomic age"? It depends, we might think, partly on which man stands at the beginning of the long causal chain ending in that inadequate dictionary entry.

Reply: If this doctrine is correct, I can treat it as a subtle sort of indexicality. The set of worlds in which a sentence σ is true in a language £ may depend on features of possible occasions of utterance of σ. One feature of a possible occasion of utterance — admittedly a more recondite feature than the time, place, or speaker — is the causal history of a dictionary entry in a speaker's mind.

As with other kinds of indexicality, we face a problem of nomenclature. Let a *meaning*$_1$ be that which an indexical language £ assigns to a sentence σ on a possible occasion o of its utterance: £(σ, o), a set of worlds on our account. Let a *meaning*$_2$ be that fixed function whereby the meaning$_1$ in £ of a sentence σ varies with its occasions of utterance. Which one is a meaning? That is unclear — and it is no clearer which one is a sense, intension, interpretation, truth-condition, or proposition.

The objection says that we sometimes cannot find the meaning$_1$ of σ on o in P by looking into the minds of members of P. Granted. But what prevents it is that the minds do not contain enough information about o: in particular, not enough information about its causal history. We have been given no reason to doubt that we can find the meaning$_2$

15

of σ in P by looking into minds; and that is all we need do to identify the indexical language used by P.

An exactly similar situation arises with more familiar kinds of indexicality. We may be unable to discover the time of an utterance of a tensed sentence by looking into minds, so we may know the meaning$_2$ of the sentence uttered in the speaker's indexical language without knowing its meaning$_1$ on the occasion in question.

Objection: It makes no sense to say that a mere string of sounds or of marks can bear a meaning or a truth-value. The proper bearers of meanings and truth-values are particular speech acts.

Reply: I do not say that a string of types of sounds or of marks, by itself, can bear a meaning or truth-value. I say it bears a meaning and truth-value relative to a language, or relative to a population. A particular speech act by itself, on the other hand, can bear a meaning and truth-value, since in most cases it uniquely determines the language that was in use on the occasion of its performance. So can a particular uttered string of vocal sounds, or a particular inscribed string of marks, since in most cases that uniquely determines the particular speech act in which it was produced, which in turn uniquely determines the language.

Objection: It is circular to give an account of meanings in terms of possible worlds. The notion of a possible world must itself be explained in semantic terms. Possible worlds are models of the analytic sentences of some language, or they are the diagrams or theories of such models.[5]

Reply: I do not agree that the notion of a possible world ought to be explained in semantic terms, or that possible worlds ought to be eliminated from our ontology and replaced by their linguistic representatives — models or whatever.

For one thing, the replacement does not work properly. Two worlds indistinguishable in the representing language will receive one and the same representative.

But more important, the replacement is gratuitous. The notion of a possible world is familiar in its own right, philosophically fruitful, and

---

[5] Possible worlds are taken as models in S. Kripke, "A Completeness Theorem in Modal Logic," *Journal of Symbolic Logic*, 24(1959):1–15; in Carnap's recent work on semantics and inductive logic, discussed briefly in secs. 9, 10, and 25 of "Replies and Systematic Expositions," *The Philosophy of Rudolf Carnap*, ed. P. Schilpp; and elsewhere. Worlds are taken as state-descriptions — diagrams of models — in Carnap's earlier work: for instance, sec. 18 of *Introduction to Semantics*. Worlds are taken as complete, consistent novels — theories of models — in R. Jeffrey, *The Logic of Decision*, sec. 12.8.

tolerably clear. Possible worlds are deemed mysterious and objectionable because they raise questions we may never know how to answer: are any possible worlds five-dimensional? We seem to think that we do not understand possible worlds at all unless we are capable of omniscience about them — but why should we think that? Sets also raise unanswerable questions, yet most of us do not repudiate sets.

But if you insist on repudiating possible worlds, much of my theory can be adapted to meet your needs. We must suppose that you have already defined truth and analyticity in some base language — that is the price you pay for repudiating possible worlds — and you want to define them in general, for the language of an arbitrary population P. Pick your favorite base language, with any convenient special properties you like: Latin, Esperanto, Begriffsschrift, Semantic Markerese, or what have you. Let's say you pick Latin. Then you may redefine a language as any function from certain strings of sound or of marks to sentences of Latin. A sentence σ of a language £ (in your sense) is true, analytic, etc., if and only if £(σ) is true, analytic, etc., in Latin.

You cannot believe in languages in my sense, since they involve possible worlds. But I can believe in languages in your sense. And I can map your languages onto mine by means of a fixed function from sentences of Latin to sets of worlds. This function is just the language Latin, in my sense. My language £ is the composition of two functions: your language £, and my language Latin. Thus I can accept your approach as part of mine.

Objection: Why all this needless and outmoded hypostasis of meanings? Our ordinary talk about meaning does not commit us to believing in any such entities as meanings, any more than our ordinary talk about actions for the sake of ends commits us to believing in any such entities as sakes.

Reply: Perhaps there are some who hypostatize meanings compulsively, imagining that they could not possibly make sense of our ordinary talk about meaning if they did not. Not I. I hypostatize meanings because I find it convenient to do so, and I have no good reason not to. There is no point in being a part-time nominalist. I am persuaded on independent grounds that I ought to believe in possible worlds and possible beings therein, and that I ought to believe in sets of things I believe in. Once I have these, I have all the entities I could ever want.

Objection: A language consists not only of sentences with their mean-

17

ings, but also of constituents of sentences — things sentences are made of — with their meanings. And if any language is to be learnable without being finite, it must somehow be determined by finitely many of its constituents and finitely many operations on constituents.

Reply: We may define a class of objects called *grammars*. A grammar Γ is a triple comprising (1) a large finite *lexicon* of *elementary constituents* paired with meanings; (2) a finite set of *combining operations* which build larger constituents by combining smaller constituents, and derive a meaning for the new constituent out of the meanings of the old ones; and (3) a *representing operation* which effectively maps certain constituents onto strings of sounds or of marks. A grammar Γ generates a function which assigns meanings to certain constituents, called *constituents in* Γ. It generates another function which assigns meanings to certain strings of sounds or of marks. Part of this latter function is what we have hitherto called a language. A grammar uniquely determines the language it generates. But a language does not uniquely determine the grammar that generates it, not even when we disregard superficial differences between grammars.

I have spoken of meanings for constituents in a grammar, but what sort of things are these? Referential semantics tried to answer that question. It was a near miss, failing because contingent facts got mixed up with the meanings. The cure, discovered by Carnap,[6] is to do referential semantics not just in our actual world but in every possible world. A meaning for a name can be a function from worlds to possible individuals; for a common noun, a function from worlds to sets; for a sentence, a function from worlds to truth-values (or more simply, the set of worlds where that function takes the value truth). Other derived categories may be defined by their characteristic modes of combination. For instance, an adjective combines with a common noun to make a compound common noun; so its meaning may be a function from common-noun meanings to common-noun meanings, such that the meaning of an adjective-plus-common-noun compound is the value of this function when given as argument the meaning of the common noun being modified. Likewise a verb phrase takes a name to make a sentence; so its meaning may be a function that takes the meaning of the name as argument to give the

---

[6] "Replies and Systematic Expositions," sec. 9.v. A better-known presentation of essentially the same idea is in S. Kripke, "Semantical Considerations on Modal Logic," *Acta Philosophica Fennica*, 16(1963):83–94.

meaning of the sentence as value. An adverb (of one sort) takes a verb phrase to make a verb phrase, so its meaning may be a function from verb-phrase meanings to verb-phrase meanings. And so on, as far as need be, to more and more complicated derived categories.[7]

If you repudiate possible worlds, an alternative course is open to you: let the meanings for constituents in a grammar be phrases of Latin, or whatever your favorite base language may be.

A grammar, for us, is a semantically interpreted grammar — just as a language is a semantically interpreted language. We shall not be concerned with what are called grammars or languages in a purely syntactic sense. My definition of a grammar is meant to be general enough to encompass transformational or phrase-structure grammars for natural language[8] (when provided with semantic interpretations) as well as systems of formation and valuation rules for formalized languages. Like my previous definition of a language, my definition of a grammar is too general: it gives a large superset of the interesting grammars.

A grammar, like a language, is a set-theoretical entity which can be discussed in complete abstraction from human affairs. Since a grammar generates a unique language, all the semantic concepts we earlier defined relative to a language £ — sentencehood, truth, analyticity, etc. — could just as well have been defined relative to a grammar Γ. We can also handle other semantic concepts pertaining to constituents, or to the constituent structure of sentences.

We can define the meaning in Γ, denotation in Γ, etc., of a subsentential constituent in Γ. We can define the meaning in Γ, denotation in Γ, etc., of a phrase: a string of sounds or of marks representing a subsentential constituent in Γ via the representing operation of Γ. We can define something we may call the *fine structure of meaning* in Γ of a sentence or phrase: the manner in which the meaning of the sentence or phrase is derived from the meanings of its constituents and the way it is built out of them. Thus we can take account of the sense in which, for instance, different analytic sentences are said to differ in meaning.

[7] See my "General Semantics," *Synthese*, 22(1970):18–67.

[8] For a description of the sort of grammars I have in mind (minus the semantic interpretation) see N. Chomsky, *Aspects of the Theory of Syntax*, and G. Harman, "Generative Grammars without Transformation Rules," *Language*, 37(1963):597–616. My "constituents" correspond to semantically interpreted deep phrase-markers, or sub-trees thereof, in a transformational grammar. My "representing operation" may work in several steps and thus subsumes both the transformational and the phonological components of a transformational grammar.

Now the objection can be restated: what ought to be called a language is what I have hitherto called a grammar, not what I have hitherto called a language. Different grammar, different language — at least if we ignore superficial differences between grammars. Verbal disagreement aside, the place I gave to my so-called languages ought to have been given instead to my so-called grammars. Why not begin by saying what it is for a grammar Γ to be used by a population P? Then we could go on to define sentencehood, truth, analyticity, etc., in P as sentencehood, truth, analyticity, etc., in whatever grammar is used by P. This approach would have the advantage that we could handle the semantics of constituents in a population in an exactly similar way. We could say that a constituent or phrase has a certain meaning, denotation, etc., in P if it has that meaning, denotation, etc., in whatever grammar is used by P. We could say that a sentence or phrase has a certain fine structure of meaning in P if it has it in whatever grammar is used by P.

Unfortunately, I know of no promising way to make objective sense of the assertion that a grammar Γ is used by a population P whereas another grammar Γ', which generates the same language as Γ, is not. I have tried to say how there are facts about P which objectively select the languages used by P. I am not sure there are facts about P which objectively select privileged grammars for those languages. It is easy enough to define truthfulness and trust in a grammar, but that will not help: a convention of truthfulness and trust in Γ will also be a convention of truthfulness and trust in Γ' whenever Γ and Γ' generate the same language.

I do not propose to discard the notion of the meaning in P of a constituent or phrase, or the fine structure of meaning in P of a sentence. To propose that would be absurd. But I hold that these notions depend on our methods of evaluating grammars, and therefore are no clearer and no more objective than our notion of a *best* grammar for a given language. For I would say that a grammar Γ is used by P if and only if Γ is a best grammar for a language £ that is used by P in virtue of a convention in P of truthfulness and trust in £; and I would define the meaning in P of a constituent or phrase, and the fine structure of meaning in P of a sentence, accordingly.

The notions of a language used by P, of a meaning of a sentence in P, and so on, are independent of our evaluation of grammars. Therefore I take these as primary. The point is not to refrain from ever saying anything that depends on the evaluation of grammars. The point is to do so only when we must, and that is why I have concentrated on languages rather than grammars.

We may meet little practical difficulty with the semantics of constituents in populations, even if its foundations are as infirm as I fear. It may often happen that all the grammars anyone might call best for a given language will agree on the meaning of a given constituent. Yet there is trouble to be found: Quine's examples of indeterminacy of reference[9] seem to be disagreements in constituent semantics between alternative good grammars for one language. We should regard with suspicion any method that purports to settle objectively whether, in some tribe, "gavagai" is true of temporally continuant rabbits or time-slices thereof. You can give their language a good grammar of either kind — and that's that.

It is useful to divide the claimed indeterminacy of constituent semantics into three separate indeterminacies. We begin with undoubted objective fact: the dependence of the subject's behavioral output on his input of sensory stimulation (both as it actually is and as it might have been) together with all the physical laws and anatomical facts that explain it. (a) This information either determines or underdetermines the subject's system of propositional attitudes: in particular, his beliefs and desires. (b) These propositional attitudes either determine or underdetermine the truth conditions of full sentences — what I have here called his language. (c) The truth conditions of full sentences either determine or undetermine the meanings of sub-sentential constituents — what I have here called his grammar.

My present discussion has been directed at the middle step, from beliefs and desires to truth conditions for full sentences. I have said that the former determine the latter — provided (what need not be the case) that the beliefs and desires of the subject and his fellows are such as to comprise a fully determinate convention of truthfulness and trust in some definite language. I have said nothing here about the determinacy of the first step; and I am inclined to share in Quine's doubts about the determinacy of the third step.

Objection: Suppose that whenever anyone is party to a convention of truthfulness and trust in any language £, his competence to be party to

[9] W. V. Quine, "Ontological Relativity," *Journal of Philosophy*, 65(1968):185–212; *Word and Object*, pp. 68–79.

that convention — to conform, to expect conformity, etc. — is due to his possession of some sort of unconscious internal representation of a grammar for £. That is a likely hypothesis, since it best explains what we know about linguistic competence. In particular, it explains why experience with some sentences leads spontaneously to expectations involving others. But on that hypothesis, we might as well bypass the conventions of language and say that £ is used by P if and only if everyone in P possesses an internal representation of a grammar for £.

Reply: In the first place, the hypothesis of internally represented grammars is not an explanation — best or otherwise — of anything. Perhaps it is *part* of some theory that best explains what we know about linguistic competence; we can't judge until we hear something about what the rest of the theory is like.

Nonetheless, I am ready enough to believe in internally represented grammars. But I am much less certain that there are internally represented grammars than I am that languages are used by populations; and I think it makes sense to say that languages might be used by populations even if there were no internally represented grammars. I can tentatively agree that £ is used by P if and only if everyone in P possesses an internal representation of a grammar for £, if that is offered as a scientific hypothesis. But I cannot accept it as any sort of analysis of "£ is used by P", since the analysandum clearly could be true although the analysans was false.

Objection: The notion of a convention of truthfulness and trust in £ is a needless complication. Why not say, straightforwardly, that £ is used by P if and only if there prevails in P a convention to bestow upon each sentence of £ the meaning that £ assigns to it? Or, indeed, that a grammar Γ of £ is used by P if and only if there prevails in P a convention to bestow upon each constituent in Γ the meaning that Γ assigns to it?

Reply: A convention, as I have defined it, is a regularity in action, or in action and belief. If that feature of the definition were given up, I do not see how to salvage any part of my theory of conventions. It is essential that a convention is a regularity such that conformity by others gives one a reason to conform; and such a reason must either be a practical reason for acting or an epistemic reason for believing. What other kind of reason is there?

Yet there is no such thing as an action of bestowing a meaning (except for an irrelevant sort of action that is performed not by language-users but by creators of language) so we cannot suppose that language-using populations have conventions to perform such actions. Neither does bestowal of meaning consist in forming some belief. Granted, bestowal of meaning is conventional in the sense that it depends on convention: the meanings would have been different if the conventions of truthfulness and trust had been different. But bestowal of meaning is not an action done in conformity to a convention, since it is not an action, and it is not a belief-formation in conformity to a convention, since it is not a belief-formation.

Objection: The beliefs and desires that constitute a convention are inaccessible mental entities, just as much as hypothetical internal representations of grammars are. It would be best if we could say in purely behavioristic terms what it is for a language £ to be used by a population P. We might be able to do this by referring to the way in which members of P would answer counterfactual questionnaires; or by referring to the way in which they would or would not assent to sentences under deceptive sensory stimulation; or by referring to the way in which they would intuitively group sentences into similarity-classes; or in some other way.

Reply: Suppose we succeeded in giving a behavioristic operational definition of the relation "£ is used by P." This would not help us to understand what it is for £ to be used by P; for we would have to understand that already, and also know a good deal of common-sense psychology, in order to check that the operational definition was a definition of what it is supposed to be a definition of. If we did not know what it meant for £ to be used by P, we would not know what sort of behavior on the part of members of P would indicate that £ was used by P.

Objection: The conventions of language are nothing more nor less than our famously obscure old friends, the rules of language, renamed.

Reply: A convention of truthfulness and trust in £ might well be called a rule, though it lacks many features that have sometimes been thought to belong to the essence of rules. It is not promulgated by any authority. It is not enforced by means of sanctions except to the extent that, because one has some sort of reason to conform, something bad may happen if one does not. It is nowhere codified and therefore is not "laid down in the course of teaching the language" or "appealed to in

23

the course of criticizing a person's linguistic performance."[10] Yet it is more than a mere regularity holding "as a rule"; it is a regularity accompanied and sustained by a special kind of system of beliefs and desires.

A convention of truthfulness and trust in £ might have as consequences other regularities which were conventions of language in their own right: specializations of the convention to certain special situations. (For instance, a convention of truthfulness in £ on weekdays.) Such derivative conventions of language might also be called rules; some of them might stand a better chance of being codified than the overall convention which subsumes them.

However, there are other so-called rules of language which are not conventions of language and are not in the least like conventions of language: for instance, "rules" of syntax and semantics. They are not even regularities and cannot be formulated as imperatives. They might better be described not as rules, but as clauses in the definitions of entities which are to be mentioned in rules: clauses in the definition of a language £, of the act of being truthful in £, of the act of stating that the moon is blue, etc.

Thus the conventions of language might properly be called rules, but it is more informative and less confusing to call them conventions.

Objection: Language is not conventional. We have found that human capacities for language acquisition are highly specific and dictate the form of any language that humans can learn and use.

Reply: It may be that there is less conventionality than we used to think: fewer features of language which depend on convention, more which are determined by our innate capacities and therefore are common to all languages which are genuine alternatives to our actual language. But there are still conventions of language; and there are still convention-dependent features of language, differing from one alternative possible convention of language to another. That is established by the diversity of actual languages. There are conventions of language so long as the regularity of truthfulness in a given language has even a single alternative.

Objection: Unless a language-user is also a set-theorist, he cannot expect his fellows to conform to a regularity of truthfulness and trust in a

[10] P. Ziff, *Semantic Analysis*, pp. 34–35.

certain language £. For to conform to this regularity is to bear a relation to a certain esoteric entity: a set of ordered pairs of sequences of sound-types or of mark-types and sets of possible worlds (or something more complicated still, if £ is a natural language with indexicality, ambiguity, and non-indicatives). The common man has no concept of any such entity. Hence he can have no expectations regarding such an entity.

Reply: The common man need not have any concept of £ in order to expect his fellows to be truthful and trusting in £. He need only have suitable particular expectations about how they might act, and how they might form beliefs, in various situations. He can tell whether any actual or hypothetical particular action or belief-formation on their part is compatible with his expectations. He expects them to conform to a regularity of truthfulness and trust in £ if any particular activity or belief-formation that would fit his expectations would fall under what we — but not he — could describe as conformity to that regularity.

It may well be that his elaborate, infinite system of potential particular expectations can only be explained on the hypothesis that he has some unconscious mental entity somehow analogous to a general concept of £ — say, an internally represented grammar. But it does not matter whether this is so or not. We are concerned only to say what system of expectations a normal member of a language-using population must have. We need not engage in psychological speculation about how those expectations are generated.

Objection: If there are conventions of language, those who are party to them should know what they are. Yet no one can fully describe the conventions of language to which he is supposedly a party.

Reply: He may nevertheless know what they are. It is enough to be able to recognize conformity and non-conformity to his convention, and to be able to try to conform to it. We know ever so many things we cannot put into words.

Objection: Use of language is almost never a rational activity. We produce and respond to utterances by habit, not as the result of any sort of reasoning or deliberation.

Reply: An action may be rational, and may be explained by the agent's beliefs and desires, even though that action was done by habit, and the agent gave no thought to the beliefs or desires which were his reason for acting. A habit may be under the agent's rational control in this sense: if that habit ever ceased to serve the agent's desires according to his beliefs, it would at once be overridden and corrected by conscious reasoning. Action done by a habit of this sort is both habitual and rational. Likewise for habits of believing. Our normal use of language is rational, since it is under rational control.

Perhaps use of language by young children is not a rational activity. Perhaps it results from habits which would not be overridden if they ceased to serve the agent's desires according to his beliefs. If that is so, I would deny that these children have yet become party to conventions of language, and I would deny that they have yet become normal members of a language-using population. Perhaps language is first acquired and afterward becomes conventional. That would not conflict with anything I have said. I am not concerned with the way in which language is acquired, only with the condition of a normal member of a language-using population when he is done acquiring language.

Objection: Language could not have originated by convention. There could not have been an agreement to begin being truthful and trusting in a certain chosen language, unless some previous language had already been available for use in making the agreement.

Reply: The first language could not have originated by an agreement, for the reason given. But that is not to say that language cannot be conventional. A convention is so-called because of the way it persists, not because of the way it originated. A convention need not originate by convention — that is, by agreement — though many conventions do originate by agreement, and others could originate by agreement even if they actually do not. In saying that language is convention-governed, I say nothing whatever about the origins of language.

Objection: A man isolated all his life from others might begin — through genius or a miracle — to use language, say to keep a diary. (This would be an accidentally private language, not the necessarily private language Wittgenstein is said to have proved to be impossible.) In this case, at least, there would be no convention involved.

Reply: Taking the definition literally, there would be no convention. But there would be something very similar. The isolated man conforms to a certain regularity at many different times. He knows at each of these times that he has conformed to that regularity in the past, and he has an interest in uniformity over time, so he continues to conform to that regularity instead of to any of various alternative regularities that would have done about as well if he had started out using them. He knows at all times that this is so, knows that he knows at all times that this is so, and so on. We might think of the situation as one in which a convention prevails in the population of different time-slices of the same man.

Objection: It is circular to define the meaning in P of sentences in terms of the beliefs held by members of P. For presumably the members of P think in their language. For instance, they hold beliefs by accepting suitable sentences of their language. If we do not already know the meaning in P of a sentence, we do not know what belief a member of P would hold by accepting that sentence.

Reply: It may be true that men think in language, and that to hold a belief is to accept a sentence of one's language. But it does not follow that belief should be analyzed as acceptance of sentences. It should not be. Even if men do in fact think in language, they might not. It is at least possible that men — like beasts — might hold beliefs otherwise than by accepting sentences. (I shall not say here how I think belief *should* be analyzed.) No circle arises from the contingent truth that a member of P holds beliefs by accepting sentences, so long as we can specify his beliefs without mentioning the sentences he accepts. We can do this for men, as we can for beasts.

Objection: Suppose a langauge £ is used by a population of inveterate liars, who are untruthful in £ more often than not. There would not be even a regularity — still less a convention, which implies a regularity — of truthfulness and trust in £.

Reply: I deny that £ is used by the population of liars. I have undertaken to follow ordinary usage only where it is determinate; and, once it is appreciated just how extraordinary the situation would have to be, I do not believe that ordinary usage is determinate in this case. There are many similarities to clear cases in which a language is used by a population, and it is understandable that we should feel some inclination to classify this case along with them. But there are many important differences as well.

Although I deny that the population of liars *collectively* uses £, I am willing to say that each liar *individually* may use £, provided that he falsely believes that he is a member — albeit an exceptional, untruthful member — of a population wherein there prevails a convention of truthfulness and trust in £. He is in a position like that of a madman who thinks he belongs to a population which uses £, and behaves accordingly, and so can be said to use £, although in reality all the other members of this £-using population are figments of his imagination.

Objection: Suppose the members of a population are untruthful in their language £ more often than not, not because they lie, but because they go in heavily for irony, metaphor, hyperbole, and such. It is hard to deny that the language £ is used by such a population.

Reply: I claim that these people *are* truthful in their language £, though they are not *literally truthful* in £. To be literally truthful in £ is to be truthful in another language related to £, a language we can call literal-£. The relation between £ and literal-£ is as follows: a good way to describe £ is to start by specifying literal-£ and then to describe £ as obtained by certain systematic departures from literal-£. This two-stage specification of £ by way of literal-£ may turn out to be much simpler than any direct specification of £.

Objection: Suppose they are often untruthful in £ because they are not communicating at all. They are joking, or telling tall tales, or telling white lies as a matter of social ritual. In these situations, there is neither truthfulness nor trust in £. Indeed, it is common knowledge that there is not.

Reply: Perhaps I can say the same sort of thing about this non-serious language use as I did about non-literal language use. That is: their seeming untruthfulness in non-serious situations is untruthfulness not in the language £ that they actually use, but only in a simplified approximation to £. We may specify £ by first specifying the approximation language, then listing the signs and features of context by which non-serious language use can be recognized, then specifying that when these signs or features are present, what would count as untruths in the approximation language do not count as such in £ itself. Perhaps they are automatically true in £, regardless of the facts; perhaps they cease to count as indicative.

Example: what would otherwise be an untruth may not be one if said by a child with crossed fingers. Unfortunately, the signs and features of context by which we recognize non-serious language use are seldom as simple, standardized, and conventional as that. While they must find a place somewhere in a full account of the phenomenon of language, it may be inexpedient to burden the specification of £ with them.

Perhaps it may be enough to note that these situations of non-serious language use must be at least somewhat exceptional if we are to have anything like a clear case of use of £; and to recall that the definition of a convention was loose enough to tolerate some exceptions. We could take the non-serious cases simply as violations — explicable and harmless ones — of the conventions of language.

There is a third alternative, requiring a modification in my theory. We may say that a *serious communication situation* exists with respect to a sentence σ of £ whenever it is true, and common knowledge between a speaker and a hearer, that (a) the speaker does, and the hearer does not, know whether σ is true in £; (b) the hearer wants to know; (c) the speaker wants the hearer to know; and (d) neither the speaker nor the hearer has other (comparably strong) desires as to whether or not the speaker utters σ. (Note that when there is a serious communication situation with respect to σ, there is one also with respect to synonyms or contradictories in £ of σ, and probably also with respect to other logical relatives in £ of σ.) Then we may say that the convention whereby P uses £ is a convention of truthfulness and trust in £ in serious communication situations. That is: when a serious communication situation exists with respect to σ, then the speaker tries not to utter σ unless it is true in £, and the hearer responds, if σ is uttered, by coming to believe that σ is true in £. If that much is a convention in P, it does not matter what goes on in other situations: they use £.

The definition here given of a serious communication resembles that of a signaling problem in *Convention*, chapter IV, the difference being that the hearer may respond by belief-formation only, rather than by what speaker and hearer alike take to be appropriate action. If this modification were adopted, it would bring my general account of language even closer to my account in *Convention* of the special case of signaling.

Objection: Truthfulness and trust cannot be a convention. What could be the alternative to uniform truthfulness — uniform untruthfulness, perhaps? But it seems that if such untruthfulness were not intended to deceive, and did not deceive, then it too would be truthfulness.

Reply: The convention is not the regularity of truthfulness and trust *simpliciter*. It is the regularity of truthfulness and trust in some particular language £. Its alternatives are possible regularities of truthfulness and trust in other languages. A regularity of uniform untruthfulness and non-trust in a language £ can be redescribed as a regularity of truthfulness and trust in a different language anti-£ complementary to £. Anti-£ has exactly the same sentences as £, but with opposite truth conditions. Hence the true sentences of anti-£ are all and only the untrue sentences of £.

There is a different regularity that we may call a regularity of truthfulness and trust *simpliciter*. That is the regularity of being truthful and trusting in whichever language is used by one's fellows. This regularity neither is a convention nor depends on convention. If any language whatever is used by a population P, then a regularity (perhaps with exceptions) of truthfulness and trust *simpliciter* prevails in P.

Objection: Even truthfulness and trust in £ cannot be a convention. One conforms to a convention, on my account, because doing so answers to some sort of interest. But a decent man is truthful in £ if his fellows are, whether or not it is in his interest. For he recognizes that he is under a moral obligation to be truthful in £: an obligation to reciprocate the benefits he has derived from others' truthfulness in £, or something of that sort. Truthfulness in £ may bind the decent man against his own interest. It is more like a social contract than a convention.

Reply: The objection plays on a narrow sense of "interest" in which only selfish interests count. We commonly adopt a wider sense. We count also altruistic interests and interests springing from one's recognition of obligations. It is this wider sense that should be understood in the definition of convention. In this wider sense, it is nonsense to think of an obligation as outweighing one's interests. Rather, the obligation provides one interest which may outweigh the other interests.

A convention of truthfulness and trust in £ is sustained by a mixture of selfish interests, altruistic interests, and interests derived from obligation. Usually all are present in strength; perhaps any one would be enough to sustain the convention. But occasionally truthfulness in £ answers only to interests derived from obligation and goes against one's selfish or even altruistic interests. In such a case, only a decent man will have an interest in remaining truthful in £. But I dare say such cases are not as common as moralists might imagine. A convention of truthfulness and trust among scoundrels might well be sustained — with occasional lapses — by selfish interests alone.

A convention persists because everyone has reason to conform if others do. If the convention is a regularity in action, this is to say that it persists because everyone prefers general conformity rather than almost-general conformity with himself as the exception. A (demythologized) social contract may also be described as a regularity sustained by a general preference for general conformity, but the second term of the preference is different. Everyone prefers general conformity over a certain state of general non-conformity called the state of nature. This general preference sets up an obligation to reciprocate the benefits derived from others' conformity, and that obligation creates an interest in conforming which sustains the social contract. The objection suggests that, among decent men, truthfulness in £ is a social contract. I agree; but there is no reason why it cannot be a social contract and a convention as well, and I think it is.

Objection: Communication cannot be explained by conventions of truthfulness alone. If I utter a sentence σ of our language £, you — expecting me to be truthful in £ — will conclude that I take σ to be true in £. If you think I am well informed, you will also conclude that probably σ is true in £. But you will draw other conclusions as well, based on your legitimate assumption that it is for some good reason that I chose to utter σ rather than remain silent, and rather than utter any of the other sentences of £ that I also take to be true in £. I can communicate all sorts of misinformation by exploiting your beliefs about my conversational purposes, without ever being untruthful in £. Communication depends on principles of helpfulness and relevance as well as truthfulness.

Reply: All this does not conflict with anything I have said. We do conform to conversational regularities of helpfulness and relevance. But these regularities are not independent conventions of language; they result from our convention of truthfulness and trust in £ together with certain general facts — not dependent on any convention — about our conversational purposes and our beliefs about one another. Since they are by-products of a convention of truthfulness and trust, it is unnecessary to mention them separately in specifying the conditions under which a language is used by a population.

Objection: Let £ be the language used in P, and let $£^-$ be some fairly rich fragment of £. That is, the sentences of $£^-$ are many but not all of the sentences of £ (in an appropriate special sense if £ is infinite); and any sentence of both has the same meaning in both. Then $£^-$ also turns out to be a language used by P; for by my definition there prevails

31

in *P* a convention of truthfulness and trust in £⁻, sustained by an interest in communication. Not one but many — perhaps infinitely many — languages are used by *P*.

Reply: That is so, but it is no problem. Why not say that any rich fragment of a language used by *P* is itself a used language?

Indeed, we will need to say such things when *P* is linguistically inhomogeneous. Suppose, for instance, that *P* divides into two classes: the learned and the vulgar. Among the learned there prevails a convention of truthfulness and trust in a language £; among *P* as a whole there does not, but there does prevail a convention of truthfulness and trust in a rich fragment £⁻ of £. We wish to say that the learned have a common language with the vulgar, but that is so only if £⁻, as well as £, counts as a language used by the learned.

Another case: the learned use £₁, the vulgar use £₂, neither is included in the other, but there is extensive overlap. Here £₁ and £₂ are to be the most inclusive languages used by the respective classes. Again we wish to say that the learned and the vulgar have a common language: in particular, the largest fragment common to £₁ and £₂. That can be so only if this largest common fragment counts as a language used by the vulgar, by the learned, and by the whole population.

I agree that we often do not count the fragments; we can speak of *the* language of *P*, meaning by this not the one and only thing that is a language used by *P*, but rather the most inclusive language used by *P*. Or we could mean something else: the union of all the languages used by substantial sub-populations of *P*, provided that some quite large fragment of this union is used by (more or less) all of *P*. Note that the union as a whole need not be used at all, in my primary sense, either by *P* or by any sub-population of *P*. Thus in my example of the last paragraph, *the* language of *P* might be taken either as the largest common fragment of £₁ and £₂ or as the union of £₁ and £₂.

Further complications arise. Suppose that half of the population of a certain town uses English, and also uses basic Welsh; while the other half uses Welsh, and also uses basic English. The most inclusive language used by the entire population is the union of basic Welsh and basic English. The union of languages used by substantial sub-populations is the union of English and Welsh, and the proviso is satisfied that some quite large fragment of this union is used by the whole population. Yet we would be reluctant to say that either of these unions is the language of the population of the town. We might say that Welsh and English are the two languages of the town, or that basic English and basic Welsh are. It is odd to call either of the two language-unions a language; though once they are called that, it is no further oddity to say that one or other of them is the language of the town. There are two considerations. First: English, or Welsh, or basic English, or basic Welsh, can be given a satisfactory unified grammar; whereas the language-unions cannot. Second: English, or Welsh, or basic Welsh, or basic English, is (in either of the senses I have explained) the language of a large population outside the town; whereas the language-unions are not. I am not sure which of the two considerations should be emphasized in saying when a language is the language of a population.

Objection: Let £ be the language of P; that is, the language that ought to count as the most inclusive language used by P. (Assume that P is linguistically homogeneous.) Let $£^+$ be obtained by adding garbage to £: some extra sentences, very long and difficult to pronounce, and hence never uttered in P, with arbitrarily chosen meanings in $£^+$. Then it seems that $£^+$ is a language used by P, which is absurd.

A sentence never uttered at all is *a fortiori* never uttered untruthfully. So truthfulness-as-usual in £ plus truthfulness-by-silence on the garbage sentences constitutes a kind of truthfulness in $£^+$; and the expectation thereof constitutes trust in $£^+$. Therefore we have a prevailing regularity of truthfulness and trust in $£^+$. This regularity qualifies as a convention in P sustained by an interest in communication.

Reply: Truthfulness-by-silence is truthfulness, and expectation thereof is expectation of truthfulness; but expectation of truthfulness-by-silence is not yet trust. Expectation of (successful) truthfulness — expectation that a given sentence will not be uttered falsely — is a necessary but not sufficient condition for trust. There is no regularity of trust in $£^+$, so far as the garbage sentences are concerned. Hence there is no convention of truthfulness and trust in $£^+$, and $£^+$ is not used by P.

For trust, one must be able to take an utterance of a sentence as evidence that the sentence is true. That is so only if one's degree of belief that the sentence will be uttered falsely is low, not only absolutely, but as a fraction of one's degree of belief — perhaps already very low — that the sentence will be uttered at all. Further, this must be so not merely because one believes in advance that the sentence is probably true: one's degree of belief that the sentence will be uttered falsely must be sub-

33

stantially lower than the product of one's degree of belief that the sentence will be uttered times one's prior degree of belief that it is false. A garbage sentence of £+ will not meet this last requirement, not even if one believes to high degrees both that it is true in £+ and that it never will be uttered.

This objection was originally made, by Stephen Schiffer, against my former view that conventions of language are conventions of truthfulness. I am inclined to think that it succeeds as a counter-example to that view. I agree that £+ is not used by P, in any reasonable sense, but I have not seen any way to avoid conceding that £+ is a possible language — it might *really* be used — and that there does prevail in P a convention of truthfulness in £+, sustained by an interest in communication. Here we have another advantage of the present account over my original one.

Objection: A sentence either is or isn't analytic in a given language, and a language either is or isn't conventionally adopted by a given population. Hence there is no way for the analytic-synthetic distinction to be unsharp. But not only can it be unsharp; it usually is, at least in cases of interest to philosophers. A sharp analytic-synthetic distinction is available only relative to particular rational reconstructions of ordinary language.

Reply: One might try to explain unsharp analyticity by a theory of degrees of convention. Conventions do admit of degree in a great many ways: by the strengths of the beliefs and desires involved, and by the fraction of exceptions to the many almost-universal quantifications in the definition of convention. But this will not help much. It is easy to imagine unsharp analyticity even in a population whose conventions of language are conventions to the highest degree in every way.

One might try to explain unsharp analyticity by recalling that we may not know whether some worlds are really possible. If a sentence is true in our language in all worlds except some worlds of doubtful possibility, then that sentence will be of doubtful analyticity. But this will not help much either. Unsharp analyticity usually seems to arise because we cannot decide whether a sentence would be true in some bizarre but clearly possible world.

A better explanation would be that our convention of language is not exactly a convention of truthfulness and trust in a single language, as I have said so far. Rather it is a convention of truthfulness and trust in whichever we please of some cluster of similar languages: languages with more or less the same sentences, and more or less the same truth-values for the sentences in worlds close to our actual world, but with increasing divergence in truth-values as we go to increasingly remote, bizarre worlds. The convention confines us to the cluster, but leaves us with indeterminacies whenever the languages of the cluster disagree. We are free to settle these indeterminacies however we like. Thus an ordinary, open-textured, imprecise language is a sort of blur of precise languages — a region, not a point, in the space of languages. Analyticity is sharp in each language of our cluster. But when different languages of our cluster disagree on the analyticity of a sentence, then that sentence is unsharply analytic among us.

Rational reconstructions have been said to be irrelevant to philosophical problems arising in ordinary, unreconstructed language. My hypothesis of conventions of truthfulness and trust in language-clusters provides a defense against this accusation. Reconstruction is not — or not always — departure from ordinary language. Rather it is selection from ordinary language: isolation of one precise language, or of a sub-cluster, out of the language-cluster wherein we have a convention of truthfulness and trust.

Objection: The thesis and the antithesis pertain to different subjects. The thesis, in which languages are regarded as semantic systems, belongs to the philosophy of artificial languages. The antithesis, in which language is regarded as part of human natural history, belongs to the philosophy of natural language.

Reply: Not so. Both accounts — just like almost any account of almost anything — can most easily be applied to simple, artificial, imaginary examples. Language-games are just as artificial as formalized calculi.

According to the theory I have presented, philosophy of language is a single subject. The thesis and antithesis have been the property of rival schools; but in fact they are complementary essential ingredients in any adequate account either of languages or of language.

35

Logic and Language:
An Examination of
Recent Criticisms of Intensionalism

To give the history of a physical principle is at the same time to make a logical analysis of it. The criticism of the intellectual processes that physics puts into play is related indissolubly to the exposition of the gradual evolution by which deduction perfects a theory and makes of it a more precise and better-ordered representation of laws revealed by observation.

Besides, the history of science alone can keep the physicist from the mad ambitions of dogmatism as well as the despair of Pyrrhonian skepticism.

Pierre Duhem

1. On the Intensionalist-Extensionalist Controversy

This paper is a defense of a version of intensionalism against certain recent objections.[1] Principle (1.1) expresses this version.[2]

(1.1)   Logical form and meaning are one and the same thing; a theory of one is a theory of the other.

Principle (1.1) implies that the relation between logic and language, as subjects of study, is that the entities or objects to which the laws of logic apply are the senses of sentences of natural language — and only indirectly the sentences themselves. From this viewpoint, logic, as a scientific discipline seeking to uncover the laws of valid inference, is concerned with stating the regularities in the behavior of semantic ob-

[1] I wish to thank the members of my seminars in the philosophy of language at M.I.T. in the spring of 1971 and the spring of 1972 for many useful discussions of the material in this essay. I also want to thank Virginia Victoria Valian for her help.

[2] This is a programmatic statement, not something that intensionalists can be expected to establish by direct argument. It is an assumption of the intensionalist approach and of the intensionalist attempt to construct a semantic theory of natural language. As such, it can only be established indirectly through the success of this approach and attempt.

jects. Semantic theory, as part of the theory of universal linguistic structure, is concerned with describing the structure of senses with sufficient depth and richness of detail to provide an account of every feature that logic requires for a complete statement of its laws. The general idea behind this approach can be put as follows: semantic theory explains why logical laws hold in terms of the semantic structure of sentences in a way similar to that in which physical theory explains macro-regularities in terms of micro-structure.[3]

This identification of senses of sentences with the objects to which the laws of logic apply may seem strange even to those who might not consider themselves extensionalists. But I think this strangeness results from an unfamiliar rather than an unacceptable way of talking: the peculiarity of such locutions as "proving (asserting, denying, etc.) the sense of a sentence" — as contrasted with the naturalness of their counterparts like "proving (asserting, denying, etc.) the proposition" — is like the peculiarity of rendering God's words at the beginning of Genesis as "Let there be electromagnetic radiation!"[4] When doing pure logic, we think of and talk about the entities to which laws of logic apply independently of their syntactic and phonetic dress in natural languages because, for purposes of such discussions, a notion of proposition which ignores linguistic connections allows us to avoid irrelevant aspects of syntax and phonology. Thus we habituate to a mode of thinking of and talking about propositions that makes the unfamiliar seem strange and unacceptable to us.

When we change perspective and focus on discussions of logical argumentation in natural language, we find that the propositions in our study of pure logic are related to sentences (of natural language) in the way that senses are related to sentences. Indeed, the correspondence is so striking that it is immediately plausible to take the situation as one calling for the theoretical identification of propositions with senses of sentences.

Consider some examples. Meaningless sentences like

(1.2)   Smelly itches think soapy memories

---

[3] Katz (1972), chap. 7, presents an example of how the semantic representation of sentences explains why a logical law holds. The example treated is why sentences formed from converse relations imply each other.

[4] I refer here to the argument in Cartwright (1962), p. 101 (repeated in various places, e.g., Pitcher (1964). A full reply to this argument can be found in Katz (1972), pp. 123–124.

37

are also taken to fail to express a proposition. That is, sentences which lack a sense are regarded as not entering into deductive connections: it is absurd to say that (1.2) implies some other proposition (not merely false). Sentences like (1.2) lack whatever it is that makes laws of logic applicable to sentences. Thus "they do not have a sense" and "they do not express a proposition" seem to state the same deficiency.

Ambiguous sentences like

(1.3)   Someone took my photograph

that is, sentences which have more than one sense, are also taken to express more than one proposition. Each separate sense of an ambiguous sentence constitutes an independent proposition for purposes of inference; that is, each is regarded as entering into distinct sets of deductive connections. Failure to treat them as entering into distinct sets of deductive connections results in (what is traditionally called) a fallacy of ambiguity. For example, the sense of (1.3) on which it means 'someone made off with a possession of mine' behaves quite differently in argumentation from the sense on which it means 'someone photographed me', and these differences reflect exactly the semantic differences between these senses. Thus a semantic difference between these senses such as that the first contrasts as in (1.4) but the second does not

(1.4)   Someone took my photograph but returned it immediately,

— i.e., (1.4) is semantically anomalous if its first clause is taken as identical with the second sense of (1.3) only[5] — directly corresponds to the logical difference that the argument from (1.3) to (1.5),

(1.5)   Someone is or was in possession of a photograph that I once had,

is valid when the premise is taken to bear the first sense but invalid when the premise is taken to bear the second sense.

Synonymous sentences like (1.3) and (1.6),

(1.6)   Someone got possession of a picture of me that was produced by a photographic process,

sentences that have a sense in common, are also taken to express the same proposition. On their shared sense, such sentences are regarded as

---

[5] With the second sense as the reading of (1.4), this sentence is meaningless in the manner of "The doctor took my temperature but returned it immediately."

functioning identically in argumentation. Synonymous sentences enter into exactly the same sets of deductive connections.

Such correspondences between semantic properties and relations and logical ones call for explanation: why is it that whenever we find meaninglessness we find that no laws of logic apply; whenever we find ambiguity we find that distinct sets of laws apply; whenever we find synonymy we find that the same laws apply; etc.? Intensionalists take this question to be of the same kind as 'Why is it that wherever we find water we find H_2O?' Accordingly, they give the same kind of answer, namely, (1.1).

Intensionalists claim that only with a level in the grammar at which senses are represented can we provide an adequate account of the aspects of sentence structure that determine the role of sentences in argumentation and provide a complete explication of the concept of a valid argument. Extensionalists deny this. The essential issue between them is thus whether the aspect of linguistic structure that makes the laws of logic applicable to sentences of natural languages can be properly explained without information about the semantic representation of sentences. Intensionalism holds that there is an independent level of semantic representation in grammars and that the facts about meaning stated at this level are indispensable to any account of the inference potentialities of sentences. Extensionalism, on the other hand, holds that logical form can be explained without appeal to meanings. Thus intensionalism bases its explanation of logical implication in natural language on the theory of meaning as well as on the theory of reference, whereas extensionalism bases its explanation simply on the theory of reference.

The theory of meaning is understood as a system of statements designed to answer a set of questions about certain relations of linguistic objects, expressions and sentences, to one another. The theory of reference is understood as a system of statements designed to answer a set of question about certain relations of linguistic objects to the world. To make these characterizations more informative, it is necessary to indicate which relations among linguistic objects are relevant to the former theory and which relations between linguistic objects and the world are relevant to the latter. Often this is done by giving examples. For instance, the relations 'x is synonymous with y', 'x is meaningful in L', 'x is ambiguous in L', and others are customarily cited as examples of the

Jerrold J. Katz

relations with which the theory of meaning concerns itself, and the relations 'x denotes y', 'x satisfies y', 'x is true of y', and others are customarily cited as examples of relations with which the theory of reference concerns itself.

In terms of this distinction between the two theories, the issue between extensionalism and intensionalism can be formulated more sharply. The extensionalist position claims that the theory of meaning is reducible to the theory of reference in the sense in which one scientific theory might reduce to another, e.g., the reduction of classical thermodynamics to statistical mechanics.[6] The extensionalist program is thus to show that the relation between the theory of meaning and the theory of reference (with perhaps some other theory such as the theory of syntax) is that of 'to-be-reduced' theory and 'reducing' theory(s), namely, that the theory of reference permits us to define the *viable* theoretical terms in the vocabulary of the theory of meaning and to derive from these definitions and the principles of the 'reducing theory' the empirical generalizations (or laws) of the 'to-be-reduced-theory'.[7]

I put the issue in terms of such a reduction because this is the most general way of stating the controversy: extensionalists include both anti-intensionalists who believe that the concepts of the theory of meaning are otiose, as well as the anti-intensionalists who believe that these concepts are viable but reducible to the concepts in the theory of reference (as well as those who believe that some semantic concepts are otiose and others reducible to the theory of reference).

This characterization takes the rejection of intensionalism (as defined above) to be central and allows wide divergence among extensionalists with respect to their reasons for this rejection. At one extreme of the extensionalist spectrum, we would find a position that takes the concepts of the theory of meaning to be disguised versions of the concepts of the theory of reference; at the other we find a position, like Quine's,

---

[6] Cf. Nagel (1961), chap. 11.

[7] Although the intensionalist position which I defend below claims that no such reduction goes through, it does not claim that there is a reduction the other way around. One might try to argue that the theory of reference reduces to the theory of meaning by contending that questions about the relations between linguistic objects and the world can be answered by analyses of the meaning of words such as "true," "refers," "names," "denotes," etc., in natural language. Intensionalism need not argue this, and I think it would be a mistake. The mistake is the same one philosophers make when they say that the meaning of "meaning," etc., in English bears directly on the theory of meaning. But this is another matter. See Katz (1972), chap. 8, sec. 5, for discussion.

40

that holds that concepts of the theory of meaning are worthless by some criterion of scientific acceptability.

2. The Roots of Quinian Philosophy of Language in Bloomfieldian Linguistics

Quine's paper "The Problem of Meaning in Linguistics"[8] and related papers of the same period have had a significant effect on recent philosophy of language. Their influence has, I think, been far greater than is generally recognized. These papers appeared at a critical moment in contemporary analytic philosophy. Prior to its appearance, philosophers of language working within the framework of analytic philosophy had essentially two approaches to the problem of meaning to choose from, one provided by ordinary language philosophy and the other by logical empiricism. The former looked at meaning as something to be understood in terms of how words are used in ordinary language, while the latter looked at it as something to be understood in terms of formal rules of inference set up in artificial languages. Both approaches were strictly philosophical, with philosophical ulterior motives (the elimination of certain forms of metaphysics). Moreover, from a scientific viewpoint, each was flawed: the former observed language without attempting to theorize about it and the latter theorized without attempting to observe it.[9]

It was one of Quine's contributions to philosophy of language to attempt to find an alternative to this choice between two unscientific approaches to a scientific subject by asking what the relevant science, in this case linguistics, had to say about meanings. He deserves credit for having been the first to become dissatisfied with the qualifications of both ordinary language philosophy and logical empiricism to pronounce on questions in the science of language. Quine was thus first to recognize the relevance of linguistics to philosophy.

But what linguistics said about meanings was what the prevailing theory at the time had to say. Since this theory was the "taxonomic theory of grammar" developed within Bloomfieldian structuralism, and since it had been worked out on the basis of the same neo-positivist ideas upon which early logical empiricism was based, the theory of language

[8] Quine (1953a), pp. 47–64; also Quine (1953b), Quine (1953c), etc.
[9] Cf. Fodor and Katz (1962) and also Katz (1966).

that Quine encountered when he turned to linguistics was strongly empiricist, behaviorist, and operationalist. What it said about meanings, therefore, was not very flattering to them.

This theory, however, must have had an enormous appeal to Quine. Not only did it enjoy the status of an undisputed, well-entrenched scientific orthodoxy, but, being empiricist, behaviorist, and operationalist, its philosophical qualifications, for Quine at least, were above reproach. For instance, compare Bloomfield's neo-positivist remarks:

Within the next generations mankind will learn that only [such terms as are translatable into the language of physical and biological science] are usable in any science. The terminology in which at present we try to speak of human affairs — the terminology of 'consciousness,' 'mind,' 'perception,' 'ideas,' and so on — in sum, the terminology of mentalism and animism — will be discarded . . .

and "If we are right, then the term 'idea' is simply a traditional obscure synonym for 'speech-form' . . ." with Quine's statement:

Now there is considerable agreement among modern linguists that the idea of an idea, the idea of the mental counterpart of a linguistic form, is worse than worthless for linguistic science. I think the behaviorists are right in holding that talk of ideas is bad business even for psychology. The evil of the idea idea is that its use . . . engenders an illusion of having explained something. And the illusion is increased by the fact that things wind up in a vague enough state to insure . . . freedom from further progress.[10]

Bloomfieldian structuralism thus easily convinced so like-minded a philosopher as Quine that he had found the scientific basis he needed to reformulate the problem of meaning in philosophy as the problem of meaning in linguistics. Taxonomic linguistics lacked only the sort of rational reconstruction that a philosopher of Quine's caliber and disposition was uniquely suited to provide.

"The Problem of Meaning in Linguistics" *is* Quine's rational reconstruction (in the philosophy of science sense) of Bloomfieldian structuralism. It improves this theory considerably, but it also reflects the theory's basic assumptions about language and mind. Because this reformulation of the problem of meaning occupies a key place for Quine's subsequent philosophizing about language and logic, these assumptions of taxonomic linguistics became assumptions of the entire extensionalist philosophy Quine was to construct, and of the work of philosophers in-

[10] See Bloomfield (1936), pp. 89, 95, and Quine (1953a), p. 48.

fluenced by it. I shall try to show that there is little or nothing to recommend these assumptions other than the taxonomic theory from which they originally came.

2.1. THE CENTRAL ASSUMPTION OF QUINIANISM

In this subsection I will argue that this is true of the central assumption of Quinianism, namely, the supposition, now widespread, that there is something wrong with the notion of meaning, that it is somehow suspect as occult or unscientific, and that we are better off not depending on it in philosophical or linguistic inquiry. I will argue that the corollary of this assumption — namely, the further supposition that a stimulus-response account of verbal behavior is preferable to a semantic account because the former is prima facie more scrutable, scientific, and progressive — also has nothing to back it up in Quine's philosophy of language other than the taxonomic theory from which it comes. Elsewhere I have tried to show that even Quine's famous thesis of the indeterminacy of radical translation, rather than establishing or supporting these assumptions, actually rests on them.[11]

By showing that these assumptions originate in Bloomfieldian structuralism and derive their entire support from this theory, we thus show that what undermines this theory of linguistic structure also undermines these assumptions. Hence, we thereby open Quinian extensionalism and other forms of extensionalism based on it to the same transformationalist criticism that deposed the taxonomic theory of language in linguistics.

Typifying Bloomfield's attitude toward meanings and ideas is:

Non-linguists (unless they happen to be physicalists) constantly forget that a speaker is making noise, and credit him, instead, with the possession of impalpable 'ideas'. It remains for linguists to show, in detail, that the speaker has no 'ideas', and that the noise is sufficient — for the speaker's words act with a trigger-effect upon the nervous systems of his speech-fellows.[12]

On the positive side, meaning for Bloomfield is best thought of not as ideas but as the stimulus features of "the situation which prompt people to utter speech" and of "the response which it calls forth in the

---

[11] See Quine (1960), p. 76, where the argument for indeterminacy comes to rest on an appeal to this assumption; I argue this in Katz (1972), pp. 283–292, and Katz (1974).

[12] Bloomfield (1936), p. 93.

hearer" (Bloomfield (1933), p. 139) — the finger on the trigger, as it were. Everyone will recognize that this conception of linguistic meaning, where stimulus-response properties of the environment replace ideas, is the same as the one Quine developed in a number of works, namely, his conception of stimulus-meaning. Compare the above remarks of Bloomfield's with Quine's remark in "The Problem of Meaning in Linguistics" (Quine (1953a), p. 60): "Synonymy of two forms is supposed vaguely to consist in an approximate likeness in the situations which evoke the two forms, and an approximate likeness in the effect of either form on the hearer." I do not mean to suggest that Quine took his conception from Bloomfield, but only that they are the same conception of how semantics should be done.

The fate of the physicalist, operationalist, and verifiability doctrines that encouraged Bloomfield to adopt the taxonomic conception of grammar is well known. Although I think their failure deprives Quine's attitude toward meaning of much of its plausibility, I am not much interested in trying to establish this here. Rather, I wish to show that Quine's central assumption seems acceptable because most philosophers still retain an essentially taxonomic conception of grammar. I want to argue that this central assumption depends completely on the Bloomfieldian conception of grammatical analysis and makes no sense outside of it. If we can show this, then the transformationalist refutation of the Bloomfieldian conception of grammar, which swept taxonomic theories out of linguistics in the early sixties,[13] refutes Quine's assumption that meanings are unacceptable entities.

To refute this assumption in this way, we must begin by looking at two things. The first is the nature of the theory of language that Quine found when he decided that a scientifically minded philosophy of language ought to consider what the science of language has to say about meaning. The second is why such a theory of language makes the notion of meaning as the idea that a word, phrase, clause, or sentence expresses seem out of place in a scientific account of the structure and use of language.

The theory of language in linguistics at this time represented the grammatical analysis of sentences as a process very similar to library cata-

---

[13] There is an enormous literature, but the following are sufficient for purposes of documenting this claim: Chomsky (1957), (1962), and (1965); also Postal (1964) and Katz (1971).

loging. The linguist's corpus of utterances, suitably recorded in the course of fieldwork, corresponds to the librarian's collection of writings and printings. Corresponding to the elements of the librarian's collection, the books, magazines, journals, letters, newspapers, etc., are the linguist's phones, the smallest segments of the utterances that represent differentiable speech sounds of the language. In both the case of library cataloging and the case of grammatical taxonomy, the classification proceeds by forming equivalence classes (appropriately labeled), at successively higher and higher levels. In grammatical analysis, at the lowest level, phones are individuated. Then they are grouped into phoneme classes on the basis of a test for assigning phones to different classes in case replacing one for the other changes one word into another. Next, phonemic stretches in utterances are classified into morphemic classes, and then word classes, which in combination are further classified into the higher syntactic categories of noun, verb, preposition, article, adverb, etc., of various phrasal and clausal types, and finally of one or another sentence type (e.g., simple-compound, declarative, interrogative, imperative, etc.). These strata of classes are not different substantively from the stratification of books, magazines, etc., into fiction and nonfiction, adult and juvenile, etc.; then into such subcategories as science-fiction, adventure, etc., science, history, art, etc.; then into such narrower groupings as mathematics, physics, chemistry, etc.; and so on down to the bibliographical equivalent of phones, namely, different editions of the same book. But, although the items and categories differ, the form of the classification is the same.

The attractiveness of such a conception of grammatical analysis is that it begins with utterances, physical disturbances in the air, which are properly material and intersubjective; each successive level of grammatical description involves nothing more than a reclassification of the components of such physical events, a re-analysis of the speech sounds described at the lowest level. This means that nothing non-physical can appear as part of the grammatical structure of sentences in such an analysis. But note that such successive analysis, from phone to sentence, also provides no place in grammatical structure for anything but sounds and relations of sounds in sequential order, no place for anything but phonological and syntactic structure. In particular, in this scheme, there is simply no place for representing the meaning of a sentence as an integral part of its grammatical structure. To put it another way: the scheme works just as well for uninterpreted, formal calculi as for natural languages. Accordingly, if one accepts this scheme for writing grammars, one must define meaning as external to grammar as is the case with Bloomfield's and Quine's definition of meaning in stimulus-response terms.

The shift from the Bloomfieldian taxonomic theory to the Chomskyian transformational theory is critical because meaning now finds a quite natural place in grammatical structure. To see why meaning can be part of grammatical structure on this theory, let us look briefly at the new features of the transformational model.

The transformational model rejects the basic claim of taxonomic theory that the elements of grammatical structure are as truly physical, public, and inter-subjective as the books of a library. Transformational theory claims that many levels of grammatical structure lie beneath the surface form of sentences and that taxonomic grammars have no means to do more than describe this surface form.[14] The reality of the structures at these "underlying" levels cannot be accounted for on the taxonomic model, which regards grammatical analysis as cataloging of the linguistic data in a corpus, but requires something akin to the physicist's theory of atomic structure on which the surface continuity of matter hides a discontinuous microstructure. To account for this underlying linguistic reality, transformational theory provides base rules and transformations that represent each underlying level of grammatical structure up to the surface, deriving the representation of each transformationally from that of the one lying just below it, until the representation of the surface form is itself derived. Thus on the transformational theory taxonomic description is treated as a special and limited case of grammatical representation.

But the rules of a grammar that define the underlying levels of the language cannot be accorded the kind of reality that is accorded to the rules of a taxonomic grammar, namely, as representations of regularities in distribution inductively generalized from the utterances in some sample of the language. Rather, the rules for representing underlying levels must be conceived of as representations of the mental principles (or brain mechanisms) inside the heads of speakers. This interpretation of grammars, as simulations of the principles by virtue of which a speaker

[14] Katz (1971).

is fluent in a natural language, immediately provides a natural place for semantic structure as part of grammatical structure. The meaning of a sentence can be taken as represented by some specific set of recursive rules for representing underlying levels of grammatical structure. These rules are in turn connected to the rules that determine the syntactic form and the phonetic shape of the sentence. They thus fit in as a stage in the sentence-generation process and as such share the psychological reality of the deeper syntactic rules. The grammar as a whole can now be thought of as a formal explication of the principles by which fluent speakers correlate sound and meaning.

Furthermore, on this view, the extensionalist can no longer take it for granted that sense or meaning are "impalpable ideas" better gotten rid of in the manner of occult notions surviving from some prescientific past. Now statements about meanings take the form of hypothetical postulations about the rules internalized by a speaker in acquiring knowledge of the grammar of a language. These postulations are employed in a theory of the grammatical competence underlying the use of sentences and lead to predictions about verbal performance that can be empirically verified, thereby confirming or disconfirming these postulations. Thus, rather than being "impalpable ideas," senses and meanings enjoy the status of hypothetical postulations in science, like the molecule or the gene. We shall return to this point in the next sections and try to show that Quinian philosophy of language and logic depend on taking it for granted that senses and meanings are "impalpable ideas."

2.2. QUINE'S UNDERESTIMATION OF THE RANGE OF SEMANTIC PROPERTIES AND RELATIONS

Not only does Quine's attitude toward the concept of meaning derive from Bloomfieldian structuralism, so does his conception of how the concepts of the theory of meaning can either be reduced to those of theories having no commitment to "occult entities" or be shown to be scientifically unacceptable. In this subsection we shall look at this aspect of the roots of Quine's anti-intensionalism.

The practice of taxonomic linguists (studying synchronic phenomena) at the time of Quine's survey divided into grammar and lexicography. The former concerned the syntactic organization and phonemic constitution of utterances and was carried out in the manner described above in our brief sketch of taxonomic grammar. The latter concerned problems of dictionary construction, and the conception of a dictionary was that of an ordinary reference or desk dictionary. Accordingly, when Quine looked to see where linguists employed the term "meaning" in their actual practice in order to determine what semantic concepts a reduction would have to handle, he came up with only two cases. First, there was the lexicographer's concept of "same in meaning" or "synonymous" used in discussions of dictionary entries, and second, the grammarian's concept of "grammatical," "significant sentence," "having a meaning," etc. On the basis of finding no other cases where linguists employed "meaning" or some other semantic notion in their scientific practice, Quine concluded: "What had been the problem of meaning boils down now to a pair of problems in which meaning is best not mentioned; one is the problem of making sense of the notion of significant sequence, and the other is the problem of making sense of the notion of synonymy."[15]

There is no argument to establish the claim that the problem of meaning reduces to these two and *only* these two cases. Quine simply assumes that the practice of linguists (undergoing rational reconstruction) is adequate in the required sense, namely, the taxonomic linguist's use of "meaning" reflects the role of this concept in a fully worked out, empirically valid theory of grammatical structure. Thus the assumption has no more to recommend it than whatever there is to recommend the Bloomfieldian framework which provided the guidelines for the practice of linguists at the time. Again, Quine assumes that what he found in linguistics could be relied on as a basis for his philosophical critique of intensionalism.

Quine's restriction of the explanatory concepts of the theory of meaning to significance and synonymy seriously underestimates the actual range of semantic properties and relations that a reduction would have to account for. To see how seriously, we should now ask whether there is any general reason for thinking that there are legitimate semantic properties and relations excluded by this restriction. Suppose one were to start as Quine does with the objective of providing a reduction of the theory of meaning, *but* without Quine's bias in favor of thinking the practice of Bloomfieldian linguists reliable. Question: would one

---

[15] Quine (1953a), p. 49.

come to Quine's conclusion that the semantic concepts that needed to be considered to achieve this objective are only those of significance and synonymy? I think not. One would quite naturally begin with an attempt to frame some general principle to circumscribe the domain of semantics, in order to be sure that no relevant property or relation is omitted. Thus one would reason somewhat as follows. Semantics, if it is about anything, is clearly about whatever answers the question 'What is meaning?'[16] A semantic theory would thus tell us what meaning is and anything that is part of this answer is part of semantic theory. Accordingly, if the question 'What is meaning?' breaks down into a set of more particular questions — as big questions in other sciences tend to do — then the concepts asked about in these special cases of the basic question will qualify as part of semantics. These concepts will be the ones that a semantic theory will have to provide definitions for, which explain the nature of the concept. Conversely, then, these concepts will be the ones that a reduction of semantics will have to handle.

The two such questions that were recognized in Bloomfieldian linguistics were:

(2.1)   What is the property of having a meaning (in L)?
(2.2)   What is the relation of sameness of meaning (in L or between L_i and L_j)?

If no antecedent philosophical or linguistic bias stops the enumeration at this point, a number of other questions, and hence properties and relations, must be accepted as equally genuine. For instance:

(2.3)   What is multiplicity of meaning or ambiguity?
(2.4)   What is similarity and difference in meaning?
(2.5)   What is analyticity?
(2.6)   What is contradictoriness?
(2.7)   What is entailment by virtue of meaning?
(2.8)   What is semantic redundancy?
(2.9)   What is antonymy?
(2.10)   What is superordination?

Before we illustrate the concepts asked about in these questions, we

---

[16] For an abstract account of sciences in terms of the questions with which they are concerned, see Bromberger (1963), and for further discussion of the approach taken here to circumscribing semantics see Katz (1972), chap. 1.

should note that the ones considered here are not the only ones that could be mentioned.[17] Nor should we expect that even a complete list of those we could now mention would exhaust the range of semantic properties and relations that an optimal semantic theory ought to define. As Fodor observes: "a science has to discover what it is about: it does so by discovering that the laws and concepts it produced in order to explain one set of phenomena can be fruitfully applied to phenomena of other sorts as well."[18]

Question (2.1) asks about the concept that explains the difference between cases like (1.2) and (1.3). Question (2.2) asks about the one that explains the relation between (1.3) and (1.6). Question (2.3) asks about the concept that explains the property of (1.3) by virtue of which it is synonymous (in a sense) with both (1.6) and (2.11):

(2.11)   Someone got possession of a photograph of mine.

Question (2.4) asks about the concept that explains the semantic similarities of the words in

(2.12)   nun, aunt, cow, sister, actress, . . .

and the dissimilarity of the words in

(2.13)   house, pain, reflection, truth, lap, . . .

Question (2.5) asks about the concept that explains why declarative sentences like

(2.14)   Nudes are naked

make only true statements. Question (2.6) asks about the concept that explains the property found in

(2.15)   My uncle made a complete recovery from his fatal illness

in the sense where the possessive pronoun is coreferential with the subject. Question (2.7) asks about the concept that explains the consequence relation in the argument from (2.16) to (2.17):

(2.16)   Socrates had a nightmare
(2.17)   Socrates had a dream

---

[17] Examples of other semantic concepts that might have been mentioned and discussed if space had permitted are 'conversion triple', 'suppressed sense', 'self-answered question', 'evasion of a question', etc. See Katz (1972).
[18] Fodor (1968), pp. 10–11.

Question (2.8) asks about the concept that explains the superfluousness of the modifier in

(2.18)   kings and queens who are monarchs

Question (2.9) asks about the concept that explains the relations between pairs like those in

(2.19)   open/close, whisper/shout, high/low, . . .

Question (2.10) asks about the concept that explains the relations between pairs like those in

(2.20)   finger/thumb, dwelling/cottage, human/child, . . .

Admittedly there is much room for argument about whether this or that concept in (2.1) to (2.10) is really semantic and also about whether this or that concept is definable in terms of others. Such questions, together with the further question of whether there are still other properties and relations that belong on the list, are ones for theory construction to settle. Nonetheless, at the very least, some such fuller, more realistic initial account of the scope of a theory of meaning ought to be considered by anyone seriously contemplating the reduction of this theory to one that makes no use of intensional notions.[19]

[19] It might be replied: "How do you know that the semantic properties and relations beyond significance and synonymy, those referred to in (2.1) to (2.10) and others that you foresee being added, are not reducible to (definable in terms of) the former two? If so, and if Quine is right in claiming to have discredited synonymy and significance, then all these other semantic properties and relations are discredited, too." This begins with the true supposition that we know little about whether further semantic properties and relations are definable in terms of significance and synonymy but draws the wrong moral from it. In advance of the necessary empirical investigation, we cannot know much about their definability. But in the discussion above I didn't purport to establish their irreducibility. My purpose was to expose the hidden assumption of the Quinian program: that significance and synonymy are the only semantic properties and relations that a reductionist needs to consider. Thus the reply concedes my point that Quine presents no argument to show all other semantic properties and relations are definable in terms of synonymy and significance. Therefore the Quinian attack on the theory of meaning lacks an essential step: it fails to account for an indefinitely large number of semantic properties and relations that have as good a claim to the status of semantic concepts as synonymy and significance and are thus equally relevant to the question of reductive elimination. Moreover, in the one case where Quine offers such a definition, Quine's definition of analyticity in terms of the substitution of synonyms in a logical truth, his definition fails because it does not capture the full notion of analyticity that emerges from an empirical study of natural language; cf. Katz (1972), chap. 4.

Jerrold J. Katz

2.3. MEANINGFULNESS

On the basis of his assumption that the only two semantic concepts with which a reduction must deal are meaningfulness and synonymy, Quine proceeds[20] to try to eliminate the theory of meaning by showing that both these concepts can be accounted for without a commitment to meanings. That is, he tries to show that neither of the uses of "meaning" found in the practice of Bloomfieldian linguistics — neither "has a meaning" nor "same in meaning" — imply a commitment to intensionalism. In the present subsection we will examine Quine's handling of the former use; in the next we will look at his handling of the latter.

Quine tries to handle "has a meaning" by the device of assimilating the concept of meaningfulness to that of well-formedness or grammaticality. Quine writes that we should

treat the context 'having a meaning' in the spirit of a single word, 'significant', and continue to turn our backs on the supposititious entities called meanings. . . .
Significance is the trait with respect to which the subject matter of linguistics is studied by the grammarian. The grammarian catalogues short forms and works out the laws of their concatenation . . .
[The problem of significant sequence] is one of the two aspects into which the problem of meaning seemed to resolve, namely, the aspect of having a meaning. The fact that this aspect of the problem is in such halfway tolerable shape accounts, no doubt, for the tendency to think of grammar as a formal, nonsemantical part of linguistics.[21]

Quine can assume that the explication of 'significance' can be so pursued without the danger of appeals to irreducible semantic notions because he supposes well-formedness is safely in the hands of taxonomic linguists like Bloomfield, Bloch, and Trager (these being the only linguists whose work is cited in "The Problem of Meaning in Linguistics").

Quine's discussion of significance is one of the high points in his rational reconstruction of Bloomfieldian linguistics: it provides a clearer, more systematic, and more philosophically sound formulation of the taxonomic conception of grammaticality than anything available in linguistics at the time. The conclusion Quine arrives at is that the notion of significant sequence is describable as applying to any sequence of morphemes that could be uttered in the language community without occasioning a bizarreness reaction on the part of native speakers. But

[20] In Quine (1953b), (1953c), (1960).
[21] Quine (1953a), pp. 49–56.

52

this explicatum for "having a meaning" is question-begging.

The explicatum will provide a noncircular reduction of the concept of meaningfulness to that of well-formedness or grammaticality only if the bizarreness reactions in question are homogeneous, that is, only if there is just one kind of linguistic bizarreness at which speakers raise an eyebrow. If such reactions to utterances are heterogeneous, that is, if these reactions ambiguously reflect both (i) judgments about the ill-formedness of concatenations of morphemes due to violations of the system of syntactic constraints and (ii) independent judgments about the conceptual incoherence of well-formed concatenations stemming from violations of semantic relations, then the attempt to reconstruct the use of "having a meaning" will break down because it involves an unexplained semantic commitment. Since there is no a priori reason to think that these reactions of speakers must be expressions of homogeneous intuitions of syntactic irregularity and no argument (either by Quine or by the linguists he is reconstructing) to establish this, there is no reason to think that there is not semantical deviance in addition to syntactic irregularity.[22]

Let us approach the question begged in Quine's account of meaningfulness from another angle. In the study of artificial, formalized languages, we distinguish between the *uninterpreted calculus* and its *semantic interpretation*. The former consists of a set of primitive symbols, a set of formation rules, a set of inference rules, and a set of axioms. The first of these, the primitive symbols, specify the formal objects that may be used to make formulas. The second tells us which of the formulas are well-formed (and which ill-formed). The well-formed formulas are the sentences, that is, the strings upon which calculations can be performed. The third of these specifications tells us what calculations can be performed, and the fourth tells us which of the well-formed formulas can initiate derivations. Thus the third and the fourth together determine a subset of the well-formed formulas, the theorems. But up to this point the notions 'sentence', 'derivation', and 'theorem' are purely formal. They have no meaning but what the rules of the calculus confer on them; they say nothing about anything.

Once an interpretation is supplied, the calculus becomes a formalized language. The sentences now make assertions about some domain

[22] And perhaps other forms of deviance such as phonological queerness, situational incongruity, etc.

53

and collectively constitute the full range of possible assertions about the domain that can be made using the formalized language. The derivations state the logical relations between sentences, and the theorems state the things about the domain that the language itself asserts as truths.

These formalized languages are artificial in the sense that they are human artifacts. Because we make them with some idea of their use, we have a well-developed idea of their semantic interpretation at the time their primitive symbols and formation rules are laid down. Thus normally logicians do not allow situations to occur where the semantic interpretation does not "properly fit," that is, where some of the sentences receive no interpretation and hence say nothing about the domain. But this could, in principle, happen. If we choose, we can construct such ill-fitting semantic interpretations by deliberately extending one or another calculus or restricting one or another standard semantic interpretation. In such an artificial language, we would have both ill-formed formulas, i.e., non-sentences, and meaningless well-formed sentences, i.e., sentences that say nothing whatever. Let us call a language "interpretationally incomplete" if there is both a set of ill-formed formulas and a set of interpretationless well-formed formulas.

The question we want to look at can now be stated as follows. Are there *natural* languages that are interpretationally incomplete? Is English interpretationally incomplete? English has a primitive basis of letters, punctuation signs, etc. It has formation rules in the form of syntactic and phonological principles of grammar and some interpretation under which sentences of English say something. Does this interpretation "fit properly" as a normal interpretation of the calculus of an artificial language does? Quine's assumption that "having a meaning" can be assimilated to the grammarian's notion of "well-formed (grammatical) string" is acceptable only if English and other natural languages are interpretationally complete.

Not only is this a non-trivial question, it is one where clear-cut evidence can be presented against an affirmative answer. Without making any claim to decide the issue, we can provide some idea of the nature of this evidence. Recent studies in transformational grammar offer examples of selectional relations that are purely semantic and others that are purely syntactic. On the former side, McCawley has shown that the selection restriction on noun phrase objects of the verb "count" is se-

54

mantic.[23] He considers the cases

(2.21)   I counted the boy
(2.22)   I counted the boys
(2.23)   I counted the crowd

where the verb is taken in the sense "determine the number of things in the collection" rather than in the sense "include in an estimation." He argues, quite plausibly, that the restriction cannot be the syntactic requirement that the direct object of "count" be plural in number, since, although (2.21) is deviant and (2.22) is not, as would be predicted on this hypothesis, (2.23) is nondeviant. McCawley thus concludes that the restriction must be the semantic requirement that the object's sense contain information that fixes its denotation as a set of things.

Alternatively, I have argued that a purely syntactic restriction on the occurrence of expressions in the sentential frame

(2.24)   A revolution is —— to occur

permits the grammar to mark the deviance of a sentence like (2.25) and to mark a sentence like (2.26) as nondeviant.[24]

(2.25)   A revolution is probable to occur
(2.26)   A revolution is likely to occur

The reasoning is as follows. There is a feature assigned to "probable" which prevents the application of the pronoun-replacement transformation. "Likely" is oppositely marked to allow pronoun replacement to apply. If this feature distinction also carried semantic information, then there would be a semantic distinction as well between "probable" and "likely." Therefore, if the feature distinction that determines the selectional relation in this case were to be semantic in some degree, it would follow that in sentences like

(2.27)   It is probable that a revolution will occur
(2.28)   It is likely that a revolution will occur
(2.29)   That a revolution will occur is probable
(2.30)   That a revolution will occur is likely
(2.31)   The occurrence of a revolution is probable
(2.32)   The occurrence of a revolution is likely

[23] McCawley (1968).
[24] Katz (1972), pp. 373–374.

"likely" and "probable" are nonsynonymous, their senses being different, at least, by the semantic information carried by the feature controlling pronoun replacement. But these occurrences of "likely" and "probable" are synonymous, and therefore this feature and the restriction it is used to state must be nonsemantic, i.e., purely syntactic.

Therefore, there are empirical grounds for thinking that English contains a set of sentences that are ill formed grammatically by virtue of their violating a purely syntactic selectional restriction and that English also contains a disjoint set of sentences that are well formed (they violate none of the former restrictions) but exhibit some degree of semantic deviance, are in some respect meaningless, by virtue of violating a semantic restriction on sense associations. If a natural language like English is interpretationally incomplete in the sense of these examples, then an acknowledged scientific use of the term "meaning" is left unexplained in Quine's attempt at reduction. Hence Quine's assimilation of "having a sense" to "well-formed (grammatical string)" begs an important question.

2.4. SYNONYMY

Quine's attempt to show that the other use of "meaning" (in the locution "same in meaning") can also be handled (and with it related notions like "analyticity") comprises the remainder of "The Problem of Meaning in Linguistics" and much of his subsequent work in the philosophy of language.

Quine's arguments against the possibility of an acceptably clear notion of synonymy in "The Problem of Meaning in Linguistics" are the only arguments of his that apply beyond particular proposals for explicating synonymy, that is, to any attempt to explicate it.[25] Quine's famous indeterminacy thesis, the thesis that there is no right choice of a radical translation no matter what the empirical evidence, itself assumes there are no such things as meanings, no propositions in the sense of whatever sentences express in common by virtue of which they are synonymous. As I have shown elsewhere, Quine's argument for the indeterminacy thesis comes to rest on the claim that one would not be

[25] For example, Quine's criticisms in (1953b) against the proposals in Carnap (1956) do not offer general reasons to reject synonymy and analyticity but only reasons to reject Carnap's explications of them. Cf. Katz (1972), chap. 6, and Katz (1974).

able to translate highly theoretical sentences like

(2.33)   Neutrinos lack mass

into "jungle language."[26] The basic reason Quine gives for their un-translatability is that there are no propositions, no language-independent, intercultural meanings. The translator of the "jungle language," faced with a sentence like (2.33) may, according to Quine, be expected to plead in extenuation that the natives lack the requisite concepts; also that they know too little physics. And he is right, except for the hint of there being some free-floating, linguistically neutral meaning which we capture, in 'Neutrinos lack mass', and the native cannot. . . . The discontinuity of radical translation tries our meanings: really sets them over against their verbal embodiments, or, more typically, finds nothing there.[27]

Furthermore, Quine's often-made claim that "we could equate a native expression with any of the disparate English terms 'rabbit', 'rabbit stage', 'undetached rabbit part', etc., and still, by compensatorily juggling the translation of numerical identity and associated particles, preserve conformity to stimulus meanings of occasion sentences"[28] also assumes there are no meanings. This claim about the existence of alternative, nonsynonymous translations might be thought easily refuted by the reply that such alternatives merely show the limitations of stimulus-response theory and that a more sophisticated theory would provide us with a conception of linguistic evidence on which the translation of "Gavagai" as "undetached rabbit part" would be a mistake. Perhaps such a conception would reveal the mistake in the form of higher confirmation for the opposite hypothesis that "Gavagai" means "Rabbit" (or in greater overall simplicity for the system containing the latter hypothesis ceteris paribus). The reply could spell out how the required linguistic evidence might be obtained in the following way. The field linguist might ask a bilingual native informant questions like 'Does "Gavagai" bear the semantic relation to "rabbit" that the first members of the following pairs

(2.34)   branch/tree, arm/body, roof/house, thumb/hand, . . .

bear to the second members?' An affirmative answer by the native informant gives evidence for one of the hypotheses under consideration,

[26] Cf. Katz (1972), chap. 6.
[27] Quine (1960), p. 76.
[28] Ibid., p. 54.

and a negative answer gives evidence for the other. To avoid this easy refutation, Quine must claim that such evidence is illegitimate on the grounds that questions of this kind ask about intuitions concerning semantic relations (such as the part-whole relation).

However, this further claim is tantamount to the classical skeptic's first refusing to accept the criteria for deciding questions in a domain and then arguing that there is no way to settle them, and so no way to justify knowledge claims in the domain. If there were a unique "right translation," the linguist could no more obtain it without the use of semantic relations than a scientist could obtain the true hypothesis without the use of inductive criteria. Synonymy is identity in meaning and two expressions or sentences are identical in meaning in case they have exactly the same semantic properties and relations. Thus disqualifying the use of semantic properties and relations a priori guarantees there can be no evidence to decide when one has chosen the right translations. Accordingly the a priori claim that there will always be nonsynonymous sentences available as equally good choices of a translation for a native sentence on the totality of dispositions to verbal behavior (including verbal behavior that comments on the semantic structure of other verbal behavior) on the part of speakers (including bilinguals) is again the claim that there is no appropriate range of language-independent meanings to be the bearers of such properties and relations and for native bilinguals to have intuitions about.

Therefore far more rests on Quine's argument against an acceptably clear formulation of the concept of synonymy than may seem to at first glance. This argument constitutes the only general criticism Quine gives of the synonymy relation. For Quine, it is critical to undermine belief in the possibility of such a formulation of this relation because the synonymy relation can provide intensionalists with a criterion for propositional identity.

The very terms in which Quine states the problem of synonymy in "The Problem of Meaning in Linguistics" reflect the Bloomfieldian framework. The conditions he adopts for a satisfactory clarification of the concept of synonymy are those that Bloomfieldian structuralists used to try to explicate concepts such as 'same phoneme' or 'same part of speech'. Quine himself is explicit on this: "So-called substitution criteria, or conditions of interchangeability, have in one form or another played central roles in modern grammar. For the synonymy problem of sentences such an approach seems more obvious still."[29] Thus, whereas Bloomfieldians tried to determine the phonemes of a language by asking questions such as 'Is this phone substitutable for that one without changing the word into another (in the manner in which such a substitution changes the word "butt" to "putt")?', Quine takes it as obvious that we ought to be able to determine synonymy relations by parallel questions at the semantic level.

But, as Quine easily shows, formulating such parallel questions in order to provide an adequate substitution criterion for synonymy does not prove easy. We are required to provide a test that counts two linguistic forms as synonymous if, and only if, one of them can be substituted for the other in a certain set of contexts leaving some special feature unchanged. Hence, any proposal of this kind for clarifying the concept of synonymy must answer two questions: first, what set of contexts should be used; and second, what features have to remain invariant under substitution of synonymous expressions. These questions were the cutting edge of Quine's criticism that the concepts of synonymy and analyticity are incapable of receiving clarification: neither, according to Quine, could be answered without something like a vicious circle. The alternatives, he argued, were either (i) a set of contexts including adverbs like "necessarily" and a feature such as 'interchangeability *salva veritate*' or (ii) a set of contexts without such terms and a feature such as 'interchangeability *salva analyticitate*'. But both presuppose "that we have already made satisfactory sense of 'analytic'. Then what are we so hard at work on right now?"[30]

This criticism persuaded many philosophers that it is the concepts of synonymy and analyticity that are at fault. This, I think, was because they were prepared, antecedently, to go along with the terms in which Quine set the problem. Now knowing the history of substitution tests in linguistics, and supposing on the authority of that science as transmitted in Quine's rational reconstruction that such tests are the proper means of clarifying a linguistic concept, they could do little else but accept Quine's diagnosis.

Once, however, we become skeptical about the authority of Bloomfieldian linguistics and recognize its connection with the framework Quine has set up for dealing with synonymy, we see that these difficulties

[29] Quine (1953a), p. 56.
[30] Quine (1963), p. 30.

Jerrold J. Katz

can be attributed equally well to the Bloomfieldian paradigm for clarifying concepts. Thus we can claim that the use of such distributional tests to determine the extension of these linguistic concepts is responsible for the difficulty of obtaining the right set of contexts and a noncircular specification of the features that remain invariant under substitution. The situation, on this viewpoint, would be like first requiring that the relation 'is the same number as' be determined by the test that two numbers N_i and N_j are the same in case $X + N_j + Y$ preserves the feature F of $X + N_i + Y$, and then arguing that the absolute (i.e., non-graded) concept 'is the same number as' cannot be made sense of because the only way in which to provide an adequate choice of F is to (circularly!) use 'is the same number as $X + N_i + Y$' or some trivial equivalent. Quine's argument would thus prove too much.

It does no good to reply here that we can talk a la Russell and Whitehead numbers as sets and explicate 'is the same number as' in terms of the notion of a 1–1 correspondence between sets. The counter-part of this reply is of no help to Quine in sustaining his criticism of synonymy. It further undermines the criticism because the counter-part is the suggestion that we can, proceeding in analogy to set theoretic accounts of arithmetic, frame a *theory* within which to explicate meanings in terms of constructs of the theory (just as numbers are explicated in terms of sets). Then, we can define 'same meaning as' in terms of an equivalence relation between such constructs (just as 'same number as' is defined in terms of the notion of a 1–1 correspondence between sets). Quine has no reply to this alternative. He cannot say that such a theory doesn't exist or that he doubts that it could because the question here is simply one of the possibility of semantics. The existence of this alternative way of making sense of the synonymy relation shows the fallaciousness of Quine's argument against an intensionalist notion of synonymy just as much as does either the fact that his argument rests on a discredited notion of the appropriate criteria for linguistic concepts or the fact that it proves too much.

Indeed, the changeover from Bloomfieldian structuralism to transformational theory directly suggests such an alternative way of dealing with the problem of explicating synonymy (and analyticity, etc.). We can try to construct a system of formal rules, as part of the grammar, as a simulation of the speaker's semantic competence: these rules will assign representations of the meaning of sentences as part of their grammatical description; semantic theory, as part of linguistic theory, will contain a definition of synonymy that says two sentences are synonymous just in case they are assigned the same semantic representation. The definition of synonymy and the definitions of other semantic properties and relations, e.g., analyticity, meaningfulness, ambiguity, etc., would be interrelated in the theory and in certain cases interdefinable too. This would not be a vicious circle, but a virtuous one, representing a typical case of the kind of interrelations among concepts that a scientific theory affords. The assignment of the same semantic representation to certain linguistic forms would not only say that they are synonymous, but by virtue of what shared conceptual content they are synonymous, since it is this information that is expressed formally in semantic representations. Accordingly, the empirical adequacy of such assignments can be determined by how well they predict the semantic properties and relations of the sentences and expressions. These predictions can be confirmed or disconfirmed by consulting the linguistic intuitions of native speakers about the same sentences and expressions.

Thus, instead of a fruitless search for an empirically sound criterion of synonymy framed in terms of an appropriate set of contexts and a feature to use as the test which is co-extensive with synonymy but is not semantically related, we can construct a system of formal semantic rules for determining the semantic properties and relations of sentences, as part of the theory of their grammatical structure. The empirical adequacy of various claims about such rules can be determined in the same way that we judge other grammatical rules or hypotheses of other theories. This option is not possible in taxonomic grammars, but then taxonomic theory has been replaced by transformational theory and transformational grammars do permit us to include meaning as part of grammatical structure.

The preceding exploration of Quine's attack on the concept of meaning and his extensionalism shows them to have their roots in the taxonomic theory of grammar and the practice of Bloomfieldian linguistics. The cloud of suspicion that now hangs over the concept of meaning is thus an anachronism. It exists only because the relevance of events that revolutionized linguistics in the course of the last decade or so have not as yet been fully appreciated in philosophy. The transformationalist revolution overthrew the fundamental assumption of taxonomic theory that all properties required to explain grammatical phenomena are present in surface structure. This forced a change in the conception of the nature of grammar from the taxonomic conception of a grammar as a catalog of linguistic data to the transformationalist conception of a grammar as a simulation of the internalized rules underlying the speaker's fluency. Since the taxonomic conception and its extreme physicalism were the only things that made it reasonable to think that meaning is not part of grammatical structure, the relevance of the transformational revolution to semantics is that there is no longer any basis for regarding the concept of meaning with suspicion. Thus the overthrow of Bloomfieldian structuralism and the establishment of the transformational theory of deep structure and its mentalistic conception of grammars make it possible to take a more critical look at some recent versions of extensionalism and at their criticisms of intensionalism.

Quinian extensionalism has a positive side as well as a negative one. So far we have been concerned exclusively with the latter, its skeptical arguments against the notion of meaning. But the positive side is equally important, since every position has to offer some theory of its own to make sense of the phenomena that it criticizes its opponents for incorrectly handling. Quinian extensionalists could not expect their position to be accepted for very long if this constructive task were ignored. Philosophers and linguists might remain convinced for a while that there is no analytic-synthetic distinction, no relation of synonymy, no concept of meaninglessness apart from ill-formedness, etc., but sooner or later acceptance of these negative claims would be undermined by the absence in the extensionalist position of any treatment of the linguistic behavior that first prompted intensionalists to invoke the notions of analyticity, synonymy, meaninglessness, etc. Sooner or later the desire to understand nature would bring about a critical re-examination of the skeptical arguments against meaning. Thus, the positive side of extensionalism supplements the negative side in a significant way: it tries to buttress the claim that Quine's skeptical arguments do not deny intuitively obvious linguistic phenomena, but only criticize certain theories about them. From the intensionalist viewpoint, then, it is important to show that extensionalist attempts to make sense of these phenomena fail.

The extensionalist position is flexible to some extent in the kind of account it can give of such linguistic phenomena. It can accommodate a number of different accounts so long as they conform to Quine's restrictions on acceptable linguistic apparatus. Quine, as we know, turned to stimulus-response theory in psychology to obtain the apparatus on which to base his account. Other extensionalists, many of whom are no more sympathetic to stimulus-response psychology than Quine's critics, have chosen other apparatus for their own constructive efforts. In the remainder of this essay I shall consider some of the more prominent Quinian extensionalists who see the constructive task before extensionalism as better founded on something other than stimulus-response psychology. I will try to present the intensionalist reply to their views. I will not consider Quine's own attempt to use stimulus-response theory, adapted from Skinner,[31] to frame an account of linguistic behavior because the criticism of this position is already available.[32]

In examining some of the more recent trends in Quinian extensionalism, we shall consider, first Davidson, who bases his form of extensionalism on Tarski's theory of truth, and later Putnam and Kripke, who base theirs on concepts about the essential nature of the things in the world as developed in empirical science (which they see replacing the meanings of intensionalism). It is one of the major themes of this study that the work of such philosophers represents far more the working out of a set of possible variations within the extensionalist approach than a set of unrelated approaches to questions about logic and language. It is one of the main claims of this study that their work, as a consequence of its dependence on the Quinian framework, rests to some extent, too, on Bloomfieldian assumptions about the nature of grammar.[33]

3. On Trying to Work Within the System: Davidson's Approach

Davidson's recent proposal concerning the development of a Tarskian theory of truth is an attempt to work within Quine's extensionalist system. Davidson shares Quine's desire to show that the theory of meaning as intentionalists conceive it is dispensable in the study of the logical structure of natural languages. Moreover, he carries over Quine's under-

---

[31] Cf. Skinner (1957).

[32] Cf. Chomsky (1959) and Chomsky (1968) for Chomsky's criticisms of Quine's conception of stimulus meaning.

[33] Harman represents a good example of how some of the Quinian arguments might be restated without the Bloomfieldian framework (cf. Harman (1973)). It is easy to show, nonetheless, that Harman's versions beg essentially the same questions as Quine's (cf. Katz (1974)).

estimation of intensional semantics. Working within the framework of Quine's rational reconstruction of Bloomfieldian structuralism, Davidson takes it for granted that there are only two notions from the theory of meaning that need to be handled: meaningfulness and synonymy. He takes the former to be something that can be safely handed over to syntax: "The main job of a modest syntax is to characterize *meaningfulness* (or sentencehood). We may have as much confidence in the correctness of such a characterization as we have in the representativeness of our sample and our ability to say when particular expressions are meaningful (sentences)."[34] The point of Davidson's proposal is to eliminate synonymy in favor of an appropriate extensional relation to use in connection with semantic explication.

Davidson adds his own underestimation of intensional semantics. This has to do with the compositional mechanism of sense combination that intensionalists from Frege on have believed necessary to explain the semantic properties and relations of syntactically complex constituents in terms of the grammar of their parts. In particular, Davidson claims that the concept of sense (or meaning) does no work in explaining how a representation of logical form can be assigned to a sentence on the basis of the syntactic structure of the sentence and the semantic structure of its constituents.

Davidson says that an intensionalist, or Fregean, answer to a request for the meaning or logical form of a sentence like

(3.1)   Theaetetus flies

"might go something like this: given the meaning of 'Theaetetus' as argument, the meaning of 'flies' yields the meaning of 'Theaetetus flies' as value. The vacuity of this answer is obvious. We wanted to know what the meaning of 'Theaetetus flies' is; it is no progress to be told that it is the meaning of 'Theaetetus flies'. This much we knew before any theory was in sight. In the bogus account just given, talk of the structure of the sentence and of the meanings of words was idle, for it played no role in producing the given description of the meaning of the sentence."[35]

There are two reasons for concluding that this is a caricature of a Fregean answer. First, the extreme simplicity of Davidson's example

---

[34] Davidson (1967), p. 308.
[35] *Ibid.*, pp. 306–307.

makes it highly implausible that his claim about the vacuity of an intensionalist analysis of (3.1) can be generalized to other, more typically complex, sentences of the language. Davidson's argument against a Frege-type semantic theory requires that the example be typical of the cases where such a theory appeals to the meanings of the constituents of sentences to account for their meaning compositionally. If the example is not typical, an intensionalist can accept what Davidson says about it as true of such extremely simple and easy-to-understand sentences, but argue that the value of the theory only reveals itself in the treatment of more complex and difficult sentences. Nothing that Davidson says supports an inference that the same criticism applies in connection with sentences of English generally. Davidson's argument is thus no better than an argument for the triviality of arithmetic that claims that for arithmetic to inform us that zero plus one is one is saying no more than what we already know on the basis of common experience with adding something and nothing.

Second, even if the example were not too simple, Davidson's criticism would fail because the semantic analysis on which the criticism is based is radically incomplete in a manner that directly undermines the criticism. A complete semantic analysis of (3.1), simple as this example is, would actually need to refer to a great deal of grammatical structure and compositional combination of meanings. No indication of this appears in Davidson's treatment only because he takes it for granted that "Theaetetus" in (3.1) can be straightforwardly taken as an argument and "flies" in (3.1) can be straightforwardly interpreted as a function. Taking this for granted begs the question at issue, since it is exactly in this area of imposing a canonical paraphrase on a sentence from a natural language that grammatical structure and word meanings play their role. Hence, by leaving out consideration of how we know that this is the correct paraphrase, it can be made to appear as if grammatical structure and word meaning play no role in determining the meaning of (3.1).

To see that grammatical structure and word meaning do play a critical role in this aspect of the compositional meaning of (3.1), it suffices to note that the following facts are left unexplained on Davidson's account of the sentence. First, the sentence cannot be rendered other than as Davidson renders it. It cannot, for example, be rendered the other way around, with "flies" as the argument and "Theaetetus" as the function. Nor can it be rendered in analogy to cases such as (3.2):

(3.2)   (a) Tsetse flies
         (b) Time flies (!)[36]

Second, (3.1) is semantically ambiguous over such senses as:

(3.3)   (a) Someone (or thing) moves through the air
         (b) Someone (or thing) flees from danger
         (c) Someone (or thing) moves swiftly

Third, (3.1) semantically entails (i.e., entails by virtue of meaning alone[37]) a sentence such as

(3.4)   Theaetetus moves while suspended above ground

in sense (a) of (3.3) but not in senses (b) or (c) of (3.3). Fourth, the occurrence of the verb in (3.1) has a suppressed sense, roughly synonymous with the phrase (3.5),[38]

(3.5)   is rapidly used up

and so (3.1) contrasts with a case like the declarative in (3.2) (b). Fifth, (3.1) is not analytic in a sense in the manner of (3.6)

(3.6)   What flies is rapidly used up

but is synthetic on each of its senses. The reason grammatical structure and word meaning seem to "play no role in producing the given description of the meaning of the sentence" is that Davidson's description fails to mention their role.

We pointed out in the previous section that the Bloomfieldian framework makes it natural to look to stimulus-response psychology for apparatus to explain the phenomena that intensionalists try to explain with the theory of meaning. But this is not the only place where extensionalists could look for such apparatus. Quine himself suggests another in comparisons between the theory of meaning and the theory of reference. The notions of the latter theory, he says,

are so very much less foggy and mysterious than the notions belonging

---

[36] Note that the ambiguity of this case, between a declarative and an imperative sense, raises the further question of why (3.1) is not ambiguous in this way. I am indebted to Mr. R. M. Harnish for this observation.

[37] This notion is explicated in Katz (1972), chap. 4.

[38] Cf. Katz (1972), chap. 3, sec. 7. Roughly, a sense of a constituent C is a suppressed sense in a sentence S just in case it is a sense of C but not one that contributes to the content of any of the senses of S. (Selection restrictions "suppress" the sense in that they prevent it from entering into some higher combination with other senses of S.)

to the theory of meaning. We have general paradigms ['——' is true-in-L if and only if——, etc.] which, though they are not definitions, yet serve to endow much clarity, in any particular application, as is enjoyed by the particular expressions of L to which we apply them. Attribution of truth in particular to 'Snow is white', for example, is every bit as clear to us as attribution of whiteness to snow. . . . See how unfavorably the notion of analytic-in-L, characteristic of the theory of meaning, compares with that of truth-in-L.[39]

Davidson pursues the alternative that Quine suggests here. Instead of following Quine in his attempt to work out substitutes for semantic concepts within stimulus-response psychology, Davidson sets out to cultivate the theory of reference, as developed in Tarskian formal semantics, as a way of treating the linguistic phenomena that intensionalists treat within the theory of meaning. Since many linguists and philosophers are unwilling to adopt either Quine's Skinnerian psychology or any other form of stimulus-response theory as a basis for their approach to problems in semantics, Davidson's attempt to construct a truth definition for natural languages comes to occupy an important place in contemporary extensionalist thought. Thus, the significance of Davidson's work in developing Quine's suggestions concerning the theory of reference is that it provides an alternative to the problematic theory of associationistic psychology on which Quine's own constructive efforts rest. Hence, we will have taken another major step toward undermining extensionalism if we can show that Davidson's alternative — whatever importance it may have for the theory of reference *per se* — provides no insight whatsoever into the linguistic phenomena that it is required to handle (as the constructive side of extensionalism).

Davidson introduces his proposal as follows:

Anxiety that we are enmeshed in the intensional springs from using the words 'means that' as filling between description of sentence and sentence, but it may be that the success of our venture depends not on the filling but on what it fills. The theory will have done its work if it provides, for every sentence s in the language under study, a matching sentence (to replace 'p') that, in some way yet to be made clear, 'gives the meaning' of s. One obvious candidate for matching sentence is just s itself, if the object language is contained in the metalanguage; otherwise a translation of s in the metalanguage. As a final bold step, let us try treating the position occupied by 'p' extensionally: to implement this, sweep away the obscure 'means that', provide the sentence that replaces

[39] Quine (1953c), pp. 137–138.

'p' with a proper sentential connective, and supply the description that replaces 's' with its own predicate.[40]

The result is the schema

(3.7)   s is T if and only if p

which is used to formulate the Davidsonian conception of a theory of meaning, namely that a theory of meaning for a language L take the form of a truth definition for L. Davidson says:

what we require of a theory of meaning for a language L is that without appeal to any (further) semantical notions it place enough restrictions on the predicate 'is T' to entail all sentences got from schema T [i.e., (3.7)] when 's' is replaced by a structural description of a sentence of L and 'p' by that sentence.[41]

Davidson himself observes that his conception of a theory of meaning counts a theory of meaning that entails

(3.8)   "Snow is white" is true if and only if snow is white

no more correct, other things being equal, than one that entails

(3.9)   "Snow is white" is true if and only if grass is green

assuming, of course, as seems reasonable, that (3.8) and (3.9) are equally certain. Davidson anticipates his critic, remarking that (3.9) "may not encourage the same confidence that a theory that entails it deserves to be called a theory of meaning," and then he goes on to justify himself against this criticism, arguing that

the grotesqueness of [(3.9)] is in itself nothing against a theory of which it is a consequence, provided the theory gives the correct results for every sentence . . . if [(3.9)] followed from a characterization of the predicate 'is true' that led to the invariable pairing of truths with truths and falsehoods with falsehoods — then there would not, I think, be anything essential to the idea of meaning that remained to be captured.[42]

This attempt at justification begs the question. Davidson, as noted above, is defending an analysis of 'gives the meaning' that claims that 'means that' can be replaced by 'is true if and only if' without the loss of anything essential to the idea of meaning. The defense, therefore, has to be that in this replacement we lose *only* "anxiety that we are en- meshed in the intensional." But as soon as the first difficulty with the

[40] Davidson (1967), p. 309.
[41] *Ibid.*
[42] *Ibid.*, p. 312.

reduction arises, the first counter-intuitive consequence, he assumes that nothing essential to the idea of meaning remains to be captured after pairing truths with truths and falsehoods with falsehoods. But this is exactly what needs to be shown. One would suppose that the mere fact that a theory that says that each sentence in (3.9) is part of the semantic analysis of the other is sufficient to establish that the theory is missing something essential to the idea of meaning, namely, an adequate account of synonymy.

Our intuition of grotesqueness in response to the suggestion that (3.9) might be as adequate a semantic analysis as (3.8), or that (3.9) might be adequate at all, is exactly an intuition that there is something essential to the idea of meaning which is lost in such a replacement. The idea that

(3.10)   Grass is green

might be relevant to giving the meaning of

(3.11)   Snow is white

is preposterous. Equally, we would consider wrongheaded any theory that claimed that any sentence whose truth we are certain of is just as relevant to the semantic analysis of (3.11) as (3.10) or any other sentence — say, some paraphrase of (3.11). How, then, can Davidson take such a theory seriously? The answer is, I think, that he is following, from his viewpoint, the reasonable principle of being content with small mercies. The viewpoint is the Quinian philosophy of language, deriving from Quine's reconstruction of Bloomfieldian structuralism. Within this viewpoint stronger constraints on semantic analysis such as that of providing the right synonymy relations are unavailable because they would make use of suspect intensional concepts. Davidson says:

What appears to the right of the biconditional in sentences of the form 's is true if and only if p' when such sentences are consequences of a theory of truth plays its role in determining the meaning of s not by *pretending* synonymy but by adding one more brush-stroke to the picture which, taken as a whole, tells *what there is to know of the meaning of s.*[43]

Davidson himself tried to soften the blow of such grotesque consequences. He writes:

It may help to reflect that [(3.9)] is acceptable, if it is, because we are

[43] *Ibid.* Italics mine.

independently sure of the truth of 'Snow is white' and 'Grass is green';
but in cases where we are unsure of the truth of a sentence, we can have
confidence in a characterization of the truth predicate only if it pairs
that sentence with one we have good reason to believe equivalent. It
would be ill advised for someone who had any doubts about the color
of snow or grass to accept a theory that yielded [(3.9)], even if his
doubts were of equal degree, unless he thought the color of the one was
tied to the color of the other.[44]

This, however, does little to help because there are infinitely many
sentences whose truth we know (and also infinitely many whose falsehood we know), e.g., "One plus one equals two," "Two plus two equals
three," etc. Since we are independently sure of their truth, we can formulate cases as grotesque as (3.9) simply by taking arbitrary pairs of
mathematical truths, should there be any doubt about the equivalence
of (3.10) and (3.11).

After first trying to make these consequences more palatable, Davidson next tries to disavow them. "It is not easy," he writes, "to see how
[(3.9)] could be party to [the enterprise of providing a characterization
of the truth predicate]."[45] That is, questions like whether (3.10) and
(3.11) have the same truth value are not the concern of a theory designed to place enough restrictions on 'is true' to derive all instances of
(3.7) in the language. Thus such a theory ought not to be considered
unacceptable because it is committed to (3.9), insofar as the theory is
not itself responsible for the premise stating the truth-values of (3.10)
and (3.11). This, however, seems a rather feeble defense. Suppose some
theory in science (say, some theory about the physiology of color perception) were, as far as we knew, to imply nothing absurdly false by
itself, but were, in conjunction with certain simple, widely accepted
truths, like snow's being white and grass's being green, to imply grotesque consequences. We would hardly accept a Davidsonian defense
which sought to get the scientific theory off the hook on the grounds
that it is not itself responsible for these simple truths. Surely, we would
reason that the grotesque consequences cannot derive from the simple
truths themselves — since they are true. Hence we would conclude that
the source of the grotesqueness must lie in the theory, which together
with them implies the grotesque consequences. Since good theories do
not have grotesque consequences when they are combined solely with

[44] *Ibid.*
[45] *Ibid.*

simple truths, Davidson's explication of 'means that' must be held fully responsible for grotesque consequences such as that (3.10) is regarded as an analysis of (3.11) — even though, as Davidson rightly points out, the characterization of the truth predicate is not responsible for the equivalence underlying (3.9). But, then, this characterization and Davidson's theory concerning the explication of 'means that' are not the same.

If certain grotesque consequences have their source in a theory, then they ought to emerge, in some way or other, from the theory alone, i.e., without conjoining the theory with some set of simple truths. Thus let us assume, for the sake of argument, that Davidson's defense does get him out of trouble in connection with examples like (3.9). Question: Can we exhibit any similarly grotesque consequences using only principles that Davidson's theory will need in order to provide the proper characterization of the truth predicate? The answer is clearly "yes." Such counter-examples immediately come to mind once we reflect that the empirical adequacy of the theory Davidson advocates requires it to imply an appropriate instance of (3.7) for each sentence of the language (English, say) and that each of these infinitely many instances of (3.7) are also sentences of the language (English). Thus, because these sentences are each consequences of the theory, the theory is responsible for the equivalence relations between them, and because these sentences are also in the language, and so in the empirical domain of the theory, the theory will have as consequences statements in which instances of (3.7) appear both within quotations on the left of 'is true if and only if' and on the right as values of p. Hence, because the theory will be responsible for (3.8) and (3.12) (a)

(3.12)   (a) "Grass is green" is true if and only if grass is green

(b) " "Grass is green" is true if and only if grass is green" is true if and only if "Snow is white" is true if and only if snow is white,

it will be responsible for the grotesque consequence (3.12) (b) and infinitely many others of the same kind. Davidson no longer even has the reply that his enterprise of characterizing the truth predicate bears no responsibility for asserting that the sentences in question have the same truth value, as he can reply in connection with (3.10) and (3.11). His enterprise, if it is to be successful generally, has got to imply both (3.8) and (3.12)(a) and to provide an analysis for (3.12)(a), and as a result,

Davidson's theory is stuck with consequences just as grotesque as (3.9).

Grotesque cases like (3.12)(b) are not all. We have not only every metalanguage version of every case like (3.9), that is, every case like (3.12)(b), but we also have cases like

(3.13) " "Grass is green" is true if and only if grass is green" is true if and only if $p \supset (p \lor q)$,

and in general all cases in which logical axioms and theorems assumed in the truth theory are paired with sentences of a natural language and with each other. What makes such cases counter-examples is exactly what causes Davidson himself to stigmatize cases like (3.9) as grotesque, namely, the purported analysis of meaning (or logical form) bears no relation to the meanings of the words and the grammar of the sentence undergoing analysis; the purported analysis fails to predict the semantic properties and relations of the sentence under analysis systematically. A close look at cases like (3.12)(b) tells us that such sentences can only predict some of the referential properties of the sentence under analysis.

Davidson's conception of a semantic theory really deserves the charge of vacuity that he falsely levels against Fregean theories of meaning. A theory that entails a biconditional of the type (3.8) in answer to a question about the meaning of a sentence like (3.11)[46] clearly merits Davidson's criticism of the Fregean theory: "We wanted to know what the meaning of 'Snow is white' is; it is no progress to be told that it is the meaning of 'Snow is white'. This much we knew before any theory was in sight."[47]

In actual practice, we judge semantic explication and synonymy on the basis of a far less liberal criterion than equivalence. No prospective traveler would consider using a Berlitz-type phrase book whose translations were arrived at on the basis of an arbitrary choice among the sentences with the same truth value as the foreign sentence. Rather, the criterion we use to decide whether the semantic explication or alleged synonym is identical to the expression or sentence in question is whether, for any property or relation like those asked about in questions like (2.1) to (2.10), one expression or sentence has the property or bears the relation

---

[46] Note in this connection that Davidson's theory cannot have apparatus for distinguishing between different biconditionals for the same sentence (i.e., different analyses), so as to rate some as better semantic analyses than others. Thus, (3.8) must be as good as the theory can do in analyzing (3.11).

[47] *Ibid.*, pp. 306–307.

if and only if the other does. For example, we would reject the pairing (3.14) (a) with (3.11)

(3.14)   (a) Zucker ist süss

          (b) Schnee is weiss

in a phrase book but accept (3.14) (b) because (3.14) (b) but not (3.14) (a) reflects the semantic properties and relations of (3.11).

Concepts of the theory of meaning like those referred to in (2.1) to (2.10) not only provide criteria for judging the adequacy of accounts of the meaning of expressions and sentences, but also provide the further constraints necessary to avoid counter-examples like those cited above.[48] Definitions of semantic properties and relations are framed in terms of formal features of semantic representations. A definition of a semantic property or relation picks out the structural configuration in semantic representations by virtue of which they represent an expression or sentence as having the property or relation. Thus the needed constraints are provided by requiring that the representations of logical form in a theory be sufficiently rich in structure that, for any sentence, its representation has the structural configurations necessary to mark its full range of semantic properties and relations on the basis of optimal definitions of them.

However, the addition of such constraints not so much avoids these counter-examples as concedes their point. The addition actually replaces the criticized theory with a far more ambitious one, particularly so in regard to ontological commitment. The following simple examples show that a theory meeting these further constraints will look nothing like Davidson's conception in any relevant respect. Consider the definition of semantic redundancy, that is, the definition that answers (2.8). It enables us to mark cases like "naked nude" as redundant on the basis of some such condition as that the semantic representation of the modifier be a proper part of the semantic representation of its head. Thus, in connection with this definition, the constraints require semantic representations to break down or analyze the senses of constituents into their component concepts, and to present these concepts as independent formal elements with relations to each other. Other definitions will entail the same consequence. For example, an adequate definition of the concept of semantic similarity, that is, one that enables us to mark

[48] And also trivializations such as Fodor's; see Fodor (1970), pp. 299–300.

73

the common semantic element in sets like (3.12), will also require that semantic representations give a componential analysis of senses. The same requirement will also be forthcoming from the definitions of analyticity and entailment. The point is that if such constraints are imposed on semantic representations, then the theory no longer tells us that the proper account of the meaning or logical form of a sentence is the sentence itself or anything that treats it or its constituents as unanalyzed terms.[49] It now tells us that the proper account is an account involving decomposition into the primitive semantic elements of the meanings of the constituents of the sentence, since to stop short of such decomposition at any point is to fail to account for some semantic property or relation of the sentence. The theory now takes on a familiar intensionalist look.

We note that Davidson's failure to appreciate the role of grammatical structure and word meaning in connection with Fregean analysis of meaning is due to his acceptance of Quine's overly weak notion of what constraints a semantic description of a sentence like (3.1) should meet. If we require a description of (3.1) to do more than to state truth conditions for the sentence in the sense of giving a sentence which is true if and only if (3.1) is, i.e., if we ask it to predict and explain the semantic properties and relations that the sentence exhibits, then the description of the meaning of (3.1) will have to account for facts such as those found unaccounted for in Davidson's caricature of a Fregean analysis. What will have to be said about the grammar of (3.1) and about the meaning of its words and the interaction between word meaning and grammar will no longer seem vacuous. Therefore, both the impression that a Fregean analysis offers little that we didn't already know and the implication that a Fregean theory of the structure of senses and sentences is of no help in answering questions about truth

---

[49] First, we note that by excluding the sentence itself, these constraints automatically provide a means of dealing with Fodor's trivialization. Second, we note that by excluding accounts that treat constituents as unanalyzed terms, these constraints offer a natural solution to a problem that troubles Davidson. He writes: "The problem is to frame a truth definition such that ' "Bardot is a good actress" is true if and only if Bardot is a good actress' — and all other sentences like it — are consequences. Obviously, 'good actress' does not mean 'good and an actress'. We might think of taking 'is a good actress' as an unanalyzed predicate. This would obliterate all connection between 'is a good actress' and 'is a good mother', and it would give us no excuse to think of 'good', in these uses, as a word or semantic element." Davidson (1967), p. 317. An approach to Davidson's problem in componential terms is found in Katz (1966), pp. 288–317.

and meaning arise because Davidson's treatment of (3.1), based upon his conception of the constraints on a theory of meaning, requires so little of semantic representations.

The problem that Davidson says is responsible for his radical move to replace 's means that p' with (3.7) warrants this extreme solution only if one makes Quine's assumptions about the viability of concepts from the theory of meaning. Davidson writes:

It looks as though we are in trouble on another count, however, for it is reasonable to expect that in wrestling with the logic of the apparently non-extensional 'means that' we will encounter problems as hard as, or perhaps identical with, the problems our theory is out to solve.

The only way I know to deal with this difficulty is simple, and radical. . . . it may be that the success of our venture depends not on the filling but on what it fills. . . . let us try treating the position occupied by 'p' extensionally: to implement this, sweep away the obscure 'means that', provide the sentence that replaces 'p' with a proper sentential connective, and supply the description that replaces 's' with its own predicate. The plausible result is [(3.7)].[50]

The "hard problems" referred to here are obviously those, referred to a page or so earlier in Davidson's paper, that allegedly doom the prospects of a compositional theory of meaning framed as syntax plus a dictionary, together with rules for projecting the meanings of the parts of sentences to assign meanings to whole sentences. According to Davidson,

The point is easily illustrated by belief sentences. Their syntax is relatively unproblematic. Yet, adding a dictionary does not touch the standard semantic problem, which is that we cannot account for even as much as the truth conditions of such sentences on the basis of what we know of the meanings of the words in them.[51]

What Davidson has in mind is clearly the difficulty raised by Mates that nobody could doubt that

(3.15)   Whoever believes that D, believes that D

but someone could doubt

(3.16)   Whoever believes that D, believes that D'

even though 'D' and 'D'' are synonymous sentences.[52] Davidson draws

---

[50] Davidson (1967), p. 309.
[51] *Ibid.*, pp. 307–308.
[52] Mates (1952), p. 215.

the conclusion that such cases are cases where we have the same component meanings in the same relations but they do not add up to the same truth conditions. But, as Church has shown, this conclusion is not at all warranted because the doubt about (3.16) can be attributed to a deficiency of linguistic knowledge rather than to a difference in the truth conditions of (3.15) and (3.16).[53] This alternative requires only that we make the intensionalist distinction between a sentence and the proposition it expresses, that is, its sense, and that we avail ourselves of the concepts of synonymy and semantic entailment (to construct a condition for substituting into opaque contexts). Since the insurmountable problem that motivates his radical extensionalism is insurmountable only if we are not allowed to make intensionalist assumptions, this problem can hardly be grounds for preferring this extensionalism to intensionalism.

4. Logical Form: Two Dogmas of Quinian Empiricism

Intensionalists understand the logical form of a sentence to consist of every property of the sentence that determines the role it plays in valid arguments. For the intensionalist, any feature of a sentence S that is part of S's grammatical structure and by virtue of which S occurs as an essential premise (or the conclusion) of a valid argument (in the standard sense of one whose conclusion must be true if its premises are)[54] is a feature of S's logical form. In Frege's terms: whatever "influences its *possible consequences*. Everything necessary for a correct inference . . ."[55] Thus, for the intensionalist, whatever semantic features of the words "nightmare" and "dream" underlie the relation between their meanings by virtue of which the argument (4.1) to (4.2) is valid is a feature of the logical form of (4.1) and (4.2):

(4.1)   Socrates had a nightmare
(4.2)   Socrates had a dream

Such features are, then, just as central to an account of the role these sentences play in valid argument as any features of the English words that express truth-functional connectives, quantifiers, and apparatus of quantification theory.

[53] Cf. Church (1954) and Katz (1972), chap. 6.
[54] In the case of arguments employing statements, cf. Katz (1972), chap. 5.
[55] Frege (1967), p. 12.

Extensionalists, following Quine, have a far narrower notion of the logical form of a sentence. For them, the features that determine the logical implications of sentences are restricted to properties of the truth-functional connectives, quantifiers, and other apparatus of quantification theory. Quine writes: "Logical implication rests wholly on how the truth functions, quantifiers, and variables stack up. It rests wholly on what we may call, in a word, the logical structure of the two sentences."[56] For an intensionalist, a sentence like

(4.3)   Nightmares are dreams

is a logical truth in that it is true by virtue of its logical structure, but for a Quinian it is not. Quine writes:

I defined a logical truth as a sentence whose truth is assured by its logical structure. . . . I have explained that what I mean by the logical structure of a sentence . . . is its composition in respect of truth functions, quantifiers, and variables. It follows that logical structure and predicates are all there is to a sentence, under the standard grammar we have adopted. Just put schematic letter 'F', 'G', etc., in place of the predicates in a sentence, and you have depicted its logical structure.[57]

This special status accorded to quantification theory in the Quinian conception of logical form is not supported by any relevant argument,[58] but rests on "a dogma" of contemporary extensionalist theories of logic and language. The article of faith is that there exists a justifiable distinction between the logical and nonlogical components of sentences, one that enables us to divide a theory of connectives and quantifiers from a theory of the meaning of nouns, verbs, adjectives, etc., that form the expressions and sentences they connect and quantify.

We find this article of faith most often expressed as a distinction between the form and content of the propositions comprising an argument. It is said that the validity of an argument, such as the classical

---

[56] Quine (1970b), p. 48.
[57] Ibid., p. 49.
[58] Quine does cite some rewards of "staying within the bounds of standard grammar" but except for the first, extensionality, where Quine is arguing from his own extensionalist position, the considerations adduced — the efficiency, elegance, completeness, etc., of quantification theory, plus the concurrence of different definitions of logical truth — are all preserved in the intensionalist position, just as the rewards of propositional logic are preserved in quantification theory. Quine's "rewards" are not relevant arguments because the intentionalist does not advocate eliminating quantification theory but only putting it in its proper place as one component of logical form.

example,

(4.4)   Every man is mortal
(4.5)   Socrates is a man
(4.6)   Socrates is mortal

does not depend on the particular choice of content words (i.e., "man," "mortal," and "Socrates" in the present case). Choose any other triple of a noun, an adjective, and another noun, and replace each occurrence in (4.4) to (4.6) of "man" with the first noun, each occurrence of "mortal" with the adjective, and each occurrence of "Socrates" with the second noun, and the resulting argument will also be valid. Accordingly, the validity of these arguments must derive from what is common to them, not from the respects in which they differ. Since they differ in their content-words and are the same in form-words, it is concluded that the validity is a matter of form as opposed to content.

On the basis of this distinction between the form and content of arguments, logic is construed as a discipline that seeks to state all and only the valid argument forms. Accordingly, the vocabulary of a system of logic has to contain both a list of particles defined in such a way as to represent aspects of the form of arguments (in the sense of the distinction in question) and a list of variables to represent parts of the content of sentences, namely, the nouns, adjectives, verbs, etc., whose particular semantic structure will be unsymbolized in the notation of the system. Therefore, the inference from (4.4) and (4.5) to (4.6) is represented as having the logical form (4.7):

(4.7)   Every x is a y, and z is an x, therefore, z is a y

In this way, the distinction between form and matter comes to be enshrined in the formal distinction between a so-called "logical vocabulary" and a so-called "extra-logical (or descriptive) vocabulary," and in the distinction between "logical axioms (or rules)" and "non-logical axioms (or rules)." The logical vocabulary consists of the "logical particles" "every," "is a," "and," "if . . . , then . . .", etc., and the extra-logical vocabulary consists of what is suppressed by the "x"'s, "y"'s, and "z"'s in renderings of arguments like (4.7).

Once such a list of particles is given, a system of logic can be constructed as a recursive representation of argument forms like (4.7) by positing logical axioms from which instances of quantificational schema expressing conditionals like

(4.8)   If every x is a y, and z is x, then z is y

follow as theorems. Once such systems are constructed, theories of logic can be framed that characterize logic as a discipline that seeks to discover all truths of the form (4.8) in which only logical particles occur essentially.[59] This, then, is how the practice of logicians becomes the Quinian theory of logical form.

Logicians can defend their practice on the grounds that concentrating on the logical structure of quantifiers and connectives has enabled them to pose manageable problems and work out precise methods for their solution. But Quine cannot defend his theory of logical form on these grounds, though at times he tends to do so,[60] because his theory is a theory about what logic is, not a proposal about the best research strategy at present. Quine must defend the extensionalist faith in the distinction between form and matter. If this distinction turns out to be an untenable dogma, the list of logical particles becomes an arbitrary collection and the theories of logic based on them also become arbitrary.

What, then, is there on which to base this distinction between form and matter, between the logical particles and the other words of a language? The only systematic answer available in contemporary philosophy is Quine's skepticism about the notion of meaning and semantic relations. Quine's view, considered above, that meanings are occult, unexplanatory, and hence scientifically unacceptable and that absolute semantic relations like synonymy are hopelessly unclear is the only thing to which one can turn to find support for the prevailing extensionalistic theory of logical form.[61]

How does Quine's skepticism support this theory? The answer is that it motivates the extensionalist's exclusion of every noun, verb, adjective, etc. — everything but quantifier terms, connectives, and pronouns — from the category 'logical', since, if Quine's skepticism is correct, none of the excluded words give rise to necessary connections; i.e., sentence connections like the relation between (4.1) and (4.2) expressed in the

---

[59] Cf. Quine (1955), p. 3.
[60] Cf. Quine (1953c), p. 138; cf. Katz (1972), preface.
[61] I say "prevailing extensionalistic" rather than "extensionalist" here because, unfortunately, even prominent intensionalists are often found endorsing this extensionalist notion of logical form, e.g., Church (1954), pp. 1–3.

conditional 'If (4.1), then (4.2)' are not necessary in the manner of logical truths.[62] Since logic is the science of necessary inferential connections, it follows on this account that nothing other than the connection contributed by the logical particles are genuine logical implications. If Quine's skepticism is wrong, on the other hand, then connections like that between (4.1) and (4.2) are also necessary connections, and nouns, verbs, adjectives, etc., make as much of a logical contribution to sentences as the logical particles. Accordingly, the allegedly "nonlogical" words cannot be distinguished from the logical particles and the distinction between form and matter collapses.

This is why the analytic-synthetic distinction has occupied so large a place in Quine's writings. If sentences like (4.3) are analytic, they express necessary truths, and so the connections between concepts like 'nightmare' and 'dream' are necessary, and the words that stand for such concepts in natural languages contribute to the logical form of sentences. On the other hand, if there is no analytic-synthetic distinction, then the connections between words like "nightmare" and "dream" are merely contingent, best explained in terms of the frequency with which they are related in experience (reinforcement, etc.) within some theory of stimulus meaning like Quine's. In this case, there is an explanation of why such connections are nonlogical, viz., they are contingent because no co-occurrence, no matter how frequently repeated or highly reinforced, can provide a basis for necessary connection.

Given the argument in section 2 of this essay, Quine's semantic skepticism and criticism of the analytic-synthetic distinction depend completely on the taxonomic theory of grammar. Thus, given the correctness of the transformational criticism of the taxonomic theory, Quine's semantic skepticism and criticism of the analytic-synthetic distinction rests on false linguistic assumptions, and the extensionalist distinction between form and matter which rests on this skepticism and criticism loses its support.

It is not hard to show that on its own this distinction is arbitrary.[63]

---

[62] The use of the term 'necessary' is not really necessary and can be dispensed with if Quinians object to its use in this discussion. They can substitute the notion 'conventional'; see Quine (1968) and also Katz and Nagel (1974), sec. 8.

[63] Arbitrary, that is, from the viewpoint of an account of the nature of logic and logical form. This is not to say that the systems of quantification theory developed in logic are not natural explications of some systematic portion of logical form. But as such they are only partial explications of 'valid argument'. One such way to interpret them more modestly within a broader theory of logical form is to take the

The argument (4.1) to (4.2) is no less valid than the argument (4.4) to (4.6) or any other argument whose validity rests on "how the truth functions, quantifiers, and variables stack up." By any appropriate construal of "validity" applicable to the latter argument, the former is valid too. For example, if the latter is said to be valid because the truth of the premises makes the truth of the conclusion necessary, then the former must be said to be valid for the same reason; if the latter is said to be valid because it is an instance of a schema that yields only truths on the substitution of terms for its variables, then the former must be said to be valid for the same reason, since any substitutions of terms for variables (the same variable, the same term) in

(4.9)   If x had a nightmare, then x had a dream

yield only truths. Furthermore, any arbitrary consistent replacement of the noun "nightmare" or "dream" by another word from the same syntactic category does not always result in another valid argument. Following the line of reasoning used to draw the form/content distinction in the case of (4.4) to (4.6), we have to conclude that "nightmare" and "dream" must appear on the list of logical particles. But, since for almost any meaningful word or expression in a natural language, one can construct an argument like (4.1) to (4.2), where arbitrary replacements of the word or expression by another from the same syntactic category does not preserve validity, the list of logical particles will include almost every meaningful word or expression in the language. Accordingly, the distinction between the logical vocabulary and the extra-logical vocabulary disappears: if we consistently apply the criterion used to make the distinction, we include every, or almost every, noun, adjective, verb, etc., in the logical vocabulary. The distinction between the logical vocabulary and the extra-logical vocabulary is shown to rest on no actual difference.

Tarski once confessed that the division of vocabularies into logical and extra-logical is essentially arbitrary. He wrote:

no objective grounds are known to me which permit us to draw a sharp boundary between [logical and extra-logical] terms. It seems to be pos-

definitions of the propositional connectives and quantifiers to be descriptions of the meaning of the corresponding words in natural language. Thus we admit that a full theory of logical form for natural language will require definitions of every other word in the language, and this means that we drop the distinction between logical and extralogical vocabularies as an absolute epistemological distinction, perhaps retaining it as a relative one, expressing the degree to which present research has provided descriptions of the meaning of the words in natural language.

sible to include among logical terms some which are usually regarded by logicians as extra-logical without running into consequences which stand in sharp contrast with ordinary usage. In the extreme case we could regard all terms of the language as logical. The concept *formal* consequence would then coincide with that of *material* consequence.[64]

Tarski adds, by way of conclusion:

Perhaps it will be possible to find important objective arguments which will enable us to justify the traditional boundary between logical and extra-logical expressions. But I consider it to be quite possible that investigations will bring no positive results in this direction, so that we shall be compelled to regard such concepts as 'logical consequence', 'analytical statement', and 'tautology' as relative concepts which must, on each occasion, be related to a definite, although in greater or less degree arbitrary, division of terms into logical and extra-logical.[65]

The final remark of the first quotation suggests that the absence of a distinction between logical particles and extra-logical terms entails the absence of a distinction between the logico-linguistic properties of words and the empirical properties of the things words refer to, such that inferences based on the former and inferences based on the latter cannot be differentiated. There are two ways Tarski's argument might be read. First, it might be taken to recognize that there is no basis for the distinction between the logical and the extra-logical terms and draw the conclusion that there is no basis for the distinction between consequence relations like that in the argument from (4.4) and (4.5) to (4.6) and consequence relations like that in the argument from (4.1) to (4.2). On this construal, Tarski's conclusion is unacceptable only to extensionalists. Second, it might be taken to draw the conclusion that there is no basis for the distinction between consequence relations like that in the argument from (4.4) and (4.5) to (4.6) or in the argument from (4.1) to (4.2) and consequence relations like that in the argument from (4.10) to (4.11):

(4.10)   This wet stuff is water
(4.11)   This wet stuff contains hydrogen, expands on freezing, and does not dissolve gold

On this construal, the conclusion of the argument would be widely regarded as unacceptable since it raises the specter of logic turning out to be unformalizable because the notion of a logical consequence depends

[64] Tarski (1956), pp. 418–419.
[65] *Ibid.*, p. 420.

on empirical facts. Accordingly, if this version of the argument were valid, we would have to think twice about giving up the distinction between the logical and extra-logical terms.

The conclusion is not, however, universally unacceptable. Quine would (at least in one mood[66]) find it acceptable because it supports his radical empiricism, his thesis that "no statement is immune from revision,"[67] and that even statements of logic can be overturned in the light of conflicts between our whole system of knowledge and belief and our experience.

The argument, on the second construal, is invalid because it has a false premise. This is the assumption that the meaning of nouns such as "water" (verbs, adjectives, and other so-called "extra-logical" terms generally) includes whatever empirical science or our accepted beliefs say about the referents of these words. If this assumption were true, Tarski's conclusion would follow. Eliminating the distinction between the logical and extra-logical vocabulary and thus putting the nouns, verbs, adjectives, etc., of a language into the category of logical words (i.e., the category of words whose meaning determines the logical consequences of sentences) would imply that inferences like the inference from (4.10) to (4.11) turn on the meanings of the words in these sentences, given the extended notion of meaning in question.

The plausibility of this premise depends on there being no way to distinguish facts about the meaning of words in the familiar lexicographical sense from empirical facts about the properties of their referents. If, in fact, there is no way, then every newly unearthed empirical finding about water (or perhaps some selection of such facts) tells us something both about water and "water."

Clearly, Quine's empiricist claim that every truth is a truth of experience and revisable in principle, as well as his criticism that the analytic-synthetic distinction is not an acceptable methodological ideal, depends significantly on this assumption that there is no adequate way to distinguish between knowledge of the meaning of words and knowledge of the things to which they refer. As Quine see it, each new scientific discovery about something introduces further information about the meaning of the term referring to it:[68]

[66] See Katz and Nagel (1974), sec. 8.
[67] Quine (1953c), p. 43.
[68] Note also that unless present languages can incorporate in the meanings of their

83

Suppose a scientist introduces a new term, for a certain substance or force. He introduces it by an act either of legislative definition or of legislative postulation. Progressing, he evolves hypotheses regarding further traits of the named substance or force. Suppose now that some such eventual hypothesis, well attested, identifies this substance or force with one named by a complex term built up of other portions of his scientific vocabulary. We all know that this new identity will figure in the ensuing developments quite on a par with the identity which first came of the act of legislative definition, if any, or on a par with the law which first came of the act of legislative postulation. Revisions, in the course of further progress, can touch any of these affirmations equally. Now I urge that scientists, proceeding thus, are not thereby slurring over any meaningful distinction. Legislative acts occur again and again; on the other hand a dichotomy of the resulting truths themselves into analytic and synthetic, truths by meaning postulate and truths by force of nature, has been given no tolerably clear meaning even as a methodological ideal.[69]

The second of the two dogmas of Quinian empiricism to which I referred in the section title is just this assertion that semantic facts about words are indistinguishable from empirical facts about their referents. It is a false dogma because, in fact, there is a criterion for distinguishing them, and thus for distinguishing changes of meaning and changes of belief about the world. The criterion takes the form of a procedure for deciding when a piece of information that something A is B belongs in the dictionary entry for a word W that refers to A and when such information belongs in an encyclopedia account of A. Given the role of a dictionary in a grammar, such a procedure decides whether the information 'is a B' is linguistic information about W that determines the logical consequences of the sentences in which W occurs or is contingent information about A that determines the material consequences of sentences in which W occurs.[70] The procedure is, then, an explanation of how to answer questions like (4.12) and (4.13):

(4.12)　　Is 'being female' part of the dictionary entry for the English

words false accounts of things which can be empirically revised in the light of new experiences and a better scientific theory, it makes no sense to say, as Quine does, that analytic sentences are falsifiable.

[69] Quine (1966a), pp. 124–125. It is interesting to note that this conception of meaning can also be traced directly back to the Bloomfieldian position that was rationally reconstructed in "The Problem of Meaning in Linguistics." In Bloomfield (1933) Bloomfield writes: "In order to give a scientifically accurate definition of meaning for every form of a language, we should have to have a scientifically accurate knowledge of everything in the speakers' world." Pp. 139–140.

[70] A material consequence being a case like (4.11)'s relation to (4.10).

word "mother"?

(4.13)   Is 'once having had a mother' part of the dictionary entry for the English word "female"?

The criterion is this:

(4.14)   (i) The answer to a question about whether a piece of information belongs in the dictionary entry for an item W is affirmative just in case there is some semantic property or relation whose extension in the language cannot be determined on the basis of the readings of the sentences in which W occurs unless these readings are projected compositionally from lexical readings for their morphemes in which the lexical reading for W contains semantic markers representing the information.

          (ii) The answer to such a question is negative just in case the extension of every semantic property and relation in the language can be determined without the information in question being represented in a lexical reading of W.

Definition (4.14) would say that the answer to (4.12) is affirmative if the inference from (4.15) to (4.16)

(4.15)   Xantippe was a mother
(4.16)   Xantippe was female

is accepted as valid and if its validity cannot be explained without one lexical reading of "mother" containing semantic markers that represent the information 'is female'.[71] Definition (4.14) would say that the answer to (4.13) is negative if, as seems the case, no semantic property or relation goes unexplained without a lexical reading for "female" containing semantic markers that represent the information 'once having a mother'. So the basis on which (4.14) decides whether a piece of information is semantic is whether or not it enters (essentially) into explanations of entailments and other semantic properties and relations. This is exactly the same basis on which we decide on definitions for propositional connectives and quantifiers. It is clear that we decide that the definition of conjunction must contain a clause saying that a statement of the form 'p and q' is true just in case both p and q are true

[71] Cf. Katz (1972), chap. 4, sec. 5.

because without such a clause, our systems of logic will be unable to explain why an argument like (4.17) to (4.15) is valid.

(4.17)   Xantippe was a mother and Socrates was a father

It is also clear that we refuse to allow definitions of truth-functional connectives such as

(4.18)   A statement of the form 'p and q' is true just in case both p and q are true and everything is self-identical.[72]

on the grounds that their further clause plays no role in explaining the validity of arguments. Although it is true of the statements over which propositional variables range, just as it is true of everything else, that each is self-identical, we do *not* ipso facto say that this information cannot be excluded from the definition of truth-functional conjunction. Likewise, although it is true of each thing that is in the extension of "female" that it once had a mother, we do *not* ipso facto have to say that this information cannot be excluded from the definition of "female." There is a perfect symmetry here: the basis on which we include or exclude information is the same whether we are defining truth-functional connectives and quantifiers or nouns, verbs, adjectives, and other so-called extra-logical terms. Therefore, either there is no such thing as logical connection of any kind, distinct from empirical regularity, or if there is, then there is an analytic-synthetic distinction and no logical-extralogical distinction.

One final point before concluding this section. The criterion we have proposed depends, basically, on two things. First, it depends on a rather general, and uncontroversial, principle of scientific methodology to the effect that we do not complicate theories with the additions that afford us no increase in explanatory or predictive power and no benefit in simplifying the system. Second, it depends on it being legitimate for us to regard the semantic properties and relations of sentences as real features requiring explanation and prediction. It is, of course, in connection with the latter that Quinians will doubtlessly challenge my claim that we have available a criterion for distinguishing between matters of lexicography and matters of extra-linguistic fact. But their side in this disagreement — that the semantic properties and relations required in

---

[72] Or, for example, "A statement of the form 'p and q' is true just in case both p and q are true or 2 + 2 = 17.

the criterion are occult and incoherent notions — represents nothing more than an application of the Bloomfieldian assumptions that transformational linguistics has shown to be empirically false.

5. If Cats Were Martian Robots, Would They Like Blue Lemons? — Some Aspects of the Work of Putnam and Kripke

Quinian extensionalism is the form that a certain version of empiricism takes when it is cast as a philosophy of language. The version is somewhat more sophisticated than Mill's (particularly in regard to its account of the certainty of logic and mathematics[73]) but less sophisticated in important respects than Hume's (principally in regard to the question of relations of ideas). Its chief tenet, an uncompromising denial of any form of necessary truth,[74] was what led Quine to criticize fellow empiricists like Carnap for allowing a place for necessary truth in their artificial languages. Quine saw that the Carnapians were especially vulnerable to an attack against their nonscientific flank and he was especially fortunate to find an empiricism like the one underlying Bloomfieldian structuralism when he turned to linguistics to compare what scientists say about language with what philosophers were saying. The most significant feature of Bloomfieldian structuralism for Quinian extensionalism was that it conceived of grammatical analysis as nothing more than raw linguistic data (transcriptions of actual speech) processed by procedures of segmentation and taxonomic classification at successively higher levels of surface phonological and syntactic structure. As already stressed, this excludes meaning from any place in the grammatical structure of sentences. This exclusion, moreover, automatically precludes using the distinction between what is a matter of grammar and what is extra-grammatical as a basis for drawing the distinction between information about the lexical meaning of words and information about their extension. Thus we are lead directly to the view that the

[73] Note Quine's own remark in the second paragraph of (1966a), pp. 100–101. He says that the truths of mathematics and logic are more remote from the experiential boundaries of our total system of knowledge and beliefs than empirical laws in the sciences. I take this to be the basis of his claim to have a better account of the certainty of logic and mathematics than Mill.

[74] Quine writes, by way of summing up his position on necessary truth (1966b), p. 56, "In principle, therefore, I see no higher or more austere necessity than natural necessity; and in natural necessity, or our attributions of it, I see only Hume's regularities . . ."

meaning of a word, insofar as we may speak this way, consists of some body of knowledge or belief about its extension.

This view, which I will refer to as "encyclopediaism,"[75] represents another approach within the Quinian system for trying to investigate the grammatical phenomena in natural language independently of the proscribed concepts of intensional semantics. We have already considered Quine's own approach based on stimulus response psychology of the Skinnerian variety and Davidson's based on his attempt to apply Tarski's work on a truth definition to sentences of a natural language. Now we will consider encyclopediaism, again in order to show that the alternatives within Quinian extensionalism do not offer us any theoretical insight into the grammatical phenomena we wish to understand. The case we shall try to make against encyclopediaism is, however, not solely that it depends on Quinian assumptions concerning dictionary meaning which derive from Bloomfieldian linguistics but that, even on its own, it offers us no conceptual apparatus for treating linguistic meaning apart from beliefs about the world and is thus unable to account for the semantic phenomena in language as adequately as intensionalism.

In this section we shall consider a form of encyclopediaism that is gaining popularity in contemporary Anglo-American philosophy. This form seeks to replace intensionalist concepts with a notion of meaning on which the meaning of words consists of information provided by an account of the essential nature of their referents. Its leading exponents are Putnam and Kripke, and each has recently criticized intensionalist positions from this standpoint.

5.1. PUTNAM

Putnam's criticisms of the possibility of semantic theory in the intensionalist tradition of Plato, Frege, and Carnap are directed primarily against my version of such a theory, which Putnam quite properly takes to carry on this tradition.[76] Putnam's encyclopediaism departs from Quine's in that Putnam finds problematic the idea of conveying the meaning of a (natural-kind) word by informing the hearer of what sen-

---

[75] This term has little to recommend it, I realize, but it does make the right connection between the present controversies and those discussed in Katz (1968).

[76] Putnam (1970), pp. 187–201.

tences about the extension of the word one finds acceptable. Putnam says:

If conveying the meaning of the word "tiger" involved conveying the totality of accepted scientific theory about tigers, or even the totality of what I believe about tigers, then it would be an impossible task. It is true that when I tell someone what a tiger is I "simply tell him certain sentences" — though not necessarily sentences I *accept*, except as descriptions of linguistically stereotypical tigers.[77]

On Putnam's version, the meaning of a word like "tiger" is given by a theory about tigers but "not the actual theory we believe about tigers." Rather, it is

an oversimplified theory which describes a, so to speak, tiger *stereotype*. It describes . . . a *normal member* of the natural kind. It is not necessary that we believe this theory . . . But it is necessary that we be aware that *this* theory is associated with the word: if our stereotype of a tiger ever changes, then the word "tiger" will have changed its meaning.[78]

The stereotype is expressed by certain facts about the referent of the word, called "core facts."[79] These give the extension of the word, as well as its intension, i.e., for Putnam, the stereotype.

Since the distinction between core and noncore facts seems suspiciously like the distinction between semantic and nonsemantic facts that would be drawn by using (4.14), one may well wonder at this point whether Putnam's departure from Quinian orthodoxy has not taken him over the line into intensionism. But Putnam makes it quite clear that this is not so. For the intensionalist, as Putnam rightly observes,[80] the meaning of a word is one definite thing, so that there is a single body of information that has to be conveyed to convey the meaning of a word. It is by virtue of being part of this body of information, as determined by (4.14), that a piece of information is semantic. For Putnam, however, there is no single set of facts to convey in giving the meaning of a word: "one must give some way of getting the extension right, but no one *particular* way *is* necessary."[81] He concludes:

Frequently it has been regarded as a defect of *dictionaries* that they are "cluttered up" with color samples, and with stray pieces of empirical

---

[77] *Ibid.*, p. 196.
[78] *Ibid.*
[79] *Ibid.*, pp. 197–200.
[80] *Ibid.*, p. 199.
[81] *Ibid.*, p. 200.

information (e.g., the atomic weight of aluminium), not sharply distinguished from "purely linguistic" information. The burden of the present discussion is that this is no defect at all, but essential to the function of conveying the core facts in each case.[82]

In this subsection, I will first consider Putnam's misunderstandings of my theory of meaning, and second consider his "counter-examples" about Martian cats and blue lemons. Putnam's misunderstandings occur mainly in his outline of semantic theory. He lists six principles. The first, that the meaning of a word is represented by a string of semantic markers, captures a feature of my theory. The second, however, does not. Putnam says that semantic markers stand for concepts which are brain processes, according to my view.[83] My view here is rather that semantic markers are neutral with respect to the members of the class of possible physical realizations of their formal structure, e.g., in brains, computers, etc. This is a strange mistake for Putnam to make, considering that in my paper dealing with the interpretation of linguistic theory[84] I base my discussion of this point on Putnam's "Minds and Machines,"[85] arguing similarly that the concepts for which semantic markers stand are indifferent with respect to isomorphic realizations of their formal structure.

Putnam's third principle errs more in vagueness than by explicit misstatement, but he exploits this vagueness to support the criticism that semantic theory contains an "internal inconsistency." Putnam says that on my theory each semantic marker expresses a linguistic universal, and that a semantic marker "stand for an *innate* notion — one in some sense or other 'built into' the human brain."[86] He thus claims that

there are internal inconsistencies in this scheme . . . For example, "seal" is given as an example of a "linguistic universal" . . . but in no theory of human evolution is contact with seals universal. Indeed, even contact with *clothing*, or with *furniture*, or with *agriculture* is by no means universal.[87]

The alleged inconsistency is that, on the one hand, every language group must have the concept of a seal (clothing, furniture, agriculture, etc.) since it is represented by a semantic marker and so innate, and

[82] *Ibid.*, p. 201.
[83] *Ibid.*, p. 193.
[84] Katz (1964b), p. 129.
[85] Putnam (1960), pp. 148–179.
[86] Putnam (1970), p. 193.
[87] *Ibid.*

yet, on the other, some language need not have the concept since there is no guarantee its speakers will make contact with the animals (artifacts, technology, etc.). But, although I am correctly represented as saying that semantic markers express linguistic universals and linguistic universals are innately determined, I am not correctly represented by Putnam's assumption about the status I accord to innately determined concepts or by his assumption that I fail to distinguish between language-universal concepts and the words that express them in particular languages. My view is that the innate mechanism for language acquisition need not have each concept that is a component of a sense in natural languages built into the brain in the form of a separate neural realization. Instead, some concepts may be so built in and others generated from them by built-in recursive rules. When, as the result of an external environmental (or internal) triggering stimulus, some selection from these rules is activated, a specific neural realization of a concept is generated. It can then be paired with a phonetic form, the pair serving as a new dictionary entry to fill an antecedent need. Accordingly, contact with the appropriate animals, artifacts, technology, etc., in the environment may be necessary for the language of a group to have a certain dictionary entry, but such contact is not necessary for the language to have a certain concept in its semantic structure or in order for the concept to be part of the innate mechanism for language acquisition.[88] Since, on this view, environmental contact is necessary only to activate and sustain the operation of the rules of the language-acquisition mechanism and to provide conditions for the pairings that serve as new dictionary entries, the alleged inconsistency disappears.[89]

The next principle is a reasonably accurate formulation of the compositional principle that the reading of a sentence is derived from the readings of its parts together with their grammatical relations, but the last two of Putnam's principles, though they appear at first acceptable formulations, are put to polemical use in an unacceptable form.

The fifth principle is that semantic theory is put forth as a scientific theory, to be justified by the success of its explanations of semantic in-

---

[88] The fact that a concept is in the language, even though there is no word for it, might be shown by finding some syntactically complex expression having the concept as its sense.

[89] Thus it is unnecessary to take Putnam's suggestion that I might retreat to the position that only some concepts, namely those that are primitive enough (for every language group to have contact with their extension), are universal.

tuitions. The problem is that Putnam goes on to interpret the notion of 'scientific theory' as an artificial "marker language" and the notion of assigning readings as that of giving translations from a natural language into such an artificial language. I shall reply to this misconception in the next section where it occurs as a more central part of an extensionalist criticism. Here I point out only that Putnam's further claim that "the string of markers associated with a word has exactly the meaning of the word" is false.[90] Semantic markers are intended as theoretical terms, and so are introduced with reference to the words they are used to define but are not themselves defined by such words.[91]

The sixth is that the semantic properties and relations of sentences can be "read off" from their readings. Putnam understands "read off" to mean "can be seen from inspection of the reading." It ought to mean "can be effectively determined on the basis of the reading and the definitions of semantic properties and relations in semantic theory." The difference between these two construals is that Putnam's neglects the role of definitions of semantic properties and relations as the basis on which the formalism of a reading is interpreted as a set of empirical predictions about the sentence to which it is assigned. This neglect, together with a simplistic conception of the relation between competence and performance, is responsible for Putnam's belief that the 'robot cat' and the 'blue lemon' examples refute the traditional intensionalist theory of meaning and my linguistic version of it.[92]

The robot cat example goes as follows. One day we find out that the things we have been calling by the term "cat" are in fact nothing more

[90] Putnam (1970), p. 194.

[91] Some confusion might arise from the fact that a semantic marker often is presented in the form of a word in parentheses. This, however, is done as a kind of mnemonic, to suggest the concept that the marker is supposed to stand for in the theory. A less benevolent semanticist might use numbers instead of words, providing the numbers were introduced with a proper explanation of the concepts they are to stand for. Thus, the fact that the semantic marker '(Artifact)' has this form is no more to be taken as grounds for saying that its meaning is the meaning of the word used in the notation than is the syntactic notation 'Noun', 'Verb', 'Determiner', etc., to be understood to mean just what the English words "noun," "verb," determiner," etc., mean.

[92] Putnam writes: "as far as general names are concerned, the only change [from the traditional theory of meaning Putnam criticized earlier in the paper] is that whereas in the traditional account each general name was associated with a list of properties, in Katz' account each general name is associated with a list of *concepts*. It follows that each counterexample to the traditional theory is at once a counterexample also to Katz' theory." See Putnam (1970), p. 194.

than robots whose movements are controlled from Mars. We are to suppose that, up to the point of the discovery that these cleverly designed, meowing automata are not animals, the word "cat" was defined in terms of animality. Thus, according to Putnam, the intensionalist is in the position of endorsing the analyticity of

(5.1)   Cats are animals

even though cats are not animals and we know this.

The blue lemon case goes as follows. Suppose we find out that the fruits that we use to make lemonade are really not normal cases of lemons, but instead diseased freaks whose undiseased counterparts, the actual normal cases of the kind, are blue. Or, suppose one day it turns out that lemons change their color to blue as the result of a gas entering the atmosphere. We are to assume that something like 'yellow when fully ripe', in the sense of yellow being the characteristic color of a normal case of the natural kind, is a defining feature of the word "lemon." We are to assume, further, that the essential nature of lemons, here understood as their physico-chemical structure, remains unchanged.[93] Thus, according to Putnam, the intensionalist is again caught in paradox. Since "lemon" is defined, in part, by the property of being yellow in color when fully ripe, the intensionalist is committed to saying that

(5.2)   (Fully ripe) lemons are yellow

is analytic even though in one case normal lemons are blue, and in the other all future lemons are blue.

Putnam is now quite sure that robot cats will still be called "cats" after the discovery that they are not animals.[94] But there was a time when he was not so sure: in the paper where Putnam first brought up this example, he said:

Once we found out that cats were created from the beginning by Martians, that they are not self-directed, that they are automata, and so on, then it is clear that we have a problem of how to speak. What is not clear is which of the available decisions should be described as the decision to keep the meaning of either word ("cat" or "animal") un-

[93] Putnam says: "What the essential nature is is not a matter of language analysis but of scientific theory construction; today we would say it was chromosome structure, in the case of lemons." *Ibid.*, p. 188.

[94] "Even if cats turn out to be robots remotely controlled from Mars we will still call them 'cats'." *Ibid.*, p. 191.

changed, and which decision should be described as the decision to change meaning.[95]

At the time Putnam agreed with Donnellan (who discussed a very similar case in a similar way in the same symposium) that the very question of whether such decisions should be called decisions to change meaning or decisions to change a belief is indeterminate. As Putnam then put it: "this question has no clear sense."[96] He went on to make a highly suggestive remark, which will serve as the basis of our reply to his present extensionalist criticisms. Putnam remarked:

Today it doesn't seem to make much difference what we say; while in the context of a developed linguistic theory it may make a difference whether we say that talking in one of these ways is changing the meaning and talking in another of these ways is keeping the meaning unchanged.[97]

To see what difference a theory can make, let us consider the Donnellan case and then apply the argument there to the cases of the robot cat and blue lemon. Donnellan coins the terms "whale$_1$", whose meaning is such that the connection between whalehood and mammality is necessary, and "whale$_2$", whose meaning is such that this connection is contingent, and then he presents his case as follows:

Against the objection [to the proposition that all whale$_1$s are mammals is a necessary truth] that if we were to discover that many of those creatures we have been calling "whales" do not have mammalian characteristics our response might well be to say that some whales are not mammals, it is alleged that this discovery might be a reason for shifting the meaning of "whale" to that of "whale$_2$". But the supposed reason for shifting of "whale" is not different in kind from reasons for shifting a belief. For it, the discovery about what we have been calling "whales" is on the hypothesis just the sort of reason which ought to lead to the belief that some whale$_2$s are not mammals. Hence it is the sort of reason which leads us to change our belief, if we have it, in the truth of 'All whale$_2$s are mammals'. Without begging the question, we cannot support the position that we now mean one thing 'whale' but something different by it in the hypothetical situation through an attempt at distinguishing reasons for changing meaning from reasons for changing belief.[98]

[95] Putnam (1962), pp. 660–661.
[96] *Ibid.*, p. 661.
[97] *Ibid.*
[98] Donnellan (1962), p. 656.

The crux of the matter here is that, although it is true that the reason in question "is not different in kind from reasons for shifting a belief," there are other reasons, based on a "developed linguistic theory," that distinguish between change of meaning and change of belief. These are provided by (4.14). Normally such reasons will take the form of arguments to the effect that the available empirical evidence about the intuitions of native speakers concerning the semantic properties and relations of sentences (in which "whale" occurs) can be most simply and revealingly predicted on the basis of the definitions of semantic properties and relations.[99] The solution in the Donnellan cases is simply that there is a difference between the reasons for saying that one meaning rather than another is the meaning of a given word and the reasons for saying that one belief rather than another is the belief we should hold about certain creatures. The former have to do with the success or failure of predictions about the semantic properties and relations of sentences, while the latter have to do with the success or failure of predictions about the biological structure and the behavior of animals.

Therefore, if, on the basis of semantic reasons, we can say that "whale" means "whale$_1$", then the sentence

(5.3)   Whales are mammals

is analytic, and anything that is a whale$_1$ is necessarily mammalian. After the discovery that "many of those creatures we have been calling 'whales' do not have mammalian characteristics," we simply say that these creatures are not whales. The semantic reasons are reasons for thinking that the term "whale" does not apply to these creatures. On the other hand, if for the same sort of reasons we can say that "whale" means "whale$_2$", then (5.3) is not analytic and anything that is a whale, if a mammal, is only contingently so. After the discovery in question, we say that some whales are not mammals (while some are), either contrary to or in accord with what we believed prior to the discovery.

Note that on this account no change of meaning need occur. One could occur, either by stipulation or by historical process. But if there is a way to decide initially whether "whale" means "whale$_1$" or "whale$_2$", then there is a way to decide well after the discovery whether a change

[99] In Katz (1967), pp. 49–50, I offer a somewhat simplified example of how predictions about the semantic properties and relations of sentences containing "whale" might allow us to decide whether the lexical reading for "whale" has a semantic marker representing mammality.

95

in meaning has occurred or shortly after it whether a change should be stipulated. In the former case, we check the hypothesis about the lexical reading of "whale" that now best predicts the semantic properties and relations of sentences in which "whale" occurs with the hypothesis that best predicted them prior to the discovery.[100] If we find them to differ in that one has the semantic marker '(Mammal)' while the other does not, we conclude that a change of meaning has occurred. In the latter case, i.e., where we are trying to decide whether to introduce a change (perhaps because a switch over to "whale₂" would give us a term to refer to the class we most wish to refer to), we can use the same form of argument. However, in this application of (4.14) we use predictions about the semantic properties and relations of sentences in which "whale" occurs to tell us what we will be committed to if we adopt the proposal to change the meaning of "whale."[101]

Putnam's robot cat case goes the way of Donnellan's. Putnam assumes that the word "cat" is defined in terms of animality. On this assumption anything that is a cat (in this sense) is necessarily an animal. Consequently, if Putnam's story about the things we have been calling cats turning out to be robots were true, it would follow simply that there never were any cats! There couldn't be, since to be a cat something must be an animal. Everyone may have been taken in by the Martians, or some of us may have had our suspicions; either way, the word "cat" ex hypothesi means "cat₁" (i.e., feline animal). Moreover, we could know this meaning on the basis of reasons that are independent of our beliefs about the world, since this knowledge would be based on predictions of the semantic intuitions of native speakers obtained prior to the discovery. That is, we would argue that a hypothesis about the dictionary entry for "cat" on which its lexical reading contains a semantic marker representing the concept of animality offers the simplest explanation of speakers' intuitions about the semantic properties and relations of the sentences in which "cat" occurs.

The point can be put another way. Putnam claims that his robot cat case is a counter-example to (5.1), and hence he argues that the intensionalist is stuck with the embarrassment of an analytic falsehood. But a

---

[100] This is a general means for determining the occurrence of a change of meaning in natural languages.

[101] This suggests a way of applying semantic theory to the development of artificial languages.

counter-example to (5.1), which is of the form (x) $(C_x \rightarrow A_x)$, is something of the form $C_b \cdot \sim A_b$. However, the robot cats of Putnam's story cannot be described in this way, since ex hypothesi they are C only if they are A. Accordingly, these objects are not x's such that $C_x \cdot \sim A_x$. Rather, on the basis of them, all that can be asserted are statements like '$R_b \cdot \sim A_b$', which are clearly not counter-examples to (5.1). Quite the contrary, Putnam's robot cat cases are counter-examples to the claim expressed by sentences like "The things we have referred to and will refer to using the word 'cat' are all of them animals," which are logically quite different from (5.1).

The only peculiarity that remains to be accounted for is how, if there never were any cats, could we have successfully referred to the automata using the word "cat"? I certainly do not doubt that a great many past utterances of sentences like

(5.4)   Some cats make more noise than others

succeeded in making true statements even on the assumption that Putnam's fantasy is fact. Interestingly, it is Donnellan who, in a more recent paper on reference,[102] provides us with the observation that suggests the lines along which an explanation of such statements can be framed. He points out that objects can be referred to successfully using terms whose meaning involves ascriptions that are literally false of the objects. For example, I can refer to the only child in the room wearing shoes under the expression

(5.5)   the child with the red shoes on

even though the color of the child's shoes is actually green, if, say, I am red-green color-blind and my audience knows this. Likewise, everyone who used the term "cat" with animality as one of its defining properties could have nonetheless referred successfully to the Martian robots insofar as everyone believed (and believed that others believed) that the objects in question are animals. After the discovery this sufficient condition for reference to the robots with the term "cat" no longer exists: the

---

[102] Donnellan (1966). Note that the explanation I provide on the basis of Donnellan's observation will strike everyone as involving an extremely implausible set of circumstances. This, however, cannot count against the explanation, since the implausibility of these circumstances is solely due to the implausibility of Putnam's story. That is, given such an implausible set of events as Putnam asks us to assume, it is no wonder that reference in these circumstances has an air of strangeness about it.

discovery is the discovery that the belief that the creatures are animals is false and presumably the discovery becomes generally known. Because further such reference is now impossible, it might be desirable to stipulate a change in the meaning of "cat" or to coin a new term, say "Marcat." Putnam feels certain that we would continue to use "cat." Maybe so, but if so, then this means we continue to use the phonetic form "cat" in application to meowing automata, either with a new meaning or on the basis of another condition for reference under false description, as, for instance, in the case of our reference to department store employees at Christmas time under the term "Santa Claus."

Putnam's claim that the robot cat example creates a difficulty for intensionalists is thus based on the mistaken supposition that intensionalists must share his simplistic conception of the relation between meaning and reference. On this conception, if $c_1, \ldots, c_n$ are the conceptual components of the meaning of a word w, then, for any object o, if w refers to o on some occasion, it does so by virtue of o's being c_1 & $\ldots$ & c_n. Given, further, that cats turn out to be Martian robots, intensionalists have to admit, according to Putnam, that analytic sentences can be used to make false statements, as when someone who is ignorant of the terrible truth and uses the word "cat" in (5.1) in reference to what are actually Martian robots thereby falsely implies that these machines are animals. But there is no reason for the intensionalist to subscribe to Putnam's conception of the relation between meaning and reference. Instead, one may take the following view. There are, roughly, two ways that declaratives like (5.1) and (5.4) can make a statement: by virtue of their presupposition (their semantic condition of statementhood, in the case of the nongeneric sense of (5.1), the condition that there are objects with the properties $c_1, \ldots, c_n$) being satisfied, and otherwise. If the utterance of an analytic sentence "Ws are P" makes a statement in the former way, then, according to this conception, the predication of P is necessarily true and cannot be false. On the other hand, if it makes a statement in the latter way, then "all (semantic) bets are off." In this case there are no implications about anything having P (although it might turn out that P is truly predicated). This distinction would have to be made even if there were no need to handle examples like the possibility of robot cats, since someone might very well succeed in making statements about anything — dogs, diamonds, or dragonflies — on the basis of a prior arrangement under which the subject word of (5.1) is used as a code word for dogs, diamonds, dragonflies, or what have you. On this view of how meaning and reference are related, then, there is no difficulty for the intensionalist in admitting that analytic sentences can be used to make false statements about Martian robots because, ex hypothesi, these uses are not ones where the presupposition of the sentence is satisfied.[103]

We turn now to the blue lemon cases. The first is where normal instances of the natural kind are blue but we have only seen abnormal yellow instances and so falsely believe that lemons are yellow. The only difference between this case and the previous one about robot cats is that in the present case two definitions for the "word" lemon are possible. These definitions can, roughly speaking, be thought of as ones that differ in the way that the definitions of the natural-kind term "quadruped" differs from "four-legged" (i.e., the former term but not the latter applies to a horse that has lost a leg). One definition expresses the sense of "fruit that is yellow when fully ripe, etc.," while the other expresses the sense of "fruit, the normal members of which are yellow when fully ripe, etc." The choice is not essential to us, since we may assume that the decision is based on linguistic evidence and inferences about its significance made in accord with (4.14).

Putnam thinks that my notion of a dictionary entry somehow excludes the representation of natural-kind concepts and he criticizes me in a number of places for too narrow a notion of a dictionary entry. He has misread the theory he is criticizing. I do not restrict dictionary entries in this manner. A lexical reading is a set of semantic markers, and nothing in the theory prevents some of them from representing natural-kind concepts. I have never said what the structure of a semantic marker would have to be like in order to represent such a concept, but this is because I have never had a clear idea of how to provide a revealing analysis of the concept of a natural kind.

I would wholly agree with Putnam that the word "lemon" is a natural-kind word and I would thus wish to write its lexical reading so as to restrict definitional attributions to something like normal (healthy?) members of the kind. But I can't convince myself that talk about normal members is much of an improvement over saying nothing.

Suppose now that linguistic evidence and inferences from it in accord

---

[103] I wish to thank Scott Soames for discussions that helped to clarify some points in my thinking on these cases.

with (4.14) establish one of the above two choices as the definition of "lemon". Thus, both (5.2) and

(5.6)   Normal lemons are yellow (when fully ripe)

are analytic. Suppose also that, as Putnam imagines, we discover that these fruits are normally blue. The evidence on the basis of which we come to make this discovery is evidence about the essential nature of this fruit; hence, on Putnam's account, it relates to the scientific facts about the physico-chemical structure of lemons (and perhaps also about the character of the lemon disease). Since, therefore, this evidence does not concern the semantic properties and relations of words, it cannot influence the choice between candidate dictionary entries for "lemon." This evidence, new and startling as it is, can have no bearing on the correctness of the definition of "lemon" already established on linguistic grounds.

Hence it cannot have a bearing on the claim that sentences (5.2) and (5.6) are analytic. Clearly, prior to the discovery that normal, healthy instances of the kind lemon are blue rather than yellow, everyone who used either of the sentences (5.2) and (5.6) had the requisite false beliefs (about the nature of lemons, about the beliefs of others, etc.). Since the reference of the subject of these sentences was a function of such beliefs, *utterances* of (5.2) and (5.6) refer to lemons and falsely predicate yellowness of them. But these are not tokens in which the presupposition of (5.2) or (5.6) is satisfied.

The same sort of explanation accounts for the fact that a *particular* sentence like

(5.7)   That lemon is sourer than the other one

sometimes made true statements and sometimes made false ones. Again, we assume everyone was possessed of the appropriate set of false beliefs, so that the subjects of the clauses in (5.7) succeed in designating some contextually specified fruits on various occasions of the use of this sentence and these utterances make statements, sometimes true and sometimes false, about their relative sourness. After the discovery, since everyone now knows there are no things that are normal members of the natural kind in question and yellow, the noun phrases in (5.7) no longer succeed in referring. The same goes for the subject of sentences like (5.2). Since their reference must now go through on the basis of meaning and true beliefs, it does not go through at all. Accordingly, these sentences no longer make statements. Formerly their subjects succeeded in referring due to widespread false beliefs and their predicates made true attributions of yellowness to their referents. Now their predicates have nothing to which to attribute yellowness, and so these sentence-uses make no statements (though this is not strictly true, since some of these utterances might be made under some code arrangements by which they make statements about the arms race, Howard Hughes, or vegetarianism).

The second blue lemon case is where normal instances of the kind were yellow but are no longer due to some causal process whereby their color is changed to blue. This case is somewhat different since here the natural kind itself undergoes change. Before the change in the natural kind, sentences like (5.2), (5.6), and (5.7) made statements without a special basis for successful reference — they refer under true descriptions. Afterward the vocabulary of the language may take one of two directions. Either the word "lemon" changes its meaning, so that the future dictionary of English represents the phonetic form "lemon" as a term of a natural kind whose normal member is blue (perhaps "lemon" might be ambiguous or the former sense might die out), or the word does not change its meaning and a new word, say "blumon," comes into the language. Which happens and why is an irrelevant matter of diachronics. After the fact we can decide, synchronically, on the basis of linguistic data and (4.14), which direction the language has taken. At the time of the discovery that lemons changed color, someone may announce the change by an appropriate use of a sentence like (5.8):

(5.8)   Lemons have changed color from yellow to blue

The dual reference, both to the yellow lemons of yesteryear and the blue lemons of today and tomorrow, goes through because everyone knows (and everyone knows everyone knows) that at this critical moment there is no word available in the language (such as "blumon") to use in making the announcement.

5.2. KRIPKE

In some respects Kripke is not happily classified in the Quinian tradition, particularly because he takes issue with Quine on the existence of necessary truths. Moreover it is not entirely clear that Kripke shares the anti-intensionalist views of Quine and Davidson. Yet he does agree with Putnam's conclusions about analyticity and he explicitly cites Putnam as the philosopher he finds most like-minded on questions about the definition of natural-kind words.[104] Most importantly, however, part of Kripke's argument, if sound, would offer direct support to extensionalism. This is the part where he contends that his arguments against the Searle-Strawson account of how proper nouns refer can be converted into a general argument against the meaningfulness of both proper and common nouns. Kripke writes:

My own view . . . regards Mill as more-or-less right about 'singular' names [proper nouns], but wrong about 'general' names [common nouns]. Perhaps some 'general' names ('foolish', 'fat', 'yellow') express properties. In significant sense, such general names as 'cow' and 'tiger' do not, unless being a cow counts trivially as a property. Certainly 'cow' and 'tiger' are not short for the conjunction of properties a dictionary would take to define them, as Mill thought.[105]

But this attempt to extend his arguments about reference to the area of meaning fails. Kripke's arguments against the Searle-Strawson account of reference provide no support for the claim that Mill is wrong about the meaning of common nouns. Nor do they even support the claim that Searle and Strawson are wrong about the meaning (as opposed to the reference) of proper nouns! I shall try to show that, on the basis of the considerations in the previous subsection, it follows that Kripke's counter-examples to the Searle-Strawson account are irrelevant to any of these questions about meaning.[106]

These counter-examples of Kripke's can be taken to show that the referent of a proper noun is not always determined on the basis of a cluster of properties K that function as uniquely identifying marks for speakers who believe correctly that the referent of the noun satisfies 'x is K'. I think that these counter-examples prove that the properties K

---

[104] Kripke (1971), p. 318.

[105] *Ibid.*, p. 322.

[106] I should point out at the outset that, in fact, I do not subscribe to the view that proper nouns are meaningful. Though I take issue with his reasons for claiming that proper nouns are not meaningful, I agree with Kripke's conclusion about them. My reasons for holding this view of proper nouns can be found in Katz (1972), pp. 380–382. Note, however, that I do not also hold that common nouns are in general meaningless. Rather, I think that Mill is right about the meaning of proper and common nouns, but I do not agree with him that words refer just by virtue of their referent's satisfying their definition. As should be clear, I think meanings determine reference but not alone nor in a direct way.

need not be uniquely identifying, and that, even when these properties are uniquely identifying, they need not be true of the referent on the occasion of use in question. But we may accept these and the other claims that make up the Kripke-Donnellan position of reference, if not as proofs, then for the sake of argument.[107]

Accepting them, we may ask whether they imply anything about the truth or falsity of statements about semantic relations such as (5.9):

(5.9)   "Homer" means "the ancient poet who wrote the Iliad and Odyssey"

Given the appropriate form for such counter-examples in this case, it would follow that someone who thought that the name "Homer" always refers by virtue of 'the ancient poet who wrote the Iliad and Odyssey' functioning as a uniquely identifying mark for speakers was wrong. But it would not also follow that someone who thought that (5.9) would be wrong. Someone who holds that proper nouns are meaningful does not need to hold in addition that their meaning is always the basis on which they refer. As we have already seen in connection with the replies to Putnam's alleged counter-examples to intensionalism, we can account for successful reference without assuming that it is directly based on meaning if we assume the appropriate beliefs on the part of the speaker and the audience.

Taking our cue from Kripke, we can imagine it turns out that Homer wrote nothing. In the case we are imagining, the Iliad and Odyssey were written by Homer's Hebrew slave Morris. (In another imaginable case, neither work was written by anyone: a large quantity of ink was spilled on a windy day and it formed the appropriate sequences of Greek letters on some nearby papyrus.) Nonetheless (5.9) could still be true. Statement (5.9) states a fact about the English language that can be verified by evidence about whether or not a dictionary entry for "Homer" based on (5.9) predicts the semantic properties and relations of sentences in which "Homer" appears better than rival entries. Clearly, given the possibility of reference under false lexical descriptions, such verification is independent of whether Morris actually wrote both the Iliad and the Odyssey and his master, Homer, took the credit. The case is exactly analogous to our previous claim against Putnam that the analyticity of

[107] Donnellan (1970), pp. 335–358.

(5.1) is entirely consistent with cats being Martian robots rather than animals. Just as we argued against Putnam's alleged counter-examples that the analyticity of (5.1) does not preclude successful references to Martian robots under the false description 'feline animal', so we can argue here that (5.9) does not preclude successful references to a Greek plagiarist under the false description "the ancient poet who wrote the Iliad and Odyssey." If it had turned out that no one wrote the Iliad and the Odyssey, that Homer found the papyrus with the inscriptions on it and "published" them under his name, then the proper noun "Homer" defined as in (5.9) would have the status of words like "witch" or "gremlin." A sentence like

(5.10)   Homer was Greek

could refer and make a true statement, just as a sentence like

(5.11)   The witch caught yesterday escaped today

could refer and make a true statement (say, in early America at Salem).[108] There are more complicated aspects of reference under false descriptions, but these do not affect my point.[109]

Kripke's observations about reference and necessity are, I think, often important and true. His mistake is to have jumped from them to a false conclusion about meaning. The reason he makes this mistake is, I think, the same as one assumption underlying Putnam's confusions concerning meaning, namely, the failure to distinguish between what might be called *lexicographical meaning*, specifications of which answer the question "What does the word (phrase, clause, or sentence) W mean in the language L?", and what might be called *scientific meaning*, specifications of which answer the question "What is the essential nature of W's, of the referent of 'W'?" I take it that Kripke wants to deny this distinction when he writes "some people think right away that there are really two concepts of metal operating here, a phenomenological one and a scientific one which then replaces it. This I reject . . ."[110] Kripke gives no independent argument against this distinction (I will return to this point below). Moreover, the distinction would appear intuitively sound in light of the

---

[108] Kripke's other example of the three-legged tiger can be handled simply by an appropriate change in the definition under consideration, namely, a change from the property of three-leggedness to the property of quadrupedness, i.e., the property of belonging to a kind whose typical member is naturally four-legged.
[109] Donnellan (1970), pp. 335–358.
[110] Kripke (1971), p. 315.

ease and naturalness with which we separate the lexicographical meaning of a noun like "wave" (i.e., "a moving ridge or swell on the surface of a liquid") from a scientific account of the essential nature of its referent. Because Kripke fails to make this distinction, what others speak of as meaning in the lexicographical sense seems to him (and Putnam) as merely a rough, and often unreliable, approximation to meaning in the scientific sense, to be replaced eventually by a scientific account of the object, event, or relation in question (i.e., by a better approximation).

On this conception of meaning, it is reasonable to think that Kripkean criticisms of the Searle-Strawson theory of referring apply also to the doctrine that proper and common nouns are meaningful in the lexicographical sense. It is, for example, reasonable to think that "gold" cannot have the property of being yellow as part of its meaning if one can imagine that there could be a general optical illusion causing everyone to see gold as yellow when it is some other color.[111] Supposing that property P cannot be part of the essential nature of something S if it is possible for S to exist without P, it follows that it cannot be part of the essential nature of gold that it is yellow. Thus, on the assumption that 'P is part of the meaning of "gold" ' is to be understood as 'P is part of the scientific account of the essential nature of gold,' the property of yellowness cannot be part of the meaning of gold.

But if we make the above distinction between the two kinds of meaning, 'P is part of the meaning of "gold" ' cannot be understood as 'P is part of the scientific account of the essential nature of gold' and the argument against the meaninglessness of proper nouns and most common nouns collapses. For the possibility that gold is not in fact yellow has no bearing on whether the property of yellowness is part of the meaning of the English word "gold"; the property of yellowness can be part of the meaning of "gold" and knowledge of this fact can be the basis on which we explain how speakers use "gold" to refer to gold when falsely believing that gold itself is yellow.

Above I said that Kripke offers no independent argument against distinguishing between lexicographical meaning (what he calls a "phenomenological" concept) and scientific meaning. There is one argument which he seems to think allows him to reject this distinction. It concerns why "light" is not synonymous with "whatever gives us visual impressions — whatever helps us to see" and why "heat" is not synonymous

---

[111] *Ibid.*, pp. 315–316.

with "whatever gives people these sensations (of heat)." The argument is roughly: "the property by which we identify [heat] originally, that of producing such and such a sensation in us, is not a necessary property [of heat as it really is] but a contingent one. This very phenomenon could have existed but due to differences in our neural structures . . . have failed to be felt as heat."[112] The conclusion is that, not being an essential property of the phenomenon, the property of having such-and-such effect on us cannot be part of the meaning (in the intensionalist's sense) and thus the only things that can be synonymous with "light" or "heat" are descriptions that involve properties that light or heat possess in every possible world.

This argument is circular. It assumes the absence of the distinction it purports to refute: in order for the argument to work we must suppose that there are not two different notions of synonymy, corresponding to the two notions of meaning distinguished above, but only one, corresponding to the notion of scientific meaning. Suppose there is both a relation 'same in lexicographical meaning', which holds between (a) and (b) in cases like (5.12) and (5.13),

(5.12)   (a) navel
          (b) belly-button
(5.13)   (a) Flattery will get you nowhere
          (b) Gratifying the vanity of others by praise or attentions
              will not get one any advantage

in addition to the relation 'necessarily the same in extension', which holds between (a) and (b) in cases like (5.14) and (5.15),

(5.14)   (a) the number two
          (b) the even prime
(5.15)   (a) Equilateral triangles are equiangular
          (b) Equiangular triangles are equilateral

where the former relation does not hold between the members of pairs like (5.14) and (5.15). On this supposition, against which the argument has nothing to say, the discovery that a component of a construction synonymous in the lexicographical sense with "light" or "heat" does not express a necessary property of light or heat can no more be grounds for rejecting the construction as synonymous in the lexicographical sense

---

[112] *Ibid.*, p. 326.

than the discovery that cats are not animals can be grounds for rejecting (5.1) as analytic. Rather than showing there is no distinction between lexicographical or phenomenological meaning and scientific meaning, Kripke's argument presupposes the absence of this distinction. If there are these two notions of synonymy there are also these two notions of meaning.

Of course, the lexicographical meaning of a word can sometimes do double duty, serving also as the scientific meaning of its referent. This seems to me to be a general feature of very early stages of scientific development. At such stages sciences use lexicographical meanings as their concepts and come to replace them with scientific meanings when phenomena arise that the former concepts cannot explain. Therefore a lexicographical meaning in its capacity as a scientific meaning might be proven inadequate by a discovery that the attributes comprising the concept are sometimes present in things we do not, and rightly do not, regard as instances of the substance in question. As Kripke points out, something yellow, a metal, soft-for-a-metal, etc. (where the "etc." is spelled out by appropriate further attributes of the lexicographical meaning of "gold"), might turn out not to have the molecular structure of gold and this discovery might make us refuse to count it as gold (from a scientific viewpoint) even though it is yellow, a metal, soft-for-a-metal, etc. But this would not show that the cluster of attributes 'yellow, a metal, soft-for-a-metal, etc.' is not the meaning of "gold" in English. We have sound scientific reasons for not classifying whales together with fish, but these do not even suggest that the attribute of fishness is not a part of the meaning of the English word "whale." These reasons do not tell us whether "whale" in English means "whale$_1$" or "whale$_2$". Other reasons, ones that qualify as semantic by (4.14), are required to determine this.

6. Another Stab at Reducing the Theory of Meaning: Lewis

Lewis's arguments[113] against intensionalism are of special interest here because they directly attack the version of intensionalism that I have tried to defend. This version is a theory about the proper form for transformational grammars, namely the view that an optimal grammar of a natural language requires a level of semantic representation at which it

[113] Lewis (1969), and also parts of Lewis (1970).

states the meaning of a sentence componentially and in so doing states its logical form. Lewis's arguments seek to refute this view by providing an extensionalist reduction (or rather the sketch of one). This attempt at reduction involves the two reduction steps mentioned in our abstract characterization of an argument to reduce the theory of meaning: first, the elimination of constructs in the 'to-be-reduced' theory's vocabulary that fail to satisfy methodology criteria; second, the demonstration that the remaining constructs and the statements of the 'to-be-reduced' theory that are expressed in terms of these constructs can be formulated, without loss of explanatory content, in the vocabulary of the 'reducing' theory.

To take the first of these steps, Lewis seeks to eliminate semantic markers and readings as viable constructs of the theory of meaning. He argues as follows:

Semantic markers are *symbols*: items in the vocabulary of an artificial language we may call *Semantic Markerese*. Semantic Interpretation by means of them amounts merely to a translation algorithm from the object language to the auxiliary language Markerese. But we can know the Markerese translation of an English sentence without knowing the first thing about the meaning of the English sentence: namely, the conditions under which it would be true. Semantics with no treatment of truth conditions is not semantics. Translation into Markerese is at best a substitute for real semantics, relying either on our tacit competence . . . as speakers of Markerese or on our ability to do real semantics at least for the one language Markerese. Translation into Latin might serve as well.[114]

Although it is true that we can know the reading that an English sentence will receive in a grammar without knowing much about its meaning, the reason this is so is not the reason that Lewis's argument requires. Someone may know that a grammar of English assigns a reading R to a sentence S in a purely formal way, say by observing that a machine constructs a derivation in which R is mapped onto S by projection rules. This way requires neither knowledge of the language nor knowledge of the theory of its grammatical structure (beyond the basics of phrase marker assignment). However, a person who does not know semantic theory cannot know what claims R makes about the meaning of S. Since semantic theory provides the intended interpretation for the semantic portion of the grammar of a natural language, the reading R

---

[114] Lewis (1970), pp. 18–19.

is merely a concatenation of meaningless symbols unless interpreted under an appropriate semantic theory. Similarly, someone may know that a chemical description of some covalent bond represents it as

$$
\begin{array}{c}
\text{H} \\
\cdot\,\cdot \\
\text{H} : \text{C} : \text{H} \\
\cdot\,\cdot \\
\text{H}
\end{array}
$$

(6.1)

without knowing what this symbolism asserts about the nature of the bond. To know this, we have to know the relevant portions of chemical theory: we have to understand that configurations of symbols like (6.1) represent electronic structures, that the element designations represent the kernel of atoms consisting of a nucleus and inner electrons, and that the dots represent paired electrons in the valence shell held jointly by two atoms. Likewise, to know what a reading claims about the semantic structure of a sentence, we have to know the relevant portion of semantic theory: we have to understand that certain semantic markers designate certain concepts and that the formal relations among the symbols in a semantic marker represent certain specific relations among the components of the concept designated by the marker.

What Lewis's argument claims, however, is not that we do not know a *theory* but that we do not know a *language*. Thus, the argument is irrelevant to his conclusions about semantic theory. The blunder in his argument is its false assumption that the theory he is discussing is simply a language.[115] Thus Lewis mistakenly construes a reading to be a "translation" into the artificial language "semantic Markerese." Because he confuses a theory with a language and thereby ignores the explanatory principles comprising semantic theory, he can go on to say that translation of sentences into Markerese is the same as translation of sentences into Latin; further, that not knowing or knowing the Markerese "translation" of an English sentence is no different from not knowing or knowing its Latin translation; and, finally, that we gain no more insight into the meaning or truth conditions of sentences from natural languages from their Markerese "translations" than we do from their

---

[115] This is the same fallacy that Putnam commits when he takes semantic theory to be an artificial language and a reading to be a translation into it. I suspect that this fallacy goes back to the general confusion between languages and theories characteristic of logical empiricism and of Quine; cf. Chomsky (1968), p. 53.

Latin translations. Lewis's conclusion that there is no point to semantic theory is not surprising, then, since there is obviously no point to constructing artificial languages that are no more relevant to problems of meaning and truth than Latin.

However, the relation between a sentence of a natural language and its reading(s) is no more translation than the relation of (6.1) to a covalent bond is translation. The terms of a translation relation are sentences, clauses, phrases, words, or morphemes from different languages. But the terms of the relation 'R is a reading of S' are in one case such linguistic objects and in the other theoretical constructions that explain what information is in the meaning of a linguistic object. Unlike translation, which simply expresses the same meaning in another language, the reading of a sentence provides part of the grammar's account of how the components of the meaning of the sentence enter into the internal phonetic-syntactic-semantic connections by virtue of which the sound and meaning of the sentence are related in the competence of speakers.

There are, moreover, obvious logical differences between 'translation of' and 'reading of'. For example, the former is symmetrical, i.e., if S_1 is a translation of S_2, then S_2 is a translation of S_1, but the latter is asymmetrical, i.e., if R is a reading of S, then S is not a reading of R.

Lewis fails to understand that a transformational grammar is designed as a theory of the native speaker's linguistic competence. He fails to understand that it explains the speaker's intuitions about grammatical structure on the basis of a hypothesis about what rules they have internalized in learning the language. Lewis might disagree with such an explanation or with the part that semantic theory is supposed to play in it, but had he understood it he could not have overlooked the glaring differences between languages and theories about their grammatical structure, such as that not knowing a language (like French, Latin, or Esperanto) disqualifies one as a competent speaker of (or translator into) that language, while not knowing its theory disqualifies someone as a competent judge on questions concerning the internalized rules underlying intuitions about its grammatical structure.

This is perhaps also the place to consider briefly a more popular version of this criticism, one resting in part on essentially the same basic confusion between languages and theories as Lewis's. "What exactly," asks John Searle,

are these 'readings'? What is the string of symbols that comes out of the semantic component supposed to *represent* or *express* in such a way as to constitute a description of the meaning of a sentence?[116] . . . Either the readings are just paraphrases, in which case the analysis is circular, or the readings consist only of lists of elements, in which case the analysis fails because of inadequacy: it cannot account for the fact that the sentence expresses a *statement*.[117]

This "dilemma" makes every mistake possible with this argument form.[118] First, it is not a real dilemma but just a disjunction because Searle provides no consideration to establish that the two alternatives he offers exhaust the possibilities. Readings might turn out (as, in fact, they do) to be a representational scheme in a theory and to be formally far richer than a mere list of elements. Thus, we are not forced to choose between the alternatives Searle cites and thereby incur their alleged bad consequences. Searle's argument is fallacious because its conclusion, that semantic theory is bad, does not follow from showing that both alternatives lead to something bad.

Second, even if the argument were a genuine dilemma, it would still be fallacious because in neither case does the bad consequence actually follow from the alternative. In the case of the former alternative, Searle tries to argue from the hypothesis that readings are paraphrases to the claim that the theory is circular. Harnish disposes of this alleged implication quite neatly:

. . . suppose that the theoretical metalanguage is some form of regimented English. Then is the analysis circular? There are two ways of taking this. First, it might be claimed that the analysis of paraphrase is *logically* circular. In this case, the definition of 'x is a paraphrase of y' would contain the expression 'is a paraphrase of' on the right side as well. However, an inspection of Katz's definition reveals no such circularity. So this could not be what is meant. Second, the theory might be said to be *epistomologically* circular in that the ability that the theory characterizes must be invoked to test the theory. But surely, if correct, this is harmless. A grammar of a language explicates the notion of 'grammaticality' for the speakers of that language, but we nevertheless *test* the theory by having fluent speakers (usually the theorist himself) judge whether the output of the grammar is grammatical or not. Indeed, when one undertakes the project of modelling another system it is hard to

[116] Searle (1972), p. 22, col. 2.
[117] *Ibid.*, col. 3.
[118] I am much endebted to Robert M. Harnish's unpublished ms. "Semantic Theory and Speech Acts: A Reply to John R. Searle" for helping me to clarify my thoughts on Searle's criticism and for supplying some of the rebuttals (see below).

111

imagine how else one could proceed to test the model. In the extreme this kind of objection could be turned into quite a bludgeon. Suppose some-one were to construct a model of the human visual system or some sig-nificant part of it. One could then argue that it cannot be tested with-out circularity. Why? because "the ability [to see the test situation] pre-supposes the very competence the [visual] theory is seeking to explain." Such an argument secures much too much, so this cannot be what is meant. So the argument for circularity does not really hold . . .[119]

In the case of the latter alternative, Searle tries to argue from the hy-pothesis that readings are lists of elements to the claim that the theory is inadequate because, as he says in the foregoing quotation, the theory "cannot account for the fact that the sentence expresses a *statement*." Harnish also disposes of this alleged implication easily:

. . . 'statement' in the above is to indicate a speech act. Now, one puz-zle with the above criticism is what sentence is 'the sentence' that Searle is talking about? Surely it is not the generic use of 'the' to refer to a class or kind of thing, because not all sentences are in any way related to stat-ing. Imperatives and interrogatives are not. Perhaps Searle was referring to the example sentence in the Katz and Fodor article which he refers the reader to and takes some examples from: "The man hits the colorful ball." But of course this *sentence* could not express an act. Perhaps Searle meant that the sentence is *used to make a statement*. Then the objection would be that Katz's theory is inadequate, if a list, because it does not account for the fact that the sentence in question is used to make a statement. But this has to be modified slightly too, if it is not to be misleadingly exclusive. For the sentence in question is used, or at least was used, to perform other speech acts as well. Indeed, Katz and Fodor used it to give an example of their semantic theory. So the ob-jection should be that Katz's theory does not account for the fact that the sentence (in question) *can be used to make a statement*. Of course, if you tell the right kind of story, almost any sentence can be used to perform almost any speech act. This makes the above inadequacy sort of thin. Presumably what Searle has in mind is that the sentence in question (and others 'like' it) can be used to make statements in *stand-ard* or *normal* circumstances, whereas it could be used to make a com-mand only in non-standard or non-normal circumstances. We all, I think, feel that there is something right about this, but it shifts a very heavy (so far unbearable) burden on the analysis of circumstances of utterance as 'standard' or 'normal' or not. Nevertheless, we must pro-ceed on the assumption that we can make this distinction at least pre-theoretically.

If this is Searle's point, then perhaps he has gained it. A quick glance

[119] Harnish, "Semantic Theory and Speech Acts," p. 6.

112

at the Katz and Fodor paper will convince one that they indeed did not account for the fact that the sentence could, under normal circumstances of utterance, be used to make a statement. So what? Determining the speech-act potential of sentences was *not* one of the abilities Katz and Fodor said they were going to try to model, nor was it one of the abilities that Searle reported them as trying to account for. Thus, Searle has not yet shown the theory to be inadequate in *its own domain*. The theory is also inadequate for accounting for stellar parallax, but I do not think we should lose any sleep over that.[120]

Third, even if Searle's argument were logically valid, it woud not be sound, since both alternatives embody false assumptions about the theory Searle is criticizing. The first alternative assumes that readings are paraphrases. This is the mistake Lewis made of confusing a language and a theory. Readings are not paraphrases but formulas in the technical notation of a theory which explicate the senses of expressions and sentences in a natural language and predict paraphrase relations. Calling them paraphrases is like calling (6.1) itself a covalent bond. The second alternative assumes that readings are lists of elements. This is simply a misrepresentation. Even on the earliest version of the theory (Katz and Fodor, 1963) readings incorporate some of the richness of truth-functional connectives, second order predication, and relational logic, while on any version from 1965 on, they have the structure of phrase markers. Calling readings lists of elements is like calling a chemical diagram like (6.1) a list of dots and letters.

Lewis's claim that my intensionalist position "leads to a semantic theory that leaves out such central semantic notions as truth and reference"[121] is like Searle's claim that semantic theory fails to account for speech acts like stating. As Lewis puts it:

The Markerese method is attractive in part just because it deals with nothing but symbols: finite combinations of entities of a familiar sort out of a finite set of elements by finitely many applications of finitely many rules. There is no risk of alarming the ontologically parsimonious. But it is just this pleasing finitude that prevents Markerese semantics from dealing with the relations between symbols and the world of non-symbols — that is, with *genuinely semantic relations*.[122]

Harnish's reply to Searle that the criticism does not show the theory to be inadequate in *its own domain* is also a fitting reply to this criticism

---

[120] *Ibid.*, p. 7.
[121] Lewis (1969), pp. 170–171.
[122] Lewis (1970), p. 19. Italics mine.

of Lewis's, since explaining truth and reference lies outside the domain of intensionalist semantics as much as do speech acts. What makes it seem as if truth and reference are semantic notions in the appropriate sense is the unfortunate ambiguity of the term "semantic" whereby the theory of reference has come to be referred to as "semantics," too. Lewis's criticisms here trade on this ambiguity.

The term is used by philosophers and logicians in two ways. On the one hand, "semantic" is used in connection with the theory of meaning to refer to relations between linguistic expressions like synonymy, ambiguity, analyticity, etc. On the other it is used in connection with the theory of reference to refer to relations between linguistic expressions and the world, like truth, denotation, satisfaction, etc. Given this ambiguity of "semantic," the question arises: In which sense is "semantic" to be understood when the term appears in Lewis's claim that my theory leaves out "such central semantic notions as truth and reference," and that "relations between symbols and the world of non-symbols" are the "genuinely semantic relations"?

Suppose "semantic" is understood to have to do with reference rather than sense. In that case the omitted notions are, without doubt, central to semantics, but the subject of semantics to which they are central is not the subject to which my theory addresses itself. On this construal, Lewis's criticism is as damaging as a criticism of syntactic theory for not directly treating the "genuinely grammatical" notions of rhyme and alliteration. Suppose, alternatively, that "semantic" is understood to have to do with sense rather than reference. In this case the term "semantic" covers the subject that my semantic theory is about, but now the omitted notions fall outside of this subject by definition. The subject, on this construal, is about concepts like those asked about in (2.1) to (2.10), concepts that concern the relations among senses of words, expressions, and sentences. Concepts like truth, reference, and satisfaction that concern the relation between such linguistic objects and the world are not central to a theory of sense and need not be handled by it. Intensionalists might be wrong in thinking that sense can be handled without the concepts of referential semantics entering centrally into the explication, but an extensionalist cannot simply assume so. Without any argument to show that these omitted concepts are not properly left out, Lewis's claim that my semantic theory has left out genuinely semantic notions begs the question.

114

A related criticism is Lewis's statement that "The markerese method . . . deals with nothing but symbols" and that this prevents it ". . . from dealing with the relations between symbols and the world of non-symbols."[123] This criticism is not simply a restatement of the criticism just dealt with. If I understand it properly, Lewis is saying further that notions like truth, reference, satisfaction, etc., are not dealt with at all in my theory because the formal representation of relations among the senses of linguistic expressions plays no role in accounting for relations between them and the world. But this criticism simply overlooks the intensionalist thesis that the formal representation of relations among senses accounts for as much about referential relations as can be accounted for without information about the various contexts of utterance. Semantic theory deals with aspects of referential properties and relations, indirectly, on the basis of what it says about the meaning relations that influence them. Because Lewis focuses exclusively on the formal representation scheme of semantic markers and readings and so ignores the part of semantic theory that provides the interpretation of this formalism, he misses what the interpretation says about the extensional relations of senses of linguistic objects.

Let us consider some examples. The semantic component of an English grammar ought to assign the same reading to grammatically distinct expressions like those in (6.2) and (6.3):

(6.2) (a) valise
       (b) traveling bag
(6.3) (a) sweat
       (b) perspiration

If it does so, it does more than merely formally mark a regularity in the judgments speakers make about their semantic intuition. By virtue of the interpretation, the assignment of the same reading implies coreferentiality for standard uses of the expressions. On this interpretation the meaning of an expression (for most expressions of a language[124]) is the information about its extension that the speaker obtains from his knowledge of the language. Therefore, if two expressions like (6.2) or (6.3)

[123] Lewis (1970), p. 19.
[124] Some words are syncategormatic. Semantic theory reconstructs their senses as operators that transform readings; cf. the discussion of "good," "not," etc., in Katz (1964c) and (1972), pp. 157–171.

115

Jerrold J. Katz

receive the same reading, this assignment says that speakers employ the same semantic information (in the sense of (4.14)) to determine their extensions, and further that in the absence of extra-linguistic information that indicates a departure from the standard use of an expression[125] the expressions must be coreferential.

By the same token the assignment of different readings to two expressions, such as those in (6.4) and (6.5)

(6.4)   (a) the even prime
         (b) the number two
(6.5)   (a) creature with a heart
         (b) creature with a kidney,

says that the possibility of non-coreferentiality is not excluded on grounds of meaning. Thus the intensionalist doctrine that sense determines reference allows for the coreferentiality in such cases, and even the necessary coreferentiality in (6.4), to be explained, respectively, in terms of mathematics and science, and so enables us to account for the difference between these cases and cases like those in (6.2) and (6.3).[126]

Clearly, then, the theory of meaning says a good deal about reference, even though the explication of referential notions is not the primary concern of this theory. But to discover what semantic representations say about referential relations it is necessary to consider their interpretation in semantic theory.

Let us now turn to the reduction step. Lewis's attempt to reduce the theory of meaning to the theory of reference accepts the legitimacy of the notions 'analytic', 'synthetic', and 'contradictory' and sets out to define them on the basis of the theory of reference. The theory of reference in this case is not, however, the same as Quine's. For Lewis, the theory is enriched with the apparatus and assumptions of modal logic. It is just this enrichment of the theory of reference that gives it whatever promise it has of providing definitions that express the content of these notions in the vocabulary of extensions. Quine's attack on inten-

---

[125] These notions might be worked out in a theory of linguistic communication that combines a grammar with an account of conversational implicature; see Katz (1972), pp. 441–450.

[126] The assignment of different readings to a single expression not only marks it as ambiguous but says that more than one extension is associated with it, requiring standard uses of the expression to select among these extensions on the basis of information about the context of utterance. Quite analogous points follow about the notion of truth.

sionalism had to argue that notions like 'analytic' are not viable constructs because a theory of reference containing no more than the apparatus of quantification theory is too restricted to define such notions. But in the case of the theory of reference supplemented with the apparatus of modal logic, as with Lewis's conception, there would seem to be a chance of reducing away intensionalism by defining 'analytic' and other such notions without appealing to senses and the like.

But Lewis's definitions are unsuccessful. He defines analyticity as truth in all possible worlds, contradictoriness as falsehood in all possible worlds, and syntheticity as truth in some possible worlds but not in others.[127] Again, Lewis's argument trades on an ambiguity. The term in question is "analytic," which originally had its sense given it by Kant. Kant called a proposition analytic just in case its subject concept contains everything in its predicate concept. The ambiguity arose when Frege introduced another sense for this term. Frege, in a curious passage in the *Grundlagen*,[128] which deals with the Kantian notion of analyticity, without explanation redefines "analytic" as follows:

If, in carrying out this process [of finding the proof of a proposition], we come only on general logical laws and on definitions, then the truth is an analytic one. . . . If, however, it is impossible to give the proof without making use of truths which are not of a general logical nature, but belong to the sphere of some special science, then the proposition is a synthetic one.[129]

This ambiguity between a 'redundant predication' sense and a 'consequence of laws of logic' sense raises the question of how we are to understand the term "analytic" appearing as the definiendum in Lewis's definition. If we understand it in Frege's sense, Lewis's definition explicates the Leibnizian distinction between truths of reason and truths of fact in terms of the distinction between deducibility from laws of logic and definitions (without appeal to premises from a special science) and deducibility from such laws and definitions together with premises from the special sciences. On this disambiguation, there is no need to take issue with Lewis's definition, since it is now irrelevant to the question of reducing the theory of meaning to the theory of reference. It can no

[127] Lewis (1969), pp. 173–202.
[128] Frege (1953), pp. 3°–4°.
[129] I do not wish to argue the historical question of whether this sense was not already latent in Kant's discussion. What matters for us is that it comes to us by virtue of Frege's influence on the philosophy of logic.

longer play a role in such a reduction because the reduction of analyticity in this sense to notions like 'truth' and 'possible world' has no implications for whether analyticity in the Kantian sense is reducible to these notions. (One could, of course, take issue with Lewis on the grounds that not all of mathematics is logic and so some things that are not analytic in Frege's sense are nonetheless true in all possible worlds.)

Hence, in order for the definition to play the role it was intended to play, we must understand its definiendum as "analytic" in Kant's sense. But on this disambiguation, Lewis's definition expresses the false claim that the defining characteristic of Kantian analyticity is truth in all possible worlds. If this claim were true, the class of sentences with predicates that attribute nothing but what is already contained in the meaning of their subject would be co-extensive with the class of sentences that are true in all possible worlds. But not every proposition that is true in all possible worlds is analytic. The sentence (6.6), (6.7), and (6.8)

(6.6)   An equilateral triangle is equiangular

(6.7)   The number two is an even prime

(6.8)   Every even number is the sum of two primes

are true in all possible worlds but are not analytic (in the present sense).

Such counter-examples might be gotten around by appealing to Frege's qualification that analytic propositions be provable without using truths that "belong to the sphere of some special science," since in the case of (6.6) to (6.8) we must appeal to geometry and arithmetic. This way out, however, will not work in cases like (6.9) and (6.10)

(6.9)   Spinsters are either living or not living

(6.10)   Something does not have both fever and chills if and only if that thing does not have fever or it does not have chills

which clearly make no appeal to a special science. These sentences are not analytic in the required sense but they are clearly true in all possible worlds.

Compare (6.9) and (6.10) with (6.11) and (6.12):

(6.11)   Spinsters are female

(6.12)   Fevers are elevations of bodily temperature above normal

Although both pairs (6.9) and (6.10), and (6.11) and (6.12), are true in all possible worlds, they are so for reasons as different as are the reasons for (6.6) to (6.8). The reason that the sentences (6.9) and

(6.10) are true in all possible worlds is that their predicates, respectively,

(6.13)   x is either living or not living

(6.14)   x does not have both fever and chills if and only if x has neither fever and chills

are universally satisfied. That is, (6.9) and (6.10) follow directly from the laws of logic

(6.15) $(x)(F_x \lor \sim F_x)$

(6.16) $(x)(\sim (G_x \& H_x) \equiv (\sim G_x \lor \sim H_x))$

without the aid of a premise belonging to a special science. On the other hand the reason that (6.11) and (6.12) are true in all possible worlds cannot be the same, since their predicates, respectively

(6.17)   x is female

(6.18)   x is an elevation of bodily temperature above normal

are not universally satisfied. Here one cannot argue that (6.11) and (6.12) follow directly from the laws of logic

(6.19) $(x)((H_x \& \sim M_x \& F_x) \supset (F_x))$

(6.20) $(x)(F_x \supset F_x)$

without the aid of a premise belonging to a special science, since both the argument that (6.11) follows directly from (6.19) and the argument that (6.12) follows directly from (6.20) require a factual premise from the special science of empirical linguistics, one that says that the meaning of the words in the sentence — "spinster" and "female" in the former case and "fever" and "elevations in bodily temperature above normal" in the latter — are such as to justify their formalization as (6.19) and (6.20).[130]

On the only construal of Lewis's definitions where they are even relevant to an attempt to reduce the theory of meaning to the theory of reference, they have to be rejected because they are too broad. Counter-examples like (6.9) and (6.10) show that Lewis's definition of 'analytic' fails to characterize the appropriate set of sentences. Instead of characterizing just the analyticities *in Kant's sense*, it characterizes a set which includes these together with analyticities *in Frege's sense*, as well as a variety of other sentences like (6.6) to (6.8) that are true in

---

[130] Frege's reference to definitions in the quotation above concerns definitions that make no truth claim, what Quine (1953b), p. 27, calls "the explicitly conventional introduction of novel notation for purposes of sheer abbreviation."

all possible worlds but for entirely different reasons. Lewis's definition thus fails to make the distinction between necessary truths (generally) and analytic truths (as a special case). Any attempt to draw distinctions of this kind within the class of necessary truths (e.g., to rule out the unwanted logical truths, etc., leaving the genuine analytic truths) would make the entire effort at reduction circular because it would require the introduction of unreduced notions from the theory of meaning. Otherwise, there would be no way to distinguish that particular propositional form that constitutes analyticity from the propositional forms of other necessary truths.[131]

Semantic theory offers an explanation of why truths like (6.11) and (6.12) are necessary even though their predicates are not universally satisfied.[132] On this account, the meaning of such sentences involves an inclusion relation between the part of the meaning that determines the thing(s) about which the assertion is made and the part that expresses the assertion about it (them). Thus a sentence is marked as 'analytic' when its reading represents the latter, the truth condition, as included within the former, the presupposition, and the satisfaction of the presupposition implies the satisfaction of the truth condition. Hence even though predicates like (6.17) and (6.18) are not universally satisfied, anything satisfying the subject of (6.11) must satisfy (6.17) and anything satisfying the subject of (6.12) must satisfy (6.18). The difference between a necessary truth of the Fregean type and one of the Kantian type is that the former is true by virtue of a relation between its truth condition and a law of logic. The latter is true by virtue of a relation, internal to the semantic structure of the proposition, between its truth conditions and its presupposition.

---

[131] This goal, stated in its most general form to include distinguishing between propositional forms so as to explain the difference between sentences that are synonymous and sentences that are merely true and false in the same sets of possible worlds, and so on, would constitute a formulation of the goal of a semantic theory in my sense. No wonder, then, that Lewis's reduction fails.

[132] See Katz (1972), chap. 4, sec. 5. I can account for the fact that there are some possible worlds in which no spinster exists without changing the substance of the argument against Lewis. The simplest way is to keep a presupposition analysis of both Kantian-type and Fregean-type analyticities but change the defining condition to 'truth in all worlds where the presupposition of the sentence is satisfied'. If there are objections to presuppositional analysis, our argument can be recast without relying on it. We can simply switch in both cases to examples that do not require such treatment, e.g., the generic sense (6.11) and (6.12) or conditionals like "If something is a spinster, it is female."

7. You Can't Have Your Cake and Eat It

In this essay I have tried to reveal the roots of contemporary extensionalism in taxonomic linguistics, to trace the main line of its development from its beginnings in Quine's attempt to rationally reconstruct Bloomfieldian structuralism through its elaboration at the hands of extensionalists such as Davidson, and to reply to various influential extensionalist criticisms of intensionalism. My basic argument was that contemporary extensionalism is simply the form Bloomfieldian structuralism takes when transposed from linguistics into philosophy and, therefore, that the assumptions of contemporary extensionalism are open to the criticisms in the transformationalist refutation of taxonomic linguistics.

Some philosophers of language are reluctant to take sides in the controversy between extensionalists and intensionalists. They want to have the virtues of both sides without the vices of either. This they think they can achieve by formulating an eclectic position. This position shares Quine's suspicion of intensional entities, his taste for a parsimonious ontology, and what I've referred to above as the "two dogmas of Quinian empiricism." At the same time they wish to draw a categorical distinction between contingent and necessary truths and to have their theorizing be relevant to natural languages. Many of these eclectics hope to construct a new theory of reference in which concepts from modal logic are used to represent the semantic structure of sentences in natural languages. By incorporating concepts of modal logic, such an expanded theory of reference permits us to retain the absolute concept of logical necessity and indeed to extend it to connections that hold by virtue of relations between words in the extra-logical vocabulary of a language; by sticking to referential properties exclusively, such a theory avoids the intensional entities — meanings, senses, etc. — against which Quine has warned for so many years. In Lewis we have already encountered one such eclectic; others are Hintikka and Montague.[133]

However, there is no such middle ground between a Quinian extensionalism that rejects meanings and necessary connections and a Fregean intensionalism that accepts them. Such a middle ground could exist only if it were possible to reduce intensional concepts to such an expanded theory of reference *in a noncircular way*. That is, the new 'reducing

[133] See Hintikka (1970), pp. 106–111, particularly footnote 26.

121

theory' must extend the absolute concept of logical necessity to connections that hold by virtue of relations among terms in the extra-logical vocabulary without appealing to meaning. In the present case, however, the notion 'possible world' (or its equivalents) depends on the full conceptual apparatus of the theory of meaning. The point can be made as follows. Kripke observes:

A possible world isn't a distant country that we are coming across, or viewing through a telescope. Generally speaking, another possible world is too far away. Even if we travel faster than light, we won't get to it. A possible world is *given by the descriptive conditions we associate with it*. What do we mean when we say 'In some other possible worlds I might not have given this lecture today'? We just imagine the situation where I didn't decide to give this lecture or decided to give it on some other day.[134]

We determine the character of a possible world by determining the meaning of the sentence used to present us with the possibility. Since the meaning of the sentence is determined by the meanings of its component words and by their syntactic structure, the "descriptive conditions," to which Kripke refers, the conditions that express a possibility, are determined by these word meanings and the syntactic relations they enter into in sentences. Thus, in either an explicit or an implicit form, an appeal to meanings is necessary with modal concepts. That is, either we explicitly recognize the semantic source of the conditions that express possible worlds by attempting to determine the features of sentences speakers use to obtain their understanding of possible worlds or we implicitly recognize it, as revealed by the fact that we rest our conception of a possible world on what we mean when we say something.

Therefore if we were to make all of the theoretical commitments of a theory of reference augmented with modal concepts fully clear, we would have to admit that the application of these concepts in natural languages depends on explications of their grammatical structure that use conceptual apparatus from the theory of meaning. The irony is that this point about the involvement of meaning with modality was well appreciated and cogently argued by extensionalists like Quine and Goodman. Appreciation of this involvement is what led them to condemn semantic notions like synonymy and modal notions like possibility as one and the same occultism. Goodman writes:

[134] Kripke (1971), p. 167.

The notion of possible entities that are not and cannot be actual is a hard one for many of us to understand or accept. And even if we do accept it, how are we to decide when there is and when there is not such a possible that satisfies one but not the other of two terms? We have already seen that we get nowhere by appealing to conceivability as a test of possibility. Can we, then, determine whether two predicates "P" and "Q" apply to the same possibles by asking whether the predicate "is a P or a Q but not both" is self-consistent? This is hardly helpful; for so long as "P" and "Q" are different predicates the compound predicate is logically self-consistent, and we have no ready means for determining whether it is otherwise self-consistent. Indeed the latter question amounts to the very question whether "P" and "Q" have the same meaning.[135]

Quine wrote:

This account [Carnap's account of analyticity or necessary truth as truth under every state-description] is an adaptation of Leibniz's "true in all posible worlds." But note that this version of analyticity serves its purpose only if the atomic statements of the language are, unlike 'John is a bachelor' and 'John is married', mutually independent. Otherwise, there would be a state-description which assigned truth to 'John is a bachelor' and 'John is married', and consequently 'No bachelors are married' would turn out synthetic rather than analytic under the proposed criterion.[136]

We can illustrate the relation between a theory of reference extended by the addition of modal concepts and the theory of meaning by considering what Montague has proposed under the title "intensional logic."[137] I do not mean to restrict my comments to his theory. I intend them to refer to the kind of theory that seems to be emerging from a tradition that beings, as far as we are concerned here, with Carnap's ideas (particularly those discussed by Quine in the above quotation) and that is being developed by many of the logicians who are contributing to the renascence of modal logic.

The theory under consideration developed on the Carnapian plan. The eclectic first constructs a formalized language that embodies certain apparatus for reference and prediction, and then tries to exploit the correspondence, such as it is, between the formalized language and a natural language. This is done in order to argue that one or another

---

[135] Goodman (1952), p. 68.
[136] Quine (1953b), p. 23. Quine further observes that "the criterion of analyticity in terms of state-descriptions serves only for languages devoid of extra-logical synonymy pairs, such as 'bachelor' and 'unmarried man' . . ."
[137] Montague (1970).

mechanism in the former is an adequate model on which to understand an aspect of the latter.[138]

The formalized language consists of an applied predicate calculus of the familiar sort and an interpretation employing modal concepts. The calculus makes the distinction between a logical vocabulary and an extralogical vocabulary that we criticized in Section 4, and so contains a set of logical particles of the standard type, a set of punctuation signs, and a set of variables, a set of individual and predicate constants, and a set of operators (including such things as symbols to be interpreted in terms of the modal concepts of necessity and possibility). The interpretation is supplied in the usual manner, except that now a so-called pragmatics may be added. In the pragmatics indexical expressions like "I," "this," "you," etc., are related to their extensions by a complex system of indexes. These indexes determine the extensions of occurrences of indexical expressions in utterances, so that changes in reference from one occasion of use to another do not lead to the consequence that sentences shift their truth value from one such occasion to the next.

I shall not be concerned with the details of this pragmatic aspect of such systems. Everything that I wish to say about the eclectic position can be said on the simplifying assumption that every referring expression has the same extension on every occasion of its use. When we look at a characteristic way of using these indexes, we see that they are nothing more than a development of the standard means for interpreting expressions in extensional systems. For example, Montague says that to specify the intension of each predicate and individual constant c of a pragmatic language (in his sense), we have

to determine, for each point of reference i, the denotation or *extension* of c with respect to i [where "i" codes "all complexes of relevant aspects of possible contexts of use"]. For example, if the points of reference were moments of time and c were the predicate constant 'is green', we should have to specify for each moment i the set of objects to be regarded as green at i. If, on the other hand, c were an individual constant, say 'the Pope', we should have to specify, for each moment i, the person regarded as the Pope at i.[139]

---

[138] Eclectics have never succeeded in making this notion of a correspondence clear enough to support arguments for the linguistic relevance of artificial languages. Cf. Chomsky (1955), pp. 36–45, and Fodor and Katz (1962). But I will not take up this line of criticism here; rather, I will return to the same point from a somewhat different angle later.

[139] Montague (1970), p. 70.

Thus we see that what we have in these systems of interpreting indexical constants is an extension of the earlier Carnapian rules for assigning extensions, typically,[140]

(7.1)   'mond' designates the moon
(7.2)   'blau' designates the property of being blue

where it is understood "for any i."

We may now see how Quine's and Goodman's point about the involvement of modal notions in intensional semantics applies in the case of "intensional logics" of the type under consideration. Roughly, their point is that the interpretation of modal operator symbols must depend on the interpretation of the predicate and individual constants in the system and these, as we see from the above, depend on the semantics of natural languages, that is, on whatever competence permits a speaker to determine the extension of terms like 'mond' and 'blau' on the basis of the meaning of expressions like the English expressions "the moon" and "the property of being blue" and definitions like (7.1) and (7.2).

Modal operator symbols represent monadic sentence-forming operators on sentences. That is, they apply to a sentence, say S_i, and form another sentence, say S_j. The relationship S_j bears to S_i is represented formally in terms of the rules of the system that express their deductive connections. Thus, if the necessity operator is understood as 'truth in all possible worlds' and the possibility operator is understood as 'truth in some possible world', then 'Necessarily S_i' must be interpreted as asserting that the truth conditions of 'S_i' are satisfied in all possible worlds and 'Possibly S_i' must be interpreted as asserting that the truth conditions of 'S_i' are satisfied in some possible world. Only in this way are the truth conditions of the modal sentence S_j properly related to the truth conditions of the sentence it operates on. For example, the sentence of a formalized language that represents the English modal sentence

(7.3)   It is necessary that squares have more sides than triangles

has to assert that the truth conditions of the embedded sentence are satisfied in every possible world. Therefore, not only do we have to know the semantic structure of the atomic statements of the language to know whether a non-atomic statement is analytic or synthetic, as Quine points

[140] Carnap (1938), p. 151.

out above, but, more generally, we have to know the semantic structure (the interpretation of the predicate and individual constants) to know what assertion is made by any modal sentence of the language. This is because the assertion made by a modal sentence S_j depends fully on the interpretation of the truth conditions of the embedded sentence S_i and these conditions depend on the interpretation of the predicate and individual constants in S_i.

Let us now take a closer look at the eclectic's position from the viewpoint of the intended interpretation of predicate and individual constants in these formalized languages. The predicate and individual constants representing words and expressions like "green," "blau," "the Pope," "the moon," etc., are interpreted on the basis of Carnapian semantical rules like (7.1) and (7.2), or some others suitably relativized to supply an extension for a particular context of use. Given that these predicates and constants are interpreted by such rules, we are told nothing about the differences between a word like "moon" and any of the indefinitely many coreferential expressions of English (and other languages), such as "the earth's satellite," "the heavenly body that has a mean distance from earth of 238,857 miles," "the direct source of the illumination that causes a thing to be moonlit," and so on. The rules merely present the designatum of the term appearing in quotes. Accordingly, the interpretation of a sentence, say

(7.4)   The Pope is fat,

is exhausted by saying that the entity designated by its subject term, as determined by a rule such as

(7.5)   'the Pope' designates the Pope,

is in the extension of its predicate, as determined by a rule such as

(7.8)   'fat' designates the fat things.

Thus these rules do not tell us what words or sentences mean, i.e., they do not tell us what condition picks out these extensions. These rules cannot tell us whether (7.4) is synonymous with (7.7) or (7.8):

(7.7)   The head of the Roman Catholic Church is fat
(7.8)   Paul the Sixth is fat

nor can they tell us about any of its other semantic properties or relations. The eclectic's intensional logic ignores the question of how to

126

specify the extension of words, expressions, and sentences (with respect to a given context). The eclectic *postulates* that a mapping of linguistic objects into sets of individuals is somehow already established.

Thus Montague's choice of the name "intensional logic" is an Orwellian euphemism: the theory it names has nothing to say about relations between the meanings of words in the extra-logical vocabulary of a language, but is made to appear as if it had something to say by virtue of what its name suggests when it is glossed in terms of the traditional and familiar extensional-intensional distinction.

From a transformationalist view of linguistic behavior, any attempt to understand mechanisms of reference and predication in natural language involves hypotheses about both the linguistic principles underlying the speaker's verbal behavior and the devices by which such principles are put into operation in real time. On this view, formalized languages will be informative about such mechanisms in natural languages to the degree to which they provide adequate simulations of the speaker's ability to use the language. The question that divides Quinian extensionalists and Fregean intensionalists is whether an adequate theory of the speaker's ability to use language has to employ concepts from the theory of meaning. Quine addresses himself to this issue by offering a simulation in terms of the Skinnerian theory of learning, revised as the theory of stimulus meaning. Some Quinians like Davidson offer other proposals based on different theories. All these efforts are clearly relevant to the issue, whatever their defects might be in regard to how well they handle semantic facts. For the eclectic position to be relevant to this issue, it must provide some account of the speaker's ability to map atomic predicates and terms into sets of things in possible worlds, not simply assume one and construct elaborate apparatus for manipulating such extensions. The eclectic who insists on having his cake and eating it too turns out to make no relevant proposal on this issue.

8. Concluding Remarks

The present essay will not settle the issue between extensionalism and intensionalism. I have no illusions on this score. But if it succeeds in causing philosophers to reexamine their suspicions about meanings and to resist their immediate inclination to prefer solutions to problems that avoid appealing to semantic relations, this essay will have achieved its purpose.

The issue between extensionalists and intensionalists will be settled in the intensionalists' favor only when they construct a semantic theory of sufficient explanatory power to establish their claim that meaning and logical form are one and the same. Thus the task ahead for intensionalists is the constructive one of building such a theory. In the course of further attempts to extend the explication of logical inference, certain demands will be made on the characterizations of the objects to which rules of logic apply. If these demands coincide with what is supplied by the semantic representations of senses of sentences, then intensionalists will have proven their claim. We must come to find that what turns out to be required to extend present formulations of logical laws so that they can account for wider and wider varieties of valid inference is just the information that empirically adequate semantic representations provide about the structure of senses of sentences.[141]

[141] See Katz (forthcoming).

Bibliography

Bloomfield, L. (1933). *Language*. New York: Henry Holt.

Bloomfield, L. (1936). "Language or Ideas?" *Language*, 12:89–95.

Bromberger, S. (1963). "A Theory about the Theory of Theory and about the Theory of Theories," in W. L. Reese, ed., *Philosophy of Science, The Delaware Seminar*. New York: Interscience.

Carnap, R. (1937). *The Logical Syntax of Language*. London: Routledge and Kegan Paul.

Carnap, R. (1938). "Foundations of Logic and Mathematics," in *International Encyclopedia of Unified Science*, vol. 1, nos. 1–5. Chicago: University of Chicago Press.

Carnap, R. (1956). *Meaning and Necessity*. Chicago: University of Chicago Press.

Cartwright, R. (1962). "Propositions," in R. J. Butler, ed., *Analytic Philosophy*. Oxford: Blackwell and Mott.

Chomsky, N. (1955). "Logical Syntax and Semantics." *Language*, 31:36–45.

Chomsky, N. (1957). *Syntactic Structures*. The Hague: Mouton.

Chomsky, N. (1959). "Review of *Verbal Behavior*." *Language*, 35:26–58.

Chomsky, N. (1962). "A Transformational Approach to Syntax," in A. A. Hill, ed. (1962), *Proceedings of the Third Texas Conference on Problems of Linguistic Analysis in English, 1958* (Austin: University of Texas). Reprinted in J. A. Fodor and J. J. Katz, eds. (1964).

Chomsky, N. (1964). *Current Issues in Linguistic Theory*. The Hague: Mouton.

Chomsky, N. (1965). *Aspects of the Theory of Syntax*. Cambridge, Mass.: M.I.T. Press.

Chomsky, N. (1966). *Cartesian Linguistics*. New York: Harper and Row.

Chomsky, N. (1968). "Quine's Empirical Assumptions." *Synthese*, 19:53–68.

Church, A. (1954). "Intensional Isomorphism and Identity of Belief." Philosophical Studies, 5:65–73.

Church, A. (1956). An Introduction to Mathematical Logic, vol. 1. Princeton, N.J.: Princeton University Press.

Davidson, D. (1967). "Truth and Meaning." Synthese, 17:304–323.

Donnellan, K. (1962). "Necessity and Criteria." Journal of Philosophy, 59:647–658.

Donnellan, K. (1966). "Reference and Definite Descriptions." Philosophical Review, 75:281–304.

Donnellan, K. (1970). "Proper Names and Identifying Descriptions." Synthese, 21:335–358.

Fodor, J. A. (1965). "Could Meaning Be an r_m?" Journal of Verbal Learning and Verbal Behavior, 4, no. 2.

Fodor, J. A. (1968). Psychological Explanation. New York: Random House.

Fodor, J. A. (1970). "Troubles about Actions." Synthese, 21:298–319.

Fodor, J. A., and J. J. Katz (1962). "What's Wrong with the Philosophy of Language." Inquiry, 5:197–237.

Fodor, J. A., and J. J. Katz, eds. (1964). The Structure of Language: Readings in the Philosophy of Language. Englewood Cliffs, N.J.: Prentice-Hall.

Frege, G. (1953). The Foundations of Arithmetic. Trans. J. L. Austin. 2nd ed. Oxford: Blackwell.

Frege, G. (1956). "The Thought: A Logical Inquiry." Mind, 65:289–311. Reprinted in E. D. Klemke, ed. (1968), Essays on Frege. Urbana: University of Illinois Press.

Frege, G. (1967). "Begriffsschrift," in J. van Heijenoort, ed., From Frege to Godel. Cambridge, Mass.: Harvard University Press.

Goodman, N. (1952). "On Likeness of Meaning," in L. Linsky, ed. (1952).

Harman, G. (1972). "Logical Form." Foundations of Language, 9:38–65.

Harman, G. (1973). Thought. Princeton: Princeton University Press.

Herzberger, H., and J. J. Katz (forthcoming). The Concept of Truth for Natural Languages.

Hintikka, J. (1970). Models for Modalities. Dordrecht, Holland: Reidel.

Katz, J. J. (1964a). "Analyticity and Contradiction in Natural Language," in J. A. Fodor and J. J. Katz, eds. (1964).

Katz, J. J. (1964b). "Mentalism in Linguistics." Language, 40:124–137.

Katz, J. J. (1964c). "Semantic Theory and the Meaning of 'Good'." Journal of Philosophy, 61:739–766.

Katz, J. J. (1965). "The Relevance of Linguistics to Philosophy." Journal of Philosophy, 62:590–602.

Katz, J. J. (1966). The Philosophy of Language. New York: Harper and Row.

Katz, J. J. (1967). "Some Remarks on Quine on Analyticity." Journal of Philosophy, 64:36–52.

Katz, J. J. (1968). "Unpalatable Recipes for Buttering Parsnips." Journal of Philosophy, 65:29–45.

Katz, J. J. (1971). The Underlying Reality of Language and Its Philosophical Import. New York: Harper and Row.

Katz, J. J. (1972). Semantic Theory. New York: Harper and Row.

Katz, J. J. (1974). "Where Things Now Stand with the Analytic-Synthetic Distinction." Synthese, 23:52–88.

Katz, J. J. (forthcoming). Propositional Structure: A Study of the Contribution of Sentence Meaning to Speech Acts.

Katz, J. J., and J. A. Fodor (1963). "The Structure of a Semantic Theory." Language, 39:170–210. Reprinted in J. A. Fodor and J. J. Katz, eds. (1964).

Katz, J. J., and P. M. Postal (1964). An Integrated Theory of Linguistic Descriptions. Cambridge, Mass.: M.I.T. Press.

Jerrold J. Katz

Katz, J. J., and R. I. Nagel (1974). "Meaning Postulates and Semantic Theory." *Foundations of Language*, 11:311–340.

Kripke, S. (1971). "Naming and Necessity," in G. Harman and D. Davidson, eds., *Semantics and Natural Language*, pp. 253–355. Dordrecht, Holland: Reidel.

Lewis, D. K. (1969). *Convention: A Philosophical Study*. Cambridge, Mass.: Harvard University Press.

Lewis, D. K. (1970). "General Semantics." *Synthese*, 22:18–67.

Linsky, L., ed. (1952). *Semantics and the Philosophy of Language*. Urbana: University of Illinois Press.

McDonald, M., ed. (1954). *Philosophy and Analysis*. New York: Philosophical Library.

Mates, B. (1952). "Synonymity," in L. Linsky, ed. (1952).

McCawley, J. D. (1968). "The Role of Semantics in a Grammar," in E. Bach and R. T. Harms, eds., *Universals in Linguistic Theory*. New York: Holt, Rinehart and Winston.

Montague, R. (1970). "Pragmatics and Intensional Logic." *Synthese*, 22:68–94.

Nagel, E. (1961). *The Structure of Science*. New York: Harcourt Brace.

Pitcher, G. (1964). *Truth*. Englewood Cliffs, N.J.: Prentice-Hall.

Postal, P. M. (1964). *Constituent Structure: A Study of Contemporary Models of Syntactic Description*. The Hague: Mouton.

Putnam, H. (1960). "Minds and Machines," in S. Hook, ed., *Dimensions of Mind*. New York: New York University Press.

Putnam, H. (1962). "It Ain't Necessarily So." *Journal of Philosophy*, 59:658–671.

Putnam, H. (1970). "Is Semantics Possible?" *Metaphilosophy*, 1:187–201.

Quine, W. V. (1953a). "The Problem of Meaning in Linguistics," in *From a Logical Point of View*. Cambridge, Mass.: Harvard University Press.

Quine, W. V. (1953b). "Two Dogmas of Empiricism," in *From a Logical Point of View*. Cambridge, Mass.: Harvard University Press.

Quine, W. V. (1953c). "Notes on the Theory of Reference," in *From a Logical Point of View*. Cambridge, Mass.: Harvard University Press.

Quine, W. V. (1955). *Mathematical Logic*. Rev. ed. Cambridge, Mass.: Harvard University Press.

Quine, W. V. (1960). *Word and Object*. Cambridge, Mass.: M.I.T. Press.

Quine, W. V. (1963). "Comments," in *Boston Studies in the Philosophy of Science*, I, pp. 97–104. Dordrecht, Holland: Reidel.

Quine, W. V. (1966a). "Carnap and Logical Truth," in *The Ways of Paradox*. New York: Random House.

Quine, W. V. (1966b). "Necessary Truth," in *The Ways of Paradox*. New York: Random House.

Quine, W. V. (1967). "On a Suggestion of Katz." *Journal of Philosophy*, 64:52–54.

Quine, W. V. (1968). "Replies." *Synthese*, 19:288–291.

Quine, W. V. (1970a). "Methodological Reflections on Current Linguistic Theory." *Synthese*, 21:386–398.

Quine, W. V. (1970b). *Philosophy of Logic*. Englewood Cliffs, N.J.: Prentice-Hall.

Searle, J. (1972). "Chomsky's Revolution in Linguistics." *New York Review of Books*, June 29.

Skinner, B. F. (1957). *Verbal Behavior*. New York: Appleton-Century-Crofts.

Tarski, A. (1952). "The Semantical Conception of Truth," in L. Linsky, ed. (1952).

Tarski, A. (1956). *Logic, Semantics, and Metamathematics*. London: Oxford University Press.

Whitehead, A. N., and B. Russell (1927). *Principia Mathematica*, vol. 2. London: Cambridge University Press.

The Meaning of "Meaning"

Language is the first broad area of human cognitive capacity for which we are beginning to obtain a description which is not exaggeratedly over-simplified. Thanks to the work of contemporary transformational linguists,[1] a very subtle description of at least some human languages is in the process of being constructed. Some features of these languages appear to be *universal*. Where such features turn out to be "species-specific" — "not explicable on some general grounds of functional utility or simplicity that would apply to arbitrary systems that serve the functions of language" — they may shed some light on the structure of mind. While it is extremely difficult to say to what extent the structure so illuminated will turn out to be a universal structure of *language*, as opposed to a universal structure of innate general learning strategies,[2] the very fact that this discussion can take place is testimony to the richness and generality of the descriptive material that linguists are beginning to provide, and also testimony to the depth of the analysis, insofar as the features that appear to be candidates for "species-specific" features of language are in no sense surface or phenomenological features of language, but lie at the level of deep structure.

The most serious drawback to all of this analysis, as far as a philosopher is concerned, is that it does not concern the meaning of words. Analysis of the deep structure of linguistic forms gives us an incomparably more powerful description of the *syntax* of natural languages than we have ever had before. But the dimension of language associated with the word "meaning" is, in spite of the usual spate of heroic if misguided attempts, as much in the dark as it ever was.

In this essay, I want to explore why this should be so. In my opinion,

---

[1] The contributors to this area are now too numerous to be listed; the pioneers were, of course, Zellig Harris and Noam Chomsky.

[2] For a discussion of this question, see my "The 'Innateness Hypothesis' and Explanatory Models in Linguistics," *Synthese*, 17(1967): 12–22.

the reason that so-called semantics is in so much worse condition than syntactic theory is that the *prescientific* concept on which semantics is based — the prescientific concept of *meaning* — is itself in much worse shape than the prescientific concept of syntax. As usual in philosophy, skeptical doubts about the concept do not at all help one in clarifying or improving the situation any more than dogmatic assertions by conservative philosophers that all's really well in this best of all possible worlds. The reason that the prescientific concept of meaning is in bad shape is not clarified by some general skeptical or nominalistic argument to the effect that meanings don't exist. Indeed, the upshot of our discussion will be that meanings don't exist in quite the way we tend to think they do. But electrons don't exist in quite the way Bohr thought they did, either. There is all the distance in the world between this assertion and the assertion that meanings (or electrons) "don't exist."

I am going to talk almost entirely about the meaning of words rather than about the meaning of sentences because I feel that our concept of word-meaning is more defective than our concept of sentence-meaning. But I will comment briefly on the arguments of philosophers such as Donald Davidson who insist that the concept of word-meaning *must* be secondary and that study of sentence-meaning must be primary. Since I regard the traditional theories about meaning as myth-eaten (notice that the topic of "meaning" is the one topic discussed in philosophy in which there is literally nothing but "theory" — literally nothing that can be labeled or even ridiculed as the "common sense view"), it will be necessary for me to discuss and try to disentangle a number of topics concerning which the received view is, in my opinion, wrong. The reader will give me the greatest aid in the task of trying to make these matters clear if he will kindly assume that *nothing* is clear in advance.

Meaning and extension. Since the Middle Ages at least, writers on the theory of meaning have purported to discover an ambiguity in the ordinary concept of meaning, and have introduced a pair of terms — *extension* and *intension*, or *Sinn* and *Bedeutung*, or whatever — to disambiguate the notion. The *extension* of a term, in customary logical parlance, is simply the set of things the term is true of. Thus, "rabbit," in its most common English sense, is true of all and only rabbits, so the extension of "rabbit" is precisely the set of rabbits. Even this notion — and it is the *least* problematical notion in this cloudy subject — has its problems, however. Apart from problems it inherits from its parent notion of *truth*, the forgoing example of "rabbit" *in its most common English sense* illustrates one such problem: strictly speaking, it is not a term, but an ordered pair consisting of a term and a "sense" (or an occasion of use, or something else that distinguishes a term in one sense from the same term used in a different sense) that has an extension. Another problem is this: a "set," in the mathematical sense, is a "yes-no" object; any given object either definitely belongs to S or definitely does not belong to S, if S is a set. But words in a natural language are not generally "yes-no": there are things of which the description "tree" is clearly true and things of which the description "tree" is clearly false, to be sure, but there are a host of borderline cases. Worse, the line between the clear cases and the borderline cases is itself fuzzy. Thus the idealization involved in the notion of *extension* — the idealization involved in supposing that there is such a thing as the set of things of which the term "tree" is true — is actually very severe.

Recently some mathematicians have investigated the notion of a *fuzzy set* — that is, of an object to which other things belong or do not belong with a given probability or to a given degree, rather than belong "yes-no." If one really wanted to formalize the notion of extension as applied to terms in a natural language, it would be necessary to employ "fuzzy sets" or something similar rather than sets in the classical sense.

The problem of a word's having more than one sense is standardly handled by treating each of the senses as a different word (or rather, by treating the word as if it carried invisible subscripts, thus: "rabbit$_1$" — animal of a certain kind; "rabbit$_2$" — coward; and as if "rabbit$_1$" and "rabbit$_2$." or whatever were different words entirely). This again involves two very severe idealizations (at least two, that is): supposing that words have discretely many senses, and supposing that the entire repertoire of senses is fixed once and for all. Paul Ziff has recently investigated the extent to which both of these suppositions distorts the actual situation in natural language;[3] nevertheless, we will continue to make these idealizations here.

Now consider the compound terms "creature with a heart" and "creature with a kidney." Assuming that every creature with a heart possesses a kidney and vice versa, the extension of these two terms is exactly the same. But they obviously differ in meaning. Supposing that there

[3] This is discussed by Ziff in *Understanding Understanding* (Ithaca: Cornell University Press, 1972), especially chap. 1.

is a sense of "meaning" in which meaning = extension, there must be another sense of "meaning" in which the meaning of a term is not its extension but something else, say the "concept" associated with the term. Let us call this "something else" the *intension* of the term. The concept of a creature with a heart is clearly a different concept from the concept of a creature with a kidney. Thus the two terms have different intension. When we say they have different "meaning," meaning = intension.

Intension and extension. Something like the preceding paragraph appears in every standard exposition of the notions "intension" and "extension." But it is not at all satisfactory. Why it is not satisfactory is, in a sense, the burden of this entire essay. But some points can be made at the very outset: first of all, what evidence is there that "extension" *is* a sense of the word "meaning"? The canonical explanation of the notions "intension" and "extension" is very much like "in one sense 'meaning' means *extension* and in the other sense 'meaning' means *meaning.*" The fact is that while the notion of "extension" is made quite precise, relative to the fundamental logical notion of *truth* (and under the severe idealizations remarked above), the notion of intension is made no more precise than the vague (and, as we shall see, misleading) notion "concept." It is as if someone explained the notion "probability" by saying: "in one sense 'probability' means *frequency,* and in the other sense it means *propensity.*" "Probability" *never* means 'frequency', and "propensity" is at least as unclear as "probability."

Unclear as it is, the traditional doctrine that the notion "meaning" possesses the extension/intension ambiguity has certain typical consequences. Most traditional philosophers thought of concepts as something *mental.* Thus the doctrine that the meaning of a term (the meaning "in the sense of intension," that is) is a concept carried the implication that meanings are mental entities. Frege and more recently Carnap and his followers, however, rebelled against this "psychologism," as they termed it. Feeling that meanings are *public* property — that the *same* meaning can be "grasped" by more than one person and by persons at different times — they identified concepts (and hence "intensions" or meanings) with abstract entities rather than mental entities. However, "grasping" these abstract entities was still an individual psychological act. None of these philosophers doubted that understanding a word (knowing its intension) was just a matter of being in a certain psychological state (somewhat in the way in which knowing how to factor numbers in one's head is just a matter of being in a certain very complex psychological state).

Secondly, the timeworn example of the two terms "creature with a kidney" and "creature with a heart" does show that two terms can have the same extension and yet differ in intension. But it was taken to be obvious that the reverse is impossible: two terms cannot differ in extension and have the same intension. Interestingly, no argument for this impossibility was ever offered. Probably it reflects the tradition of the ancient and medieval philosophers who assumed that the concept corresponding to a term was just a conjunction of predicates, and hence that the concept corresponding to a term must *always* provide a necessary and sufficient condition for falling into the extension of the term.[4] For philosophers like Carnap, who accepted the verifiability theory of meaning, the concept corresponding to a term provided (in the ideal case, where the term had "complete meaning") a *criterion* for belonging to the extension (not just in the sense of "necessary and sufficient condition," but in the strong sense of *way of recognizing* if a given thing falls into the extension or not). Thus these positivistic philosophers were perfectly happy to retain the traditional view on this point. So theory of meaning came to rest on two unchallenged assumptions:

(I) That knowing the meaning of a term is just a matter of being in a certain psychological state (in the sense of "psychological state" in which states of memory and psychological dispositions are "psychological states"; no one thought that knowing the meaning of a word was a

---

[4] This tradition grew up because *the* term whose analysis provoked all the discussion in medieval philosophy was the term "God," and the term "God" was thought to be defined through the conjunction of the terms "Good," "Powerful," "Omniscient," etc. — the so-called "Perfections." There was a problem, however, because God was supposed to be a Unity, and Unity was thought to exclude His essence's being complex in any way — i.e., "God" was defined through a conjunction of terms, but God (without quotes) could not be the logical product of properties, nor could He be the unique thing exemplifying the logical product of two or more *distinct* properties, because even this highly abstract kind of "complexity" was held to be incompatible with His perfection of Unity. This is a theological paradox with which Jewish, Arabic, and Christian theologians wrestled for centuries (e.g., the doctrine of the Negation of Privation in Maimonides and Aquinas). It is amusing that theories of contemporary interest, such as conceptualism and nominalism, were first proposed as solutions to the problem of predication in the case of God. It is also amusing that the favorite model of definition in all of this theology — the conjunction-of-properties model — should survive, at least through its consequences, in philosophy of language until the present day.

135

continuous state of consciousness, of course).

(II) That the meaning of a term (in the sense of "intension") determines its extension (in the sense that sameness of intension entails sameness of extension).

I shall argue that these two assumptions are not jointly satisfied by *any* notion, let alone any notion of meaning. The traditional concept of meaning is a concept which rests on a false theory.

"Psychological state" and methodological solipsism. In order to show this, we need first to clarify the traditional notion of a psychological state. In one sense a state is simply a two-place predicate whose arguments are an individual and a time. In this sense, *being five feet tall, being in pain, knowing the alphabet,* and even *being a thousand miles from Paris* are all states. (Note that the *time* is usually left implicit or "contextual"; the full form of an atomic sentence of these predicates would be "*x* is five feet tall at time *t,*" "*x* is in pain at time *t,*" etc.) In science, however, it is customary to restrict the term state to properties which are defined in terms of the parameters of the individual which are fundamental from the point of view of the given science. Thus, being five feet tall is a state (from the point of view of physics); being in pain is a state (from the point of view of mentalistic psychology, at least); knowing the alphabet might be a state (from the point of view of cognitive psychology), although it is hard to say; but being a thousand miles from Paris would *not* naturally be called a *state.* In one sense, a psychological state is simply a state which is studied or described by psychology. In this sense it may be trivially true that, say, *knowing the meaning of the word "water"* is a "psychological state" (viewed from the standpoint of cognitive psychology). But this is not the sense of psychological state that is at issue in the above assumption (I).

When traditional philosophers talked about psychological states (or "mental" states), they made an assumption which we may call the assumption of methodological solipsism. This assumption is the assumption that no psychological state, properly so called, presupposes the existence of any individual other than the subject to whom that state is ascribed. (In fact, the assumption was that no psychological state presupposes the existence of the subject's *body* even: if *P* is a psychological state, properly so called, then it must be logically possible for a "disembodied mind" to be in *P.*) This assumption is pretty explicit in Descartes, but it is implicit in just about the whole of traditional philosophical psychology. Making this assumption is, of course, adopting a *restrictive program* — a program which deliberately limits the scope and nature of psychology to fit certain mentalistic preconceptions or, in some cases, to fit an idealistic reconstruction of knowledge and the world. Just *how* restrictive the program is, however, often goes unnoticed. Such common or garden variety psychological states as *being jealous* have to be reconstructed, for example, if the assumption of methodological solipsism is retained. For, in its ordinary use, *x is jealous of y* entails that y exists, and *x is jealous of y's regard for z* entails that both y and z exist (as well as x, of course). Thus *being jealous* and *being jealous of someone's regard for someone else* are not psychological states permitted by the assumption of methodological solipsism. (We shall call them "psychological states in the wide sense" and refer to the states which are permitted by methodological solipsism as "psychological states in the narrow sense.") The reconstruction required by methodological solipsism would be to reconstrue *jealousy* so that I can be jealous of my own hallucinations, or of figments of my imagination, etc. Only if we assume that psychological states in the narrow sense have a significant degree of causal closure (so that restricting ourselves to psychological states in the narrow sense will facilitate the statement of psychological *laws*) is there any point to engaging in this reconstruction, or in making the assumption of methodological solipsism. But the three centuries of failure of mentalistic psychology is tremendous evidence against this procedure, in my opinion.

Be that as it may, we can now state more precisely what we claimed at the end of the preceding section. Let A and B be any two terms which differ in extension. By assumption (II) they must differ in meaning (in the sense of "intension"). By assumption (I), *knowing the meaning of A* and *knowing the meaning of B* are psychological states *in the narrow sense* — for this is how we shall construe assumption (I). *But these psychological states must determine the extension of the terms A and B just as much as the meanings ("intensions") do.*

To see this, let us try assuming the opposite. Of course, there cannot be two terms A and B such that *knowing the meaning of A* is the same state as *knowing the meaning of B* even though A and B have different extensions. For *knowing the meaning of A* isn't just "grasping the intension" of A, whatever that may come to; it is also knowing that the

137

"intension" that one has "grasped" *is* the intension of A. (Thus, someone who knows the meaning of "wheel" presumably "grasps the intension" of its German synonym "Rad"; but if he doesn't know that the
"intension" in question is the intension of *Rad*, he isn't said to "know
the meaning of *Rad*.") If A and B are different terms, then *knowing
the meaning of A* is a different state from *knowing the meaning of B*
whether the meanings of A and B be themselves the same or different.
But by the same argument, if I_1 and I_2 are different *intensions* and A is
a term, then *knowing that I_1 is the meaning of A* is a different psychological state from *knowing that I_2 is the meaning of A*. Thus, there cannot be two different logically possible worlds L_1 and L_2 such that,
say, Oscar is in the *same* psychological state (in the narrow sense) in
L_1 and L_2 (in all respects), but in L_1 Oscar understands A as having the
meaning I_1 and in L_2 Oscar understands A as having the meaning I_2.
(For, if there were, then in L_1 Oscar would be in the psychological state
knowing that I_1 is the meaning of A and in L_2 Oscar would be in the
psychological state *knowing that I_2 is the meaning of A*, and these are
different and even — assuming that A has just *one* meaning for Oscar
in each world — incompatible psychological states in the narrow sense.)

In short, if S is the sort of psychological state we have been discussing
— a psychological state of the form *knowing that I is the meaning of
A*, where *I* is an "intension" and A is a term — then the *same* necessary
and sufficient condition for falling into the extension of A "works" in
every logically possible world in which the speaker is in the psychological state S. For the state S determines the intension I, and by assumption (II) the intension amounts to a necessary and sufficient condition
for membership in the *extension*.

If our interpretation of the traditional doctrine of intension and extension is fair to Frege and Carnap, then the whole psychologism/Pla-
tonism issue appears somewhat a tempest in a teapot, as far as meaning-
theory is concerned. (Of course, it is a very important issue as far as
general philosophy of mathematics is concerned.) For even if meanings are "Platonic" entities rather than "mental" entities on the Frege-
Carnap view, "grasping" those entities is presumably a psychological
state (in the narrow sense). Moreover, the psychological state uniquely
determines the "Platonic" entity. So whether one takes the "Platonic"
entity or the psychological state as the "meaning" would appear to be
somewhat a matter of convention. And taking the psychological state to

138

be the meaning would hardly have the consequence that Frege feared, that meanings would cease to be public. For psychological states are "public" in the sense that different people (and even people in different epochs) can be in the same psychological state. Indeed, Frege's argument against psychologism is only an argument against identifying concepts with mental particulars, not with mental entities in general.

The "public" character of psychological states entails, in particular, that if Oscar and Elmer understand a word A differently, then they must be in different psychological states. For the state of knowing the intension of A to be, say I is the same state whether Oscar or Elmer be in it. Thus two speakers cannot be in the same psychological state in all respects and understand the term A differently; the psychological state of the speaker determines the intension (and hence, by assumption (II), the extension) of A.

It is this last consequence of the joint assumptions (I),(II) that we claim to be false. We claim that it is possible for two speakers to be in exactly the same psychological state (in the narrow sense), even though the extension of the term A in the idiolect of the one is different from the extension of the term A in the idiolect of the other. Extension is not determined by psychological state.

This will be shown in detail in later sections. If this is right, then there are two courses open to one who wants to rescue at least one of the traditional assumptions: to give up the idea that psychological state (in the narrow sense) determines intension, or to give up the idea that intension determines extension. We shall consider these alternatives later.

Are meanings in the head? That psychological state does not determine extension will now be shown with the aid of a little science fiction. For the purpose of the following science-fiction examples, we shall suppose that somewhere in the galaxy there is a planet we shall call Twin Earth. Twin Earth is very much like Earth; in fact, people on Twin Earth even speak English. In fact, apart from the differences we shall specify in our science-fiction examples, the reader may suppose that Twin Earth is exactly like Earth. He may even suppose that he has a Doppelgänger — an identical copy — on Twin Earth, if he wishes, although my stories will not depend on this.

Although some of the people on Twin Earth (say, the ones who call themselves "Americans" and the ones who call themselves "Canadians" and the ones who call themselves "Englishmen," etc.) speak English, there are, not surprisingly, a few tiny differences which we will now describe between the dialects of English spoken on Twin Earth and Standard English. These differences themselves depend on some of the peculiarities of Twin Earth.

One of the peculiarities of Twin Earth is that the liquid called "water" is not H_2O but a different liquid whose chemical formula is very long and complicated. I shall abbreviate this chemical formula simply as XYZ. I shall suppose that XYZ is indistinguishable from water at normal temperatures and pressures. In particular, it tastes like water and it quenches thirst like water. Also, I shall suppose that the oceans and lakes and seas of Twin Earth contain XYZ and not water, that it rains XYZ on Twin Earth and not water, etc.

If a spaceship from Earth ever visits Twin Earth, then the supposition at first will be that "water" has the same meaning on Earth and on Twin Earth. This supposition will be corrected when it is discovered that "water" on Twin Earth is XYZ, and the Earthian spaceship will report somewhat as follows:

"On Twin Earth the word 'water' means XYZ."

(It is this sort of use of the word "means" which accounts for the doctrine that extension is one sense of "meaning," by the way. But note that although "means" does mean something like *has as extension in* this example, one would *not* say

"On Twin Earth the meaning of the word 'water' is XYZ"

unless, possibly, the fact that "water is XYZ" was known to every adult speaker of English on Twin Earth. We can account for this in terms of the theory of meaning we develop below; for the moment we just remark that although the verb "means" sometimes means "has as extension," the nominalization "meaning" *never* means "extension.")

Symmetrically, if a spaceship from Twin Earth ever visits Earth, then the supposition at first will be that the word "water" has the same meaning on Twin Earth and on Earth. This supposition will be corrected when it is discovered that "water" on Earth is H_2O, and the Twin Earthian spaceship will report:

"On Earth[5] the word 'water' means H_2O."

---

[5] Or rather, they will report: "On Twin Earth (*the Twin Earthian name for Terra* — H.P.) the word 'water' means H_2O."

Note that there is no problem about the extension of the term "water." The word simply has two different meanings (as we say): in the sense in which it is used on Twin Earth, the sense of water$_{TE}$, what we call "water" simply isn't water; while in the sense in which it is used on Earth, the sense of water$_E$, what the Twin Earthians call "water" simply isn't water. The extension of "water" in the sense of water$_E$ is the set of all wholes consisting of H_2O molecules, or something like that; the extension of water in the sense of water$_{TE}$ is the set of all wholes consisting of XYZ molecules, or something like that.

Now let us roll the time back to about 1750. At that time chemistry was not developed on either Earth or Twin Earth. The typical Earthian speaker of English did not know water consisted of hydrogen and oxygen, and the typical Twin Earthian speaker of English did not know "water" consisted of XYZ. Let Oscar$_1$ be such a typical Earthian English speaker, and let Oscar$_2$ be his counterpart on Twin Earth. You may suppose that there is no belief that Oscar$_1$ had about water that Oscar$_2$ did not have about "water." If you like, you may even suppose that Oscar$_1$ and Oscar$_2$ were exact duplicates in appearance, feelings, thoughts, interior monologue, etc. Yet the extension of the term "water" was just as much H_2O on Earth in 1750 as in 1950; and the extension of the term "water" was just as much XYZ on Twin Earth in 1750 as in 1950. Oscar$_1$ and Oscar$_2$ understood the term "water" differently in 1750 *although they were in the same psychological state*, and although, given the state of science at the time, it would have taken their scientific communities about fifty years to discover that they understood the term "water" differently. Thus the extension of the term "water" (and, in fact, its "meaning" in the intuitive preanalytical usage of that term) is *not* a function of the psychological state of the speaker by itself.

But, it might be objected, why should we accept it that the term "water" had the same extension in 1750 and in 1950 (on both Earths)? The logic of natural-kind terms like "water" is a complicated matter, but the following is a sketch of an answer. Suppose I point to a glass of water and say "this liquid is called water" (or "this is called water," if the marker "liquid" is clear from the context). My "ostensive definition" of water has the following empirical presupposition: that the body of liquid I am pointing to bears a certain sameness relation (say, *x is the same liquid as* y, or *x is the same$_L$ as* y) to most of the stuff I and other speakers in my linguistic community have on other occasions called

141

"water." If this presupposition is false because, say, I am without knowing it pointing to a glass of gin and not a glass of water, then I do not intend my ostensive definition to be accepted. Thus the ostensive definition conveys what might be called a defeasible necessary and sufficient condition: the necessary and sufficient condition for being water is bearing the relation same$_L$ to the stuff in the glass; but this is the necessary and sufficient condition only if the empirical presupposition is satisfied. If it is not satisfied, then one of a series of, so to speak, "fallback" conditions becomes activated.

The key point is that the relation same$_L$ is a *theoretical* relation: whether something is or is not the same liquid as *this* may take an indeterminate amount of scientific investigation to determine. Moreover, even if a "definite" answer has been obtained either through scientific investigation or through the application of some "common sense" test, the answer is *defeasible*: future investigation might reverse even the most "certain" example. Thus, the fact that an English speaker in 1750 might have called XYZ "water," while he or his successors would not have called XYZ water in 1800 or 1850 does not mean that the "meaning" of "water" changed for the average speaker in the interval. In 1750 or in 1850 or in 1950 one might have pointed to, say, the liquid in Lake Michigan as an example of "water." What changed was that in 1750 we would have mistakenly thought that XYZ bore the relation same$_L$ to the liquid in Lake Michigan, while in 1800 or 1850 we would have known that it did not (I am ignoring the fact that the liquid in Lake Michigan was only dubiously water in 1950, of course).

Let us now modify our science-fiction story. I do not know whether one can make pots and pans out of molybdenum; and if one can make them out of molybdenum, I don't know whether they could be distinguished easily from aluminum pots and pans. (I don't know any of this even though I have acquired the word "molybdenum.") So I shall suppose that molybdenum pots and pans *can't* be distinguished from aluminum pots and pans save by an expert. (To emphasize the point, I repeat that this could be true for all I know, and a fortiori it could be true for all I know by virtue of "knowing the meaning" of the words *aluminum and molybdenum.*) We will now suppose that molybdenum is as common on Twin Earth as aluminum is on Earth, and that aluminum is as rare on Twin Earth as molybdenum is on Earth. In particular, we shall assume that "aluminum" pots and pans are made of molybdenum on Twin Earth. Finally, we shall assume that the words "alumi-num" and "molybdenum" are *switched* on Twin Earth: "aluminum" is the name of *molybdenum* and "molybdenum" is the name of *aluminum*.

This example shares some features with the previous one. If a space-ship from Earth visited Twin Earth, the visitors from Earth probably would not suspect that the "aluminum" pots and pans on Twin Earth were not made of aluminum, especially when the Twin Earthians *said* they were. But there is one important difference between the two cases. An Earthian metallurgist could tell very easily that "aluminum" was molybdenum, and a Twin Earthian metallurgist could tell equally easily that aluminum was "molybdenum." (The shudder quotes in the preceding sentence indicate Twin Earthian usages.) Whereas in 1750 no one on either Earth or Twin Earth could have distinguished water from "wa-ter," the confusion of aluminum with "aluminum" involves only a part of the linguistic communities involved.

The example makes the same point as the preceding one. If $Oscar_1$ and $Oscar_2$ are standard speakers of Earthian English and Twin Earthian English respectively, and neither is chemically or metallurgically sophisti-cated, then there may be no difference at all in their psychological state when they use the word "aluminum"; nevertheless we have to say that "aluminum" has the extension *aluminum* in the idiolect of $Oscar_1$ and the extension *molybdenum* in the idiolect of $Oscar_2$. (Also we have to say that $Oscar_1$ and $Oscar_2$ mean different things by "aluminum," that "aluminum" has a different meaning on Earth than it does on Twin Earth, etc.) Again we see that the psychological state of the speaker does *not* determine the extension (or the "meaning," speaking preana-lytically) of the word.

Before discussing this example further, let me introduce a *non-science-fiction* example. Suppose you are like me and cannot tell an elm from a beech tree. We still say that the extension of "elm" in my idiolect is the same as the extension of "elm" in anyone else's, viz., the set of all elm trees, and that the set of all beech trees is the extension of "beech" in *both* of our idiolects. Thus "elm" in my idiolect has a different exten-sion from "beech" in your idiolect (as it should). Is it really credible that this difference in extension is brought about by some difference in our *concepts*? My *concept* of an elm tree is exactly the same as my con-cept of a beech tree (I blush to confess). (This shows that the identifi-

143

cation of meaning "in the sense of intension" with *concept* cannot be correct, by the way.) If someone heroically attempts to maintain that the difference between the extension of "elm" and the extension of "beech" in *my* idiolect is explained by a difference in my psychological state, then we can always refute him by constructing a "Twin Earth" example — just let the words "elm" and "beech" be switched on Twin Earth (the way "aluminum" and "molybdenum" were in the previous example). Moreover, suppose I have a *Doppelgänger* on Twin Earth who is molecule for molecule "identical" with me (in the sense in which two neckties can be "identical"). If you are a dualist, then also suppose my *Doppelgänger* thinks the same verbalized thoughts I do, has the same sense data, the same dispositions, etc. It is absurd to think *his* psychological state is one bit different from mine: yet he "means" *beech* when he says "elm" and I "mean" *elm* when I say elm. Cut the pie any way you like, "meanings" just ain't in the *head*!

A socio-linguistic hypothesis. The last two examples depend upon a fact about language that seems, surprisingly, never to have been pointed out: that there is *division of linguistic labor*. We could hardly use such words as "elm" and "aluminum" if no one possessed a way of recognizing elm trees and aluminum metal; but not everyone to whom the distinction is important has to be able to make the distinction. Let us shift the example: consider *gold*. Gold is important for many reasons: it is a precious metal, it is a monetary metal, it has symbolic value (it is important to most people that the "gold" wedding ring they wear *really* consist of gold and not just *look* gold), etc. Consider our community as a "factory": in this "factory" some people have the "job" of *wearing gold wedding rings*, other people have the "job" of selling gold wedding rings, still other people have the job of *telling whether or not something is really gold*. It is not at all necessary or efficient that everyone who wears a gold ring (or a gold cuff link, etc.), or discusses the "gold standard," etc., engage in buying and selling gold. Nor is it necessary or efficient that everyone who buys and sells gold be able to tell whether or not something is really gold in a society where this form of dishonesty is uncommon (selling fake gold) and in which one can easily consult an expert in case of doubt. And it is *certainly* not necessary or efficient that everyone who has occasion to buy or wear gold be able to tell with any reliability whether or not something is really gold.

The foregoing facts are just examples of mundane division of labor (in a wide sense). But they engender a division of linguistic labor: everyone to whom gold is important for any reason has to acquire the word "gold"; but he does not have to acquire the *method* of *recognizing* if something is or is not gold. He can rely on a special subclass of speakers. The features that are generally thought to be present in connection with a general name — necessary and sufficient conditions for membership in the extension, ways of recognizing if something is in the extension ("criteria"), etc. — are all present in the linguistic community *considered as a collective body*; but that collective body divides the "labor" of knowing and employing these various parts of the "meaning" of "gold."

This division of linguistic labor rests upon and presupposes the division of non-linguistic labor, of course. If only the people who know how to tell if some metal is really gold or not have any reason to have the word "gold" in their vocabulary, then the word "gold" will be as the word "water" was in 1750 with respect to that subclass of speakers, and the other speakers just won't acquire it at all. And some words do not exhibit any division of linguistic labor: "chair," for example. But with the increase of division of labor in the society and the rise of science, more and more words begin to exhibit this kind of division of labor. "Water," for example, did not exhibit it at all prior to the rise of chemistry. Today it is obviously necessary for every speaker to be able to recognize water (reliably under normal conditions), and probably every adult speaker even knows the necessary and sufficient condition "water is H_2O," but only a few adult speakers could distinguish water from liquids which superficially resembled water. In case of doubt, other speakers would rely on the judgment of these "expert" speakers. Thus the way of recognizing possessed by these "expert" speakers is also, through them, possessed by the collective linguistic body, even though it is not possessed by each individual member of the body, and in this way the most recherché fact about water may become part of the *social* meaning of the word while being unknown to almost all speakers who acquire the word.

It seems to me that this phenomenon of division of linguistic labor is one which it will be very important for sociolinguistics to investigate. In connection with it, I should like to propose the following hypothesis:

HYPOTHESIS OF THE UNIVERSALITY OF THE DIVISION OF LINGUISTIC LABOR:

145

Every linguistic community exemplifies the sort of division of linguistic labor just described, that is, possesses at least some terms whose associated "criteria" are known only to a subset of the speakers who acquire the terms, and whose use by the other speakers depends upon a structured cooperation between them and the speakers in the relevant subsets.

It would be of interest, in particular, to discover if extremely primitive peoples were sometimes exceptions to this hypothesis (which would indicate that the division of linguistic labor is a product of social evolution), or if even they exhibit it. In the latter case, one might conjecture that division of labor, including linguistic labor, is a fundamental trait of our species.

It is easy to see how this phenomenon accounts for some of the examples given above of the failure of the assumptions (I), (II). Whenever a term is subject to the division of linguistic labor, the "average" speaker who acquires it does not acquire anything that fixes its extension. In particular, his individual psychological state *certainly* does not fix its extension; it is only the sociolinguistic state of the collective linguistic body to which the speaker belongs that fixes the extension.

We may summarize this discussion by pointing out that there are two sorts of tools in the world: there are tools like a hammer or a screwdriver which can be used by one person; and there are tools like a steamship which require the cooperative activity of a number of persons to use. Words have been thought of too much on the model of the first sort of tool.

Indexicality and rigidity.[6] The first of our science-fiction examples — "water" on Earth and on Twin Earth in 1750 — does not involve division of linguistic labor, or at least does not involve it in the same way the examples of "aluminum" and "elm" do. There were not (in our story, anyway) any "experts" on water on Earth in 1750, nor any experts on "water" on Twin Earth. (The example *can* be construed as involving division of labor *across time*, however. I shall not develop this method of treating the example here.) The example *does* involve

---

[6] The substance of this section was presented at a series of lectures I gave at the University of Washington (Summer Institute in Philosophy) in 1968, and at a lecture at the University of Minnesota (at the conference out of which this volume originated).

things which are of fundamental importance to the theory of reference and also to the theory of necessary truth, which we shall now discuss.

There are two obvious ways of telling someone what one means by a natural-kind term such as "water" or "tiger" or "lemon." One can give him a so-called ostensive definition — "this (liquid) is water"; "this (animal) is a tiger"; "this (fruit) is a lemon"; where the parentheses are meant to indicate that the "markers" liquid, animal, fruit, may be either explicit or implicit. Or one can give him a description. In the latter case the description one gives typically consists of one or more markers together with a stereotype[7] — a standardized description of features of the kind that are typical, or "normal," or at any rate stereo-typical. The central features of the stereotype generally are criteria — features which in normal situations constitute ways of recognizing if a thing belongs to the kind or, at least, necessary conditions (or probabilistic necessary conditions) for membership in the kind. Not all criteria used by the linguistic community as a collective body are included in the stereotype, and in some cases the stereotype may be quite weak. Thus (unless I am a very atypical speaker), the stereotype of an elm is just that of a common deciduous tree. These features are indeed necessary conditions for membership in the kind (I mean "necessary" in a loose sense; I don't think "elm trees are deciduous" is analytic), but they fall far short of constituting a way of recognizing elms. On the other hand, the stereotype of a tiger does enable one to recognize tigers (unless they are albino, or some other atypical circumstance is present), and the stereotype of a lemon generally enables one to recognize lemons. In the extreme case, the stereotype may be just the marker: the stereotype of molybdenum might be just that molybdenum is a metal. Let us consider both of these ways of introducing a term into someone's vocabulary.

Suppose I point to a glass of liquid and say "this is water," in order to teach someone the word "water." We have already described some of the empirical presuppositions of this act, and the way in which this kind of meaning-explanation is defeasible. Let us now try to clarify further how it is supposed to be taken.

In what follows, we shall take the notion of "possible world" as primitive. We do this because we feel that in several senses the notion makes

<hr />

[7] See my "Is Semantics Possible," Metaphilosophy, 1, no. 3 (July 1970).

147

sense and is scientifically important even if it needs to be made more precise. We shall assume further that in at least some cases it is possible to speak of the same individual as existing in more than one possible world.[8] Our discussion leans heavily on the work of Saul Kripke, although the conclusions were obtained independently.

Let W_1 and W_2 be two possible worlds in which I exist and in which this glass exists and in which I am giving a meaning explanation by pointing to this glass and saying "this is water." (We do *not* assume that the liquid in the glass is the same in both worlds.) Let us suppose that in W_1 the glass is full of H_2O and in W_2 the glass is full of XYZ. We shall also suppose that W_1 is the *actual* world and that XYZ is the stuff typically called "water" in the world W_2 (so that the relation between English speakers in W_1 and English speakers in W_2 is exactly the same as the relation between English speakers on Earth and English speakers on Twin Earth). Then there are two theories one might have concerning the meaning of "water."

(1) One might hold that "water" was *world-relative* but *constant* in meaning (i.e., the word has a *constant relative meaning*). On this theory, "water" *means the same* in W_1 and W_2; it's just that water is H_2O in W_1 and water is XYZ in W_2.

(2) One might hold that water is H_2O in all worlds (the stuff called "water" in W_2 isn't water), but "water" doesn't have the same meaning in W_1 and W_2.

If what was said before about the Twin Earth case was correct, then (2) is clearly the correct theory. When I say "*this* (liquid) is water," the "this" is, so to speak, a *de re* "this" — i.e., the force of my explanation is that "water" is whatever bears a certain equivalence relation (the relation we called "same$_L$" above) to the piece of liquid referred to as "this" *in the actual world*.

We might symbolize the difference between the two theories as a "scope" difference in the following way. On theory (1), the following is true:

(1') (For every world W) (For every x in W) (x is water $\equiv x$ bears same$_L$ to the entity referred to as "this" in W)

while on theory (2):

[8] This assumption is not actually needed in what follows. What *is* needed is that the same *natural kind* can exist in more than one possible world.

(2') (For every world W) (For every x in W) (x is water $\equiv x$ bears same$_L$ to the entity referred to as "this" in the actual world W_1).

(I call this a "scope" difference because in (1') "the entity referred to as 'this' " is within the scope of "For every world W" — as the qualifying phrase "in W" makes explicit, whereas in (2') "the entity referred to as 'this' " means "the entity referred to as 'this' in the actual world," and has thus a reference independent of the bound variable "W.")

Kripke calls a designator "rigid" (in a given sentence) if (in that sentence) it refers to the same individual in every possible world in which the designator designates. If we extend the notion of rigidity to substance names, then we may express Kripke's theory and mine by saying that the term "water" is rigid.

The rigidity of the term "water" follows from the fact that when I give the ostensive definition "this (liquid) is water" I intend (2') and not (1').

We may also say, following Kripke, that when I give the ostensive definition "this (liquid) is water," the demonstrative "this" is rigid.

What Kripke was the first to observe is that this theory of the meaning (or "use," or whatever) of the word "water" (and other natural-kind terms as well) has startling consequences for the theory of necessary truth.

To explain this, let me introduce the notion of a cross-world relation. A two-term relation R will be called cross-world when it is understood in such a way that its extension is a set of ordered pairs of individuals not all in the same possible world. For example, it is easy to understand the relation same height as as a cross-world relation: just understand it so that, e.g., if x is an individual in a world W_1 who is five feet tall (in W_1) and y is an individual in W_2 who is five feet tall (in W_2), then the ordered pair x,y belongs to the extension of same height as. (Since an individual may have different heights in different possible worlds in which that same individual exists, strictly speaking it is not the ordered pair x,y that constitutes an element of the extension of same height as, but rather the ordered pair x-in-world-W_1, y-in-world-W_2.)

Similarly, we can understand the relation same$_L$ (same liquid as) as a cross-world relation by understanding it so that a liquid in world W_1

which has the same important physical properties (in W_1) that a liquid in W_2 possesses (in W_2) bears same$_L$ to the latter liquid.

Then the theory we have been presenting may be summarized by saying that an entity x, in an arbitrary possible world, is *water* if and only if it bears the relation same$_L$ (construed as a cross-world relation) to the stuff we call "water" in the *actual* world.

Suppose, now, that I have not yet discovered what the important physical properties of water are (in the actual world) — i.e., I don't yet know that water is H_2O. I may have ways of *recognizing* water that are successful (of course, I may make a small number of mistakes that I won't be able to detect until a later stage in our scientific development) but not know the microstructure of water. If I agree that a liquid with the superficial properties of "water" but a different microstructure *isn't really water*, then my ways of recognizing water (my "operational defi-nition," so to speak) cannot be regarded as an analytical specification of *what it is to be* water. Rather, the operational definition, like the osten-sive one, is simply a way of pointing out a standard — pointing out the stuff *in the actual world* such that for x to be water, in *any* world, is for x to bear the relation same$_L$ to the *normal* members of the class of *local* entities that satisfy the operational definition. "Water" on Twin Earth is not water, even if it satisfies the operational definition, be-cause it doesn't bear same$_L$ to the *local* stuff that satisfies the opera-tional definition, and local stuff that satisfies the operational definition but has a microstructure different from rest of the local stuff that sat-isfies the operational definition isn't water either, because it doesn't bear same$_L$ to the *normal* examples of the local "water."

Suppose, now, that I discover the microstructure of water — that wa-ter is H_2O. At this point I will be able to say that the stuff on Twin Earth that I earlier *mistook* for water isn't really water. In the same way, if you describe not another planet in the actual universe, but an-other possible universe in which there is stuff with the chemical for-mula XYZ which passes the "operational test" for *water*, we shall have to say that that stuff isn't water but merely XYZ. You will not have de-scribed a possible world in which "water is XYZ," but merely a possible world in which there are lakes of XYZ, people drink XYZ (and not water), or whatever. In fact, once we have discovered the nature of wa-ter, nothing counts as a possible world in which water doesn't have that nature. Once we have discovered that water (in the actual world) is

H_2O, nothing counts as a possible world in which water isn't H_2O. In particular, if a "logically possible" statement is one that holds in some "logically possible world," it isn't logically possible that water isn't H_2O.

On the other hand, we can perfectly well imagine having experiences that would convince us (and that would make it rational to believe that) water isn't H_2O. In that sense, it is conceivable that water isn't H_2O. It is conceivable but it isn't logically possible! Conceivability is no proof of logical possibility.

Kripke refers to statements which are rationally unrevisable (assuming there are such) as *epistemically necessary*. Statements which are true in all possible worlds he refers to simply as necessary (or sometimes as "metaphysically necessary"). In this terminology, the point just made can be restated as: a statement can be (metaphysically) necessary and epistemically contingent. Human intuition has no privileged access to metaphysical necessity.

Since Kant there has been a big split between philosophers who thought that all necessary truths were analytic and philosophers who thought that some necessary truths were synthetic a priori. But none of these philosophers thought that a (metaphysically) necessary truth could fail to be a priori: the Kantian tradition was as guilty as the empiricist tradition of equating metaphysical and epistemic necessity. In this sense Kripke's challenge to received doctrine goes far beyond the usual empiricism/Kantianism oscillation.

In this paper our interest is in theory of meaning, however, and not in theory of necessary truth. Points closely related to Kripke's have been made in terms of the notion of *indexicality*.[9] Words like "now," "this," "here," have long been recognized to be *indexical*, or *token-reflexive* — i.e., to have an extension which varied from context to context or token to token. For these words no one has ever suggested the traditional theory that "intension determines extension." To take our Twin Earth example: if I have a *Doppelgänger* on Twin Earth, then when I think "I have a headache," he thinks "I have a headache." But the extension of the particular token of "I" in his verbalized thought is himself (or his unit class, to be precise), while the extension of the token of "I" in *my* verbalized thought is *me* (or my unit class, to be precise). So the same word, "I," has two different extensions in two different idio-

[9] These points were made in my 1968 lectures at the University of Washington and the University of Minnesota.

lects; but it does not follow that the concept I have of myself is in any way different from the concept my *Doppelgänger* has of himself.

Now then, we have maintained that indexicality extends beyond the *obviously* indexical words and morphemes (e.g., the tenses of verbs). Our theory can be summarized as saying that words like "water" have an unnoticed indexical component: "water" is stuff that bears a certain similarity relation to the water *around here*. Water at another time or in another place or even in another possible world has to bear the relation same$_L$ to our "water" *in order to be water*. Thus the theory that (1) words have "intensions," which are something like concepts associated with the words by speakers; and (2) intension determines extension — this theory cannot be true of natural-kind words like "water" for the same reason it cannot be true of obviously indexical words like "I."

The theory that natural-kind words like "water" are indexical leaves it open, however, whether to say that "water" in the Twin Earth dialect of English has the same *meaning* as "water" in the Earth dialect and a different extension (which is what we normally say about "I" in different idiolects), thereby giving up the doctrine that "meaning (intension) determines extension"; or to say, as we have chosen to do, that difference is extension is *ipso facto* a difference in meaning for natural-kind words, thereby giving up the doctrine that meanings are concepts, or, indeed, mental entities of *any* kind.

It should be clear, however, that Kripke's doctrine that natural-kind words are rigid designators and our doctrine that they are indexical are but two ways of making the same point. We heartily endorse what Kripke says when he writes:

Let us suppose that we do fix the reference of a name by a description. Even if we do so, we do not then make the name synonymous with the description, but instead we use the name rigidly to refer to the object so named, even in talking about counterfactual situations where the thing named would not satisfy the description in question. Now, this is what I think is in fact true for those cases of naming where the reference is fixed by description. But, in fact, I also think, contrary to most recent theorists, that the reference of names is rarely or almost never fixed by means of description. And by this I do not just mean what Searle says: "It's not a single description, but rather a cluster, a family of properties that fixes the reference." I mean that properties in this sense are not used at all.[10]

[10] See Kripke's "Identity and Necessity," in M. Munitz, ed., *Identity and Individuation* (New York: New York University Press, 1972), p. 157.

Let's be realistic. I wish now to contrast my view with one which is popular, at least among students (it appears to arise spontaneously). For this discussion, let us take as our example of a natural-kind word the word *gold*. We will not distinguish between "gold" and the cognate words in Greek, Latin, etc. And we will focus on "gold" in the sense of gold in the solid state. With this understood, we maintain: "gold" has not changed its *extension* (or not changed it significantly) in two thousand years. Our methods of *identifying* gold have grown incredibly sophisticated. But the extension of χρυσὸς in Archimedes' dialect of Greek is the same as the extension of *gold* in my dialect of English.

It is possible (and let us suppose it to be the case) that just as there were pieces of metal which could not have been determined *not* to be gold prior to Archimedes, so there were or are pieces of metal which could not have been determined *not* to be gold in Archimedes' day, but which we can distinguish from gold quite easily with modern techniques. Let X be such a piece of metal. Clearly X does not lie in the extension of "gold" in standard English; my view is that it did not lie in the extension of χρυσὸς in Attic Greek, either, although an ancient Greek would have *mistaken* X for gold (or, rather, χρυσὸς).

The alternative view is that "gold" *means* whatever satisfies the *contemporary* "operational definition" of *gold*. "Gold" a hundred years ago meant whatever satisfied the "operational definition" of *gold* in use a hundred years ago; "gold" now means whatever satisfies the operational definition of *gold* in use in 1973; and χρυσὸς meant whatever satisfied the operational definition of χρυσὸς in use *then*.

One common motive for adopting this point of view is a certain skepticism about *truth*. On the view I am advocating, when Archimedes asserted that something was gold (χρυσὸς) he was not just saying that it had the superficial characteristics of gold (in exceptional cases, something may belong to a natural kind and *not* have the superficial characteristics of a member of that natural kind, in fact); he was saying that it had the same general *hidden structure* (the same "essence," so to speak) as any normal piece of local gold. Archimedes would have said that our hypothetical piece of metal X was gold, but he would have been *wrong*. But *who's to say* he would have been wrong?

The obvious answer is: *we are* (using the best theory available today). For most people either the question (*who's to say?*) has bite, and our answer has no bite, or our answer has bite and the question has no bite. Why is this?

The reason, I believe, is that people tend either to be strongly anti-realistic or strongly realistic in their intuitions. To a strongly antirealistic intuition it makes little sense to say that what is in the extension of Archimedes' term χρυσὸς is to be determined using our theory. For the antirealist does not see our theory and Archimedes' theory as two approximately correct descriptions of some fixed realm of theory-independent entities, and he tends to be skeptical about the idea of "convergence" in science — he does not think our theory is a *better* description of the *same* entities that Archimedes was describing. But if our theory is *just* our theory, then to use *it* in deciding whether or not X lies in the extension of χρυσὸς is just as arbitrary as using Neanderthal theory to decide whether or not X lies in the extension of χρυσὸς. The only theory that it is *not* arbitrary to use is the one the speaker himself subscribes to.

The trouble is that for a strong antirealist *truth* makes no sense except as an intra-theoretic notion.[11] The antirealist can use truth intra-theoretically in the sense of a "redundancy theory"; but he does not have the notions of truth and reference available *extra-theoretically*. But *extension is tied to the notion of truth*. The extension of a term is just what the term is *true of*. Rather than try to retain the notion of extension via an awkward operationalism, the antirealist should reject the notion of extension as he does the notion of truth (in any extra-theoretic sense). Like Dewey, for example, he can fall back on a notion of "warranted assertibility" instead of truth (relativized to the scientific method, if he thinks there is a *fixed* scientific method, or to the best methods available at the time, if he agrees with Dewey that the scientific method itself evolves). Then he can say that "X is gold (χρυσὸς)" was warrantedly assertible in Archimedes' time and is not warrantedly assertible today (indeed, this is a *minimal* claim, in the sense that it represents the minimum that the realist and the antirealist can agree on); but the assertion that X was in the extension of χρυσὸς will be rejected as meaningless, like the assertion that "X is gold (χρυσὸς)" was true.

It is well known that narrow operationalism cannot successfully ac-

[11] For a discussion of this point, see my "Explanation and Reference," in G. Pearce and P. Maynard, eds., *Conceptual Change* (Dordrecht: Reidel, 1973).

count for the actual use of scientific or common-sense terms. Loosened versions of operationalism, like Carnap's version of Ramsey's theory, agree with if they do not account for actual scientific use (mainly because the loosened versions agree with any possible use!), but at the expense of making the communicability of scientific results a *miracle*. It is beyond question that scientists use terms as if the associated criteria were not *necessary and sufficient conditions*, but rather *approximately* correct characterizations of some world of theory-independent entities, and that they talk as if later theories in a mature science were, in general, *better* descriptions of the *same* entities that earlier theories referred to. In my opinion the hypothesis that this is *right* is the only hypothesis that can account for the communicability of scientific results, the closure of acceptable scientific theories under first-order logic, and many other features of the scientific method.[12] But it is not my task to argue this here. My point is that if we are to use the notions of truth and extension in an extra-theoretic way (i.e., to regard those notions as defined for statements couched in the languages of theories other than our own), then we should accept the realist perspective to which those notions belong. The doubt about whether we can say that X does not lie in the extension of "gold" as *Jones* used it is the *same* doubt as the doubt whether it makes sense to think of Jones's statement that "X is gold" as *true or false* (and not just "warrantedly assertible for Jones and not warrantedly assertible for us"). To square the notion of truth, which is essentially a realist notion, with one's antirealist prejudices by adopting an untenable theory of meaning is no progress.

A second motive for adopting an extreme operationalist account is a dislike of unverifiable hypotheses. At first blush it may seem as if we are saying that "X is gold (χρυσός)" was false in Archimedes' time although Archimedes could not *in principle* have known that it was false. But this is not exactly the situation. The fact is that there are a host of situations that we can describe (using the very theory that tells us that X isn't gold) in which X would have behaved quite unlike the rest of the stuff Archimedes classified as gold. Perhaps X would have separated into two different metals when melted, or would have had different conductivity properties, or would have vaporized at a different tempera-

---

[12] For an illuminating discussion of just these points, see R. Boyd's "Realism and Scientific Epistemology" (unpublished; draft circulated by the author, Cornell University Department of Philosophy).

155

ture, or whatever. If we had performed the experiments with Archimedes watching, he might not have known the theory, but he would have been able to check the empirical regularity that "X behaves differently from the rest of the stuff I classify as χρυσὸς in several respects." Eventually he would have concluded that "X may not be gold."

The point is that even if something satisfies the criteria used at a given time to identify gold (i.e., to recognize if something is gold), it may behave differently in one or more situations from the rest of the stuff that satisfies the criteria. This may not prove that it isn't gold, but it puts the hypothesis that it may not be gold in the running, even in the absence of theory. If, now, we had gone on to inform Archimedes that gold has such and such a molecular structure (except for X), and that X behaved differently because it had a different molecular structure, is there any doubt that he would have agreed with us that X isn't gold? In any case, to worry because things may be true (at a given time) that can't be verified (at that time) seems to me ridiculous. On any reasonable view there are surely things that are true and can't be verified at any time. For example, suppose there are infinitely many binary stars. Must we be able to verify this, even in principle? [13]

So far we have dealt with metaphysical reasons for rejecting our account. But someone might disagree with us about the empirical facts concerning the intentions of speakers. This would be the case if, for instance, someone thought that Archimedes (in the Gedankenexperiment described above) would have said: "it doesn't matter if X does act differently from other pieces of gold; X is a piece of gold, because X has such-and-such properties and that's all it takes to be gold." While, indeed, we cannot be certain that natural-kind words in ancient Greek had the properties of the corresponding words in present-day English, there cannot be any serious doubt concerning the properties of the latter. If we put philosophical prejudices aside, then I believe that we know perfectly well that no operational definition does provide a necessary and sufficient condition for the application of any such word. We may

[13] See my "Logical Positivism and the Philosophy of Mind," in P. Achinstein, *The Legacy of Logical Positivism* (Baltimore: Johns Hopkins Press, 1969); and also my "Degree of Confirmation and Inductive Logic," in P. A. Schilpp, ed., *The Philosophy of Rudolf Carnap* (La Salle, Ill.: Open Court, 1962), and my "Probability and Confirmation" (broadcast for the Voice of America Philosophy of Science Series, Spring 1963; reprinted in A. Danto and S. Morgenbesser, eds., *Philosophy of Science Today* (New York: Basic Books, 1967).

give an "operational definition," or a cluster of properties, or whatever, but the intention is never to "make the name *synonymous* with the description." Rather "we use the name *rigidly*" to refer to whatever things share the *nature* that things satisfying the description normally possess.

Other senses. What we have analyzed so far is the predominant sense of natural-kind words (or, rather, the predominant *extension*). But natural-kind words typically possess a number of senses. (Ziff has even suggested that they possess a *continuum* of senses.)

Part of this can be explained on the basis of our theory. To be water, for example, is to bear the relation same$_L$ to certain things. But what is the relation same$_L$?

x bears the relation same$_L$ to y just in case (1) x and y are both liquids, and (2) x and y agree in important physical properties. The term "liquid" is itself a natural-kind term that I shall not try to analyze here. The term "property" is a broad-spectrum term that we have analyzed in previous papers. What I want to focus on now is the notion of *importance.* Importance is an interest-relative notion. Normally the "important" properties of a liquid or solid, etc., are the ones that are *structurally* important: the ones that specify what the liquid or solid, etc., is ultimately made out of — elementary particles, or hydrogen and oxygen, or earth, air, fire, water, or whatever — and how they are arranged or combined to produce the superficial characteristics. From this point of view the important characteristic of a typical bit of water is consisting of H_2O. But it may or may not be important that there are impurities; thus, in one context "water" may mean *chemically pure* water, while in another it may mean the stuff in Lake Michigan. And structure may sometimes be unimportant; thus one may sometimes refer to XYZ as water if one is *using* it as water. Again, normally it is important that water is in the liquid state; but sometimes it is unimportant, and one may refer to a single H_2O molecule as water, or to water vapor as water ("water in the air").

Even senses that are so far out that they have to be regarded as a bit "deviant" may bear a definite relation to the core sense. For example, I might say "did you see the lemon," meaning the *plastic* lemon. A less deviant case is this: we discover "tigers" on Mars. That is, they look just like tigers, but they have a silicon-based chemistry instead of a carbon-based chemistry. (A remarkable example of parallel evolution!)

157

Are Martian "tigers" tigers? It depends on the context.

In the case of this theory, as in the case of any theory that is orthogonal to the way people have thought about something previously, misunderstandings are certain to arise. One which has already arisen is the following: a critic has maintained that the *predominant* sense of, say, "lemon" is the one in which anything with (a sufficient number of) the superficial characteristics of a lemon is a lemon. The same critic has suggested that having the hidden structure — the genetic code — of a lemon is necessary to being a lemon only when "lemon" is used as a term of *science*. Both of these contentions seem to me to rest on a misunderstanding, or, perhaps, a pair of complementary misunderstandings.

The sense in which literally *anything* with the superficial characteristics of a lemon is necessarily a lemon, far from being the dominant one, is extremely deviant. In that sense something would be a lemon if it looked and tasted like a lemon, even if it had a silicon-based chemistry, for example, or even if an electron-microscope revealed it to be a *machine*. (Even if we include growing "like a lemon" in the superficial characteristics, this does not exclude the silicon lemon, if there are "lemon" trees on Mars. It doesn't even exclude the machine-lemon; maybe the tree is a machine too!

At the same time the sense in which to be a lemon something has to have the genetic code of a lemon is *not* the same as the technical sense (if there is one, which I doubt). The technical sense, I take it, would be one in which "lemon" was *synonymous* with a description which *specified* the genetic code. But when we said (to change the example) that to be *water* something has to be H_2O we did not mean, as we made clear, that the *speaker* has to *know* this. It is only by confusing *metaphysical* necessity with *epistemological* necessity that one can conclude that, if the (metaphysically necessary) truth-condition for being water is being H_2O, then "water" must be synonymous with H_2O — in which case it is certainly a term of science. And similarly, even though the predominant sense of "lemon" is one in which to be a lemon something has to have the genetic code of a lemon (I believe), it does not follow that "lemon" is synonymous with a description which specifies the genetic code explicitly or otherwise.

The mistake of thinking that there is an important sense of "lemon" (perhaps the predominant one) in which to have the superficial characteristics of a lemon is at least *sufficient* for being a lemon is more plausible if among the superficial characteristics one includes *being cross-fertile with lemons*. But the characteristic of being cross-fertile with lemons presupposes the notion of being a lemon. Thus, even if one can obtain a sufficient condition in *this* way, to take this as inconsistent with the characterization offered here is question-begging. Moreover, the characterization in terms of *lemon*-presupposing "superficial characteristics" (like being cross-fertile with *lemons*) gives no truth-condition which would enable us to decide which objects in other possible worlds (or which objects a million years ago, or which objects a million light years from here) are lemons. (In addition, I don't think this characterization, question-begging as it is, is *correct*, even as a sufficient condition. I think one could invent cases in which something which was not a lemon was cross-fertile with lemons and looked like a lemon, etc.)

Again, one might try to rule out the case of the machine-lemon (lemon-machine?) which "grows" on a machine-tree (tree-machine?) by saying that "growing" is not really *growing*. That is right; but it's right because *grow* is a natural-kind verb, and precisely the sort of account we have been presenting applies to *it*.

Another misunderstanding that should be avoided is the following: to take the account we have developed as implying that the members of the extension of a natural-kind word necessarily *have* a common hidden structure. It could have turned out that the bits of liquid we call "water" had *no* important common physical characteristics except the superficial ones. In that case the necessary and sufficient condition for being "water" would have been possession of sufficiently many of the superficial characteristics.

Incidentally, the last statement does not imply that water could have failed to have a hidden structure (or that water could have been anything but H_2O). When we say that it could have *turned out* that water had no hidden structure, what we mean is that a liquid with no hidden structure (i.e., many bits of different liquids, with nothing in common except superficial characteristics) could have looked like water, tasted like water, and have filled the lakes, etc., that are actually full of water. In short, we could have been in the same epistemological situation with respect to a liquid with no hidden structure as we were actually with

159

respect to water at one time. Compare Kripke on the "lectern made of ice."[14]

There are, in fact, almost continuously many cases. Some diseases, for example, have turned out to have no hidden structure (the only thing the paradigm cases have in common is a cluster of symptoms), while others have turned out to have a common hidden structure in the sense of an etiology (e.g., tuberculosis). Sometimes we still don't know; there is a controversy still raging about the case of multiple sclerosis.

An interesting case is the case of *jade*. Although the Chinese do not recognize a difference, the term "jade" applies to two minerals: jadeite and nephrite. Chemically, there is a marked difference. Jadeite is a combination of sodium and aluminum. Nephrite is made of calcium, magnesium, and iron. These two quite different microstructures produce the same unique textural qualities!

Coming back to the Twin Earth example, for a moment, if H_2O and XYZ had both been plentiful on Earth, then we would have had a case similar to the jadeite/nephrite case: it would have been correct to say that there were *two kinds of "water."* And instead of saying that "the stuff on Twin Earth turned out not to really be water," we would have to say "it turned out to be the XYZ *kind of water.*"

To sum up: if there is a hidden structure, then generally it determines what it is to be a member of the natural kind, not only in the actual world, but in all possible worlds. Put another way, it determines what we can and cannot counterfactually suppose about the natural kind ("water could have all been vapor?" yes/"water could have been XYZ?" no). But the local water, or whatever, may have two or more hidden structures — or so many that "hidden structure" becomes irrelevant, and superficial characteristics become the decisive ones.

Other words. So far we have only used natural-kind words as examples, but the points we have made apply to many other kinds of words as well. They apply to the great majority of all nouns, and to other parts of speech as well.

Let us consider for a moment the names of artifacts — words like "pencil," "chair," "bottle," etc. The traditional view is that these words are certainly defined by conjunctions, or possibly clusters, of properties.

[14] See Kripke's "Identity and Necessity."

Anything with all of the properties in the conjunction (or sufficiently many of the properties in the cluster, on the cluster model) is necessarily a *pencil, chair, bottle,* or whatever. In addition, some of the properties in the cluster (on the cluster model) are usually held to be *necessary* (on the conjunction-of-properties model, *all* of the properties in the conjunction are necessary). *Being an artifact* is supposedly necessary, and belonging to a kind with a certain standard purpose — e.g., "pencils are artifacts," and "pencils are standardly intended to be written with" are supposed to be necessary. Finally, this sort of necessity is held to be *epistemic* necessity — in fact, analyticity.

Let us once again engage in science fiction. This time we use an example devised by Rogers Albritton. Imagine that we someday discover that *pencils are organisms.* We cut them open and examine them under the electron microscope, and we see the almost invisible tracery of nerves and other organs. We spy upon them, and we see them spawn, and we see the offspring grow into full-grown pencils. We discover that these organisms are not imitating other (artifactual) pencils — there are not and never were any pencils except these organisms. It is strange, to be sure, that there is *lettering* on many of these organisms — e.g., BONDED Grants DELUXE made in U.S.A. No. 2 — perhaps they are intelligent organisms, and this is their form of camouflage. (We also have to explain why no one ever attempted to manufacture pencils, etc., but this is clearly a possible world, in some sense).

If this is conceivable, and I agree with Albritton that it is, then it is epistemically possible that *pencils could turn out to be organisms.* It follows that *pencils are artifacts* is not epistemically necessary in the strongest sense and, a fortiori, not analytic.

Let us be careful, however. Have we shown that there is a possible world in which pencils are organisms? I think not. What we have shown is that there is a possible world in which certain organisms are the *epistemic counterparts* of pencils (the phrase is Kripke's). To return to the device of Twin Earth: imagine this time that pencils on Earth are just what we think they are, artifacts manufactured to be written with, while "pencils" on Twin Earth are organisms a la Albritton. Imagine, further, that this is totally unsuspected by the Twin Earthians — they have exactly the beliefs about "pencils" that we have about pencils. When we discovered this, we would not say: "some pencils are organisms." We would be far more likely to say: "the things on Twin Earth that pass for pencils aren't really pencils. They're really a species of organism."

Suppose now the situation to be as in Albritton's example both on Earth and on Twin Earth. Then we would say "pencils are organisms." Thus, whether the "pencil-organisms" on Twin Earth (or in another possible universe) are really *pencils* or not is a function of whether or not the *local* pencils are organisms or not. If the local pencils are just what we think they are, then a possible world in which there are pencil-organisms is *not* a possible world in which *pencils are organisms*; there are *no* possible worlds in which pencils are organisms in this case (which is, of course, the actual one). That pencils are artifacts *is* necessary in the sense of true in all possible worlds — metaphysically necessary. But it doesn't follow that it's epistemically necessary.

It follows that "pencil" is not *synonymous* with any description — not even loosely synonymous with a *loose* description. When we use the word "pencil," we intend to refer to whatever has the same *nature* as the normal examples of the local pencils in the actual world. "Pencil" is just as *indexical* as "water" or "gold."

In a way, the case of pencils turning out to be organisms is complementary to the case we discussed some years ago[15] of cats turning out to be robots (remotely controlled from Mars). In his contribution to the present volume, Katz argues that we misdescribed this case: that the case should rather be described as its *turning out that there are no cats in this world*. Katz admits that we might say "Cats have turned out not to be animals, but robots"; but he argues that this is a semantically deviant sentence which is glossed as "the things I am referring to as 'cats' have turned out not to be animals, but robots." Katz's theory is bad linguistics, however. First of all, the explanation of how it is we can say "Cats are robots" is simply an all-purpose explanation of how we can say *anything*. More important, Katz's theory predicts that "Cats are robots" is *deviant*, while "There are no cats in the world" is non-deviant, in fact standard, in the case described. Now then, I don't deny that there *is* a case in which "There are not (and never were) any cats in the world" would be standard: we might (speaking epistemically) discover that we have been suffering from a collective hallucination. ("Cats" are like pink elephants.) But in the case I described, "Cats

[15] See my "It Ain't Necessarily So," *Journal of Philosophy*, 59(1962):658–671.

have turned out to be robots remotely controlled from Mars" is surely nondeviant, and "There are no cats in the world" is highly deviant.

Incidentally, Katz's account is not only bad linguistics; it is also bad as a rational reconstruction. The reason we *don't* use "cat" as synonymous with a description is surely that we know enough about cats to know that they do have a hidden structure, and it is good scientific methodology to use the name to refer rigidly to the things that possess that hidden structure, and not to whatever happens to satisfy some description. Of course, if we *knew* the hidden structure we could frame a description in terms of *it*; but we don't at this point. In this sense the use of natural-kind words reflects an important fact about our relation to the world: we know that there are kinds of things with common hidden structure, but we don't yet have the knowledge to describe all those hidden structures.

Katz's view has more plausibility in the "pencil" case than in the "cat" case, however. We think we *know* a necessary and sufficient condition for being a *pencil*, albeit a vague one. So it is possible to make "pencil" synonymous with a loose description. We *might* say, in the case that "pencils turned out to be organisms" *either* "Pencils have turned out to be organisms" or "There are no pencils in the world" — i.e., we might use "pencil" either as a natural-kind word or as a "one-criterion" word.[16]

On the other hand, we might doubt that there are any true one-criterion words in natural language, apart from stipulative contexts. Couldn't it turn out that pediatricians aren't doctors but Martian spies? Answer "yes," and you have abandoned the synonymy of "pediatrician" and "doctor specializing in the care of children." It seems that there is a strong tendency for words which are introduced as "one-criterion" words to develop a "natural kind" sense, with all the concomitant rigidity and indexicality. In the case of artifact-names, this natural-kind sense seems to be the predominant one.

(There is a joke about a patient who is on the verge of being discharged from an insane asylum. The doctors have been questioning him for some time, and he has been giving perfectly sane responses. They

---

[16] The idea of a "one-criterion" word, and a theory of analyticity based on this notion, appears in my "The Analytic and the Synthetic," in H. Feigl and G. Maxwell, eds., *Minnesota Studies in the Philosophy of Science*, vol. 3 (Minneapolis: University of Minnesota Press, 1962).

decide to let him leave, and at the end of the interview one of the doctors inquires casually, "What do you want to be when you get out?" "A teakettle." The joke would not be intelligible if it were literally inconceivable that a person could be a teakettle.)

There are, however, words which retain an almost pure one-criterion character. These are words whose meaning derives from a transformation: *hunter = one who hunts.*

Not only does the account given here apply to most nouns, but it also applies to other parts of speech. Verbs like "grow," adjectives like "red," etc., all have indexical features. On the other hand, some syncategorematic words seem to have more of a one-criterion character. "Whole," for example, can be explained thus: *The army surrounded the town* could be true even if the A division did not take part. *The whole army surrounded the town* means every part of the army (of the relevant kind, e.g., the A Division) took part in the action signified by the verb.[17]

Meaning. Let us now see where we are with respect to the notion of meaning. We have now seen that the extension of a term is not fixed by a concept that the individual speaker has in his head, and this is true both because extension is, in general, determined *socially* — there is division of linguistic labor as much as of "real" labor — and because extension is, in part, determined *indexically*. The extension of our terms depends upon the actual nature of the particular things that serve as paradigms,[18] and this actual nature is not, in general, fully known to the speaker. Traditional semantic theory leaves out only two contributions to the determination of extension — the contribution of society and the contribution of the real world!

We saw at the outset that meaning cannot be identified with extension. Yet it cannot be identified with "intension" either, if intension is something like an individual speaker's *concept*. What are we to do?

There are two plausible routes that we might take. One route would be to retain the identification of meaning with concept and pay the price of giving up the idea that meaning determines extension. If we followed this route, we might say that "water" has the same *meaning*

---

[17] This example comes from an analysis by Anthony Kroch, in his doctoral dissertation, M.I.T. Department of Linguistics, 1974.

[18] I *don't* have in mind the Flewish notion of "paradigm" in which any paradigm of a K is *necessarily* a K (in reality).

on Earth and on Twin Earth, but a different *extension*. (Not just a different *local* extension but a different *global* extension. The XYZ on Twin Earth isn't in the extension of the tokens of "water" that I utter, but it is in the extension of the tokens of "water" that my *Doppelgänger* utters, and this isn't just because Twin Earth is far away from me, since molecules of H_2O are in the extension of the tokens of "water" that I utter no matter how far away from me they are in space and time. Also, what I can counterfactually suppose water to be is different from what my *Doppelgänger* can counterfactually suppose "water" to be.) While this is the correct route to take for an *absolutely* indexical word like "I," it seems incorrect for the words we have been discussing. Consider "elm" and "beech," for example. If these are "switched" on Twin Earth, then surely we would *not* say that "elm" has the same meaning on Earth and Twin Earth, even if my *Doppelgänger's* stereotype of a beech (or an "elm," as he calls it) is identical with my stereotype of an elm. Rather, we would say that "elm" in my *Doppelgänger's* idiolect means *beech*. For this reason, it seems preferable to take a different route and identify "meaning" with an ordered pair (or possibly an ordered *n-tuple*) of entities, *one of which is the extension*. (The other components of the, so to speak, "meaning vector" will be specified later.) Doing this makes it trivially true that *meaning determines extension* (i.e., difference in extension is ipso facto difference in meaning), but totally abandons the idea that if there is a difference in the meaning my *Doppelgänger* and I assign to a word, then there *must* be some difference in our concepts (or in our psychological state). Following this route, we can say that my *Doppelgänger* and I *mean something different* when we say "elm," but this will not be an assertion about our psychological states. All this means is that the tokens of the word he utters have a different extension than the tokens of the word I utter; but this difference in extension is not a reflection of any difference in our individual linguistic competence considered in isolation.

If this is correct, and I think it is, then the traditional problem of meaning splits into two problems. The first problem is to account for the *determination of extension*. Since, in many cases, extension is determined socially and not individually, owing to the division of linguistic labor, I believe that this problem is properly a problem for sociolinguistics. Solving it would involve spelling out in detail exactly how the division of linguistic labor works. The so-called "causal theory of reference," introduced by Kripke for proper names and extended by us to natural-kind words and physical-magnitude terms,[19] falls into this province. For the fact that, in many contexts, we assign to the tokens of a name that I utter whatever referent we assign to the tokens of the same name uttered by the person from whom I acquired the name (so that the reference is transmitted from speaker to speaker, starting from the speakers who were present at the "naming ceremony," even though no fixed *description* is transmitted) is simply a special case of social cooperation in the determination of reference.

The other problem is to describe *individual competence*. Extension may be determined socially, in many cases, but we don't assign the standard extension to the tokens of a word W uttered by Jones *no matter how* Jones uses W. Jones has to have some particular ideas and skills in connection with W in order to play his part in the linguistic division of labor. Once we give up the idea that individual competence has to be so strong as to actually determine extension, we can begin to study it in a fresh frame of mind.

In this connection it is instructive to observe that nouns like "tiger" or "water" are very different from proper names. One can use the proper name "Sanders" correctly without knowing anything about the referent except that he is called "Sanders" — and even that may not be correct. ("Once upon a time, a very long time ago now, about last Friday, Winnie-the-Pooh lived in a forest all by himself under the name of Sanders.") But one cannot use the word tiger correctly, save *per accidens*, without knowing a good deal about tigers, or at least about a certain conception of tigers. In this sense concepts *do* have a lot to do with meaning.

Just as the study of the first problem is properly a topic in sociolinguistics, so the study of the second problem is properly a topic in psycholinguistics. To this topic we now turn.

Stereotypes and communication. Suppose a speaker knows that "tiger" has a set of physical objects as its extension, but no more. If he possesses normal linguistic competence in other respects, then he could use "tiger" in *some* sentences: for example, "tigers have mass," "tigers take up space," "give me a tiger," "is that a tiger?" etc. Moreover, the *socially determined* extension of "tiger" in these sentences would be

[19] In my "Explanation and Reference," in Pearce and Maynard, eds., *Conceptual Change.*

the standard one, i.e., the set of tigers. Yet we would not count such a speaker as "knowing the meaning" of the word *tiger*. Why not?

Before attempting to answer this question, let us reformulate it a bit. We shall speak of someone as having *acquired* the word "tiger" if he is able to use it in such a way that (1) his use passes muster (i.e., people don't say of him such things as "he doesn't know what a tiger is," "he doesn't know the meaning of the word 'tiger,'" etc.); and (2) his total way of being situated in the world and in his linguistic community is such that the socially determined extension of the word "tiger" in his idiolect is the set of tigers. Clause (1) means, roughly, that speakers like the one hypothesized in the preceding paragraph don't count as having acquired the word "tiger" (or whichever). We might speak of them, in some cases, as having *partially acquired* the word; but let us defer this for the moment. Clause (2) means that speakers on Twin Earth who have the same linguistic habits as we do, count as having acquired the word "tiger" only if the extension of "tiger" in their idiolect is the set of tigers. The burden of the preceding sections of this paper is that it does *not* follow that the extension of "tiger" in Twin Earth dialect (or idiolects) is the set of tigers merely because their linguistic habits are the same as ours; the nature of Twin Earth "tigers" is also relevant. (If Twin Earth organisms have a silicon chemistry, for example, then their "tigers" aren't really tigers, even if they look like tigers, although the linguistic habits of the lay Twin Earth speaker exactly correspond to those of Earth speakers.) Thus clause (2) means that in this case we have decided to say that Twin Earth speakers have not acquired our word "tiger" (although they have acquired another word with the same spelling and pronunciation).

Our reason for introducing this way of speaking is that the question "does he know the meaning of the word 'tiger'?" is biased in favor of the theory that acquiring a word is coming to possess a thing called its "meaning." Identify this thing with a concept, and we are back at the theory that a sufficient condition for acquiring a word is associating it with the right concept (or, more generally, being in the right psychological state with respect to it) — the very theory we have spent all this time refuting. So, henceforth, we will "acquire" words, rather than "learn their meaning."

We can now reformulate the question with which this section began. The use of the speaker we described does not pass muster, although it

167

is not such as to cause us to assign a nonstandard extension to the word "tiger" in his idiolect. Why doesn't it pass muster?

Suppose our hypothetical speaker points to a snowball and asks, "is that a tiger?" Clearly there isn't much point in talking tigers with *him*. Significant communication requires that people know something of what they are talking about. To be sure, we hear people "communicating" every day who clearly know nothing of what they are talking about; but the sense in which the man who points to a snowball and asks "is that a tiger" doesn't know anything about tigers is so far beyond the sense in which the man who thinks that Vancouver is going to win the Stanley Cup, or that the Vietnam War was fought to help the South Vietnamese, doesn't know what he is talking about as to boggle the mind. The problem of people who think that Vancouver is going to win the Stanley Cup, or that the Vietnam War was fought to help the South Vietnamese, is one that obviously cannot be remedied by the adoption of linguistic conventions; but not knowing what one is talking about in the second, mind-boggling sense can be and is prevented, near enough, by our conventions of language. What I contend is that speakers are *required* to know something about (stereotypical) tigers in order to count as having acquired the word "tiger"; something about elm trees (or, anyway, about the stereotype thereof) to count as having acquired the word "elm"; etc.

This idea should not seem too surprising. After all, we do not permit people to drive on the highways without first passing some tests to determine that they have a *minimum* level of competence; and we do not dine with people who have not learned to use a knife and fork. The linguistic community too has its minimum standards, with respect both to syntax and to "semantics."

The nature of the required minimum level of competence depends heavily upon both the culture and the topic, however. In our culture speakers are required to know what tigers look like (if they acquire the word "tiger," and this is virtually obligatory); they are not required to know the fine details (such as leaf shape) of what an elm tree looks like. English speakers are *required by their linguistic community* to be able to tell tigers from leopards; they are not required to be able to tell elm trees from beech trees.

This could easily have been different. Imagine an Indian tribe, call it the Cheroquoi, who have words, say *uhaba'* and *wa'arabi* for elm trees and beech trees respectively, and who make it obligatory to know the difference. A Cheroquoi who could not recognize an elm would be said not to know what an uhaba' is, not to know the meaning of the word "uhaba'" (perhaps, not to know the word, or not to have the word), just as an English speaker who had no idea that tigers are striped would be said not to know what a tiger is, not to know the meaning of the word "tiger" (of course, if he at least knows that tigers are large felines we might say he knows part of the meaning, or partially knows the meaning), etc. Then the translation of "uhaba'" as "elm" and "wa'-arabi" as "beech" would, on our view, be only approximately correct. In this sense there is a real difficulty with radical translation,[20] but this is not the abstract difficulty that Quine is talking about.[21]

What stereotypes are. I introduced the notion of a "stereotype" in my lectures at the University of Washington and at the Minnesota Center for the Philosophy of Science in 1968. I will not review all the argumentation from the subsequently published "Is Semantics Possible" in the present essay, but I do want to introduce the notion again and to answer some questions that have been asked about it.

In ordinary parlance a "stereotype" is a conventional (frequently malicious) idea (which may be wildly inaccurate) of what an X looks like or acts like or is. Obviously, I am trading on some features of the ordinary parlance. I am not concerned with malicious stereotypes (save where the language itself is malicious); but I am concerned with conventional ideas, which may be inaccurate. I am suggesting that just such a conventional idea is associated with "tiger," with "gold," etc., and, moreover, that this is the sole element of truth in the "concept" theory.

On this view someone who knows what "tiger" means (or, as we have decided to say instead, has acquired the word "tiger") is required to know that stereotypical tigers are striped. More precisely, there is one stereotype of tigers (he may have others) which is required by the linguistic community as such; he is required to have this stereotype, and to know (implicitly) that it is obligatory. This stereotype must in-

---

[20] The term is due to Quine (in Word and Object); it signifies translation without clues from either shared culture or cognates.

[21] For a discussion of the supposed impossibility of uniquely correct radical translation, see my "The Refutation of Conventionalism" (forthcoming in Noûs and also, in a longer version, in a collection edited by M. Munitz to be published by New York University Press under the title Semantics and Philosophy).

clude the feature of stripes if his acquisition is to count as successful.

The fact that a feature (e.g. stripes) is included in the stereotype associated with a word X does not mean that it is an analytic truth that all Xs have that feature, nor that most Xs have that feature, nor that all normal Xs have that feature, nor that some Xs have that feature.[22] Three-legged tigers and albino tigers are not logically contradictory entities. Discovering that our stereotype has been based on nonnormal or unrepresentative members of a natural kind is not discovering a logical contradiction. If tigers lost their stripes they would not thereby cease to be tigers, nor would butterflies necessarily cease to be butterflies if they lost their wings.

(Strictly speaking, the situation is more complicated than this. It is possible to give a word like "butterfly" a sense in which butterflies would cease to be butterflies if they lost their wings — through mutation, say. Thus one can find a sense of "butterfly" in which it is analytic that "butterflies have wings." But the most important sense of the term, I believe, is the one in which the wingless butterflies would still be butterflies.)

At this point the reader may wonder what the value to the linguistic community of having stereotypes is, if the "information" contained in the stereotype is not necessarily correct. But this is not really such a mystery. Most stereotypes do in fact capture features possessed by paradigmatic members of the class in question. Even where stereotypes go wrong, the way in which they go wrong sheds light on the contribution normally made by stereotypes to communication. The stereotype of gold, for example, contains the feature *yellow* even though chemically pure gold is nearly white. But the gold we see in jewelry is typically yellow (due to the presence of copper), so the presence of this feature in the stereotype is even useful in lay contexts. The stereotype associated with *witch* is more seriously wrong, at least if taken with existential import. Believing (with existential import) that witches enter into pacts with Satan, that they cause sickness and death, etc., facilitates communication only in the sense of facilitating communication internal to witch-theory. It does not facilitate communication in any situation in which what is needed is more agreement with the world than agreement with the theory of other speakers. (Strictly speaking, I am speaking of

[22] This is argued in "Is Semantics Possible?"

the stereotype as it existed in New England three hundred years ago; today that witches aren't *real* is itself part of the stereotype, and the baneful effects of witch-theory are thereby neutralized.) But the fact that our language has *some* stereotypes which impede rather than facilitate our dealings with the world and each other only points to the fact that we aren't infallible beings, and how could we be? The fact is that we could hardly communicate successfully if most of our stereotypes weren't pretty accurate as far as they go.

The "operational meaning" of stereotypes. A trickier question is this: how far is the notion of stereotype "operationally definable." Here it is necessary to be extremely careful. Attempts in the physical sciences to *literally* specify operational definitions for terms have notoriously failed; and there is no reason the attempt should succeed in linguistics when it failed in physics. Sometimes Quine's arguments against the possibility of a theory of meaning seem to reduce to the demand for operational definitions in linguistics; when this is the case the arguments should be ignored. But it frequently happens that terms do have operational definitions not in the actual world but in idealized circumstances. Giving these "operational definitions" has heuristic value, as idealization frequently does. It is only when we mistake operational definition for more than convenient idealization that it becomes harmful. Thus we may ask what is the "operational meaning" of the statement that a word has such and such a stereotype, without supposing that the answer to this question counts as a theoretical account of what it is to be a stereotype.

The theoretical account of what it is to be a stereotype proceeds in terms of the notion of *linguistic obligation*; a notion which we believe to be fundamental to linguistics and which we shall not attempt to explicate here. What it means to say that being striped is part of the (linguistic) stereotype of "tiger" is that it is *obligatory* to acquire the information that stereotypical tigers are striped if one acquires the word "tiger," in the same sense of "obligatory" as that it is obligatory to indicate whether one is speaking of lions in the singular or lions in the plural when one speaks of lions in English. To describe an idealized experimental test of this hypothesis is not difficult. Let us introduce a person whom we may call the linguist's *confederate*. The confederate will be (or pretend to be) an adult whose command of English is generally excellent, but who for some reason (raised in an alien culture?

brought up in a monastery?) has totally failed to acquire the word "tiger." The confederate will say the word "tiger" or, better yet, point to it (as if he wasn't sure how to pronounce it), and ask "what does this word mean?" or "what is this?" or some such question. Ignoring all the things that go wrong with experiments in practice, what our hypothesis implies is that informants should typically tell the confederate that tigers are, inter alia, striped.

Instead of relying on confederates, one might expect the linguist to study children learning English. But children learning their native language aren't taught it nearly as much as philosophers suppose; they learn it but they aren't taught it, as Chomsky has emphasized. Still, children do sometimes ask such questions as "what is a tiger?" and our hypothesis implies that in these cases too informants should tell them, inter alia, that tigers are striped. But one problem is that the informants are likely to be parents, and there are the vagaries of parental time, temper, and attention to be allowed for.

It would be easy to specify a large number of additional "operational" implications of our hypothesis, but to do so would have no particular value. The fact is that we are fully competent speakers of English ourselves, with a devil of a good sense of what our linguistic obligations are. Pretending that we are in the position of Martians with respect to English is not the route to methodological clarity; it was, after all, only when the operational approach was abandoned that transformational linguistics blossomed into a handsome science.

Thus if anyone were to ask me for the meaning of "tiger," I know perfectly well what I would tell him. I would tell him that tigers were feline, something about their size, that they are yellow with black stripes, that they (sometimes) live in the jungle, and are fierce. Other things I might tell him too, depending on the context and his reason for asking; but the above items, save possibly for the bit about the jungle, I would regard it as *obligatory* to convey. I don't have to experiment to know that this is what I regard it as obligatory to convey, and I am sure that approximately this is what other speakers regard it as obligatory to convey too. Of course, there is some variation from idiolect to idiolect; the feature of having stripes (apart from figure-ground relations, e.g., are they black stripes on a yellow ground, which is the way I see them, or yellow stripes on a black ground?) would be found in all normal idiolects, but some speakers might regard the information that tigers (stereo-

172

typically) inhabit jungles as obligatory, while others might not. Alternatively, some features of the stereotype (big-cat-hood, stripes) might be regarded as obligatory, and others as optional, on the model of certain syntactical features. But we shall not pursue this possibility here.

Quine's "Two Dogmas" revisited. In "Two Dogmas of Empiricism" Quine launched a powerful and salutory attack on the currently fashionable analytic-synthetic distinction. The distinction had grown to be a veritable philosophical man-eater: analytic equaling necessary equaling unrevisable in principle equaling whatever truth the individual philosopher wished to explain away. But Quine's attack itself went too far in certain respects; some limited class of analytic sentences can be saved, we feel.[23] More importantly, the attack was later construed, both by Quine himself and by others, as implicating the whole notion of meaning in the downfall of the analytic-synthetic distinction. While we have made it clear that we agree that the traditional notion of meaning has serious troubles, our project in this paper is constructive, not destructive. We come to revise the notion of meaning, not to bury it. So it will be useful to see how Quine's arguments fare against our revision.

Quine's arguments against the notion of analyticity can basically be reduced to the following: that no behavioral significance can be attached to the notion. His argument (again simplifying somewhat) was that there are, basically, only two candidates for a behavioral index of analyticity, and both are totally unsatisfactory, although for different reasons. The first behavioral index is centrality: many contemporary philosophers call a sentence analytic if, in effect, some community (say, Oxford dons) holds it immune from revision. But, Quine persuasively argues, maximum immunity from revision is no exclusive prerogative of analytic sentences. Sentences expressing fundamental laws of physics (e.g. the conservation of energy) may well enjoy maximum behavioral immunity from revision, although it would hardly be customary or plausible to classify them as analytic. Quine does not, however, rely on the mere implausibility of classifying all statements that we are highly reluctant to give up as analytic; he points out that "immunity from revision" is, in the actual history of science, a matter of degree. There is no such thing, in the actual practice of rational science, as absolute immunity from revision. Thus to identify analyticity with immunity from revision would alter the notion in two fundamental ways: analyticity

[23] See "The Analytic and the Synthetic."

would become a matter of degree, and there would be no such thing as an absolutely analytic sentence. This would be such a departure from the classical Carnap-Ayer-et al. notion of analyticity that Quine feels that if *this* is what we mean to talk about, then it would be less misleading to introduce a different term altogether, say, *centrality*.

The second behavioral index is *being called "analytic."* In effect, some philosophers take the hallmark of analyticity to be that trained informants (say, Oxford dons) *call* the sentence analytic. Variants of this index are: that the sentence be deducible from the sentences in a finite list at the top of which someone who bears the ancestral of the graduate-student relation to Carnap has printed the words "Meaning Postulate"; that the sentence be obtainable from a theorem of logic by substituting synonyms for synonyms. The last of these variants looks promising, but Quine launches against it the question, "what is the criterion of synonymy?" One possible criterion might be that words W_1 and W_2 are synonymous if and only if the biconditional $(x)(x$ is in the extension of $W_1 \equiv x$ is in the extension of $W_2)$ is *analytic*; but this leads us right back in a circle. Another might be that words W_1 and W_2 are synonymous if and only if trained informants *call* them synonymous; but this is just our second index in a slightly revised form. A promising line is that words W_1 and W_2 are synonymous if and only if W_1 and W_2 are interchangeable (i.e., the words can be switched) *salva veritate* in all contexts of a suitable class. But Quine convincingly shows that this proposal too leads us around in a circle. Thus the second index reduces to this: a sentence is analytic if either it or some expression, or sequence of ordered pairs of expressions, or set of expressions, related to the sentence in certain specified ways, lies in a class to all the members of which trained informants apply a certain *noise*: either the *noise* ANALYTIC, or the *noise* MEANING POSTULATE, or the *noise* SYNONYMOUS. Ultimately, this proposal leaves "analytic," etc., *unexplicated noises.*

Although Quine does not discuss this explicitly, it is clear that taking the intersection of the two unsatisfactory behavioral indexes would be no more satisfactory; explicating the analyticity of a sentence as consisting in centrality *plus* being called ANALYTIC is just saying that the analytic sentences are a subclass of the central sentences without in any way telling us wherein the exceptionality of the subclass consists. In effect, Quine's conclusion is that analyticity is either centrality misconceived or it is nothing.

In spite of Quine's forceful argument, many philosophers have gone on abusing the notion of analyticity, often confusing it with a supposed highest degree of centrality. Confronted with Quine's alternatives, they have elected to identify analyticity with centrality, and to pay the price — the price of classifying such obviously synthetic-looking sentences as "space has three dimensions" as analytic, and the price of undertaking to maintain the view that there is, after all, such a thing as absolute unrevisability in science in spite of the impressive evidence to the contrary. But this line can be blasted by coupling Quine's argument with an important argument of Reichenbach's.

In his book *The Theory of Relativity and A Priori Knowledge*, Reichenbach showed that there exists a set of principles (see p. 31) each of which Kant would have regarded as synthetic a priori, but whose conjunction is incompatible with the principles of special relativity and general covariance. (These include normal induction, the continuity of space, and the euclidean character of space.) A Kantian can consistently hold on to euclidean geometry come what may; but then experience may force him to give up normal induction or the continuity of space. Or he may hold on to normal induction and the continuity of space come what may; but then experience may force him to give up euclidean geometry (this happens in the case that physical space is not even homeomorphic to any euclidean space). In his article in *Albert Einstein, Philosopher-Scientist*, Reichenbach gives essentially the same argument in a slightly different form.

Applied to our present context, what this shows is that there are principles such that philosophers fond of the overblown notion of analyticity, and in particular philosophers who identify analyticity with (maximum) unrevisability, would classify them as analytic, but whose conjunction has testable empirical consequences. Thus either the identification of analyticity with centrality must be given up once and for all, or one must give up the idea that analyticity is closed under conjunction, or one must swallow the unhappy consequence that an analytic sentence can have testable empirical consequences (and hence that an *analytic* sentence might turn out to be *empirically false*).

It is no accident, by the way, that the sentences that Kant would have classified as synthetic a priori would be classified by these latter-day empiricists as analytic; their purpose in bloating the notion of analyticity was precisely to dissolve Kant's problem by identifying a prioricity with analyticity and then identifying analyticity in turn with truth by convention. (This last step has also been devastatingly criticized by Quine, but discussion of it would take us away from our topic.)

Other philosophers have tried to answer Quine by distinguishing between *sentences* and *statements*: all *sentences* are revisable, they agree, but some *statements* are not. Revising a sentence is not changing our mind about the statement formerly expressed by that sentence just in case the sentence (meaning the syntactical object together with its meaning) after the revision is, in fact, not synonymous with the sentence prior to the revision, i.e., just in case the revision is a case of meaning change and not change of theory. But (1) this reduces at once to the proposal to explicate analyticity in terms of synonymy; and (2) if there is one thing that Quine has decisively contributed to philosophy, it is the realization that meaning change and theory change cannot be sharply separated. We do not agree with Quine that meaning change cannot be defined at all, but it does not follow that the dichotomy "meaning change or theory change" is tenable. Discovering that we live in a non-euclidean world *might* change the meaning of "straight line" (this would happen in the — somewhat unlikely — event that something like the parallels postulate was part of the stereotype of straightness); but it would not be a *mere* change of meaning. In particular it would not be a change of *extension*: thus it would not be right to say that the parallels postulate was "true in the former sense of the words." From the fact that giving up a sentence S would involve meaning change, it does not follow that S is *true*. Meanings may not fit the world; and meaning change can be forced by empirical discoveries.

Although we are not, in this paper, trying to explicate a notion of analyticity, we are trying to explicate a notion that might seem closely related, the notion of meaning. Thus it might seem that Quine's arguments would also go against our attempt. Let us check this out.

On our view there is a perfectly good sense in which being striped is part of the meaning of "tiger." But it does not follow, on our view, the "tigers are striped" is analytic. If a mutation occurred, all tigers might be albinos. Communication presupposes that I have a stereotype of tigers which includes stripes, and that you have a stereotype of tigers which includes stripes, and that I know that your stereotype includes stripes, and that you know that my stereotype includes stripes, and that you know that I know . . . (and so on, a la Grice, forever). But it does not presuppose that any particular stereotype be *correct*, or that the majority of our stereotypes remain correct forever. Linguistic obligatoriness is not supposed to be an index of unrevisability or even of truth; thus we can hold that "tigers are striped" is part of the meaning of "tiger" without being trapped in the problems of analyticity.

Thus Quine's arguments against identifying analyticity with centrality are not arguments against identifying a feature's being "part of the meaning" of X with its being obligatorily included in the stereotype of X. What of Quine's "noise" argument?

Of course, evidence concerning what people *say*, including explicit metalinguistic remarks, is important in "semantics" as it is in syntax. Thus, if a speaker points to a *clam* and asks "is that a tiger?" people are likely to guffaw. (When they stop laughing) they might say "he doesn't know the meaning of 'tiger,'" or "he doesn't know what tigers are." Such comments can be helpful to the linguist. But we are not *defining* the stereotype in terms of such comments. To say that being "big-cat-like" is part of the meaning of tiger is not merely to say that application of "tiger" to something which is not big-cat-like (and also not a tiger) would provoke certain *noises*. It is to say that speakers acquire the information that "tigers are (stereotypically) big-cat-like" as they acquire the word "tiger" and that they feel an obligation to guarantee that those to whom they teach the use of the word do likewise. Information about the minimum skills required for entry into the linguistic community is significant information; no circularity of the kind Quine criticized appears here.

Radical translation. What our theory does not do, by itself at any rate, is solve Quine's problem of "radical translation" (i.e., translation from an alien language/culture). We cannot translate our hypothetical Cheroquoi into English by matching stereotypes, just because finding out what the stereotype of, say, *wa'arabi* is involves translating Cheroquoi utterances. On the other hand, the constraint that each word in Cheroquoi should match its image in English under the translation-function as far as stereotype is concerned (or approximately match, since in many cases exact matching may not be attainable), places a severe *constraint* on the translation-function. Once we have succeeded in translating the basic vocabulary of Cheroquoi, we can start to elicit stereotypes, and these will serve both to constrain future translations and to check the internal correctness of the piece of the translation-function already constructed.

Even where we can determine stereotypes (relative, say, to a tentative translation of "basic vocabulary"), these do not suffice, in general, to determine a unique translation. Thus the German words *Ulme* and *Buche* have the same stereotype as *elm*; but *Ulme* means "elm" while *Buche* means "beech." In the case of German, the fact that *Ulme* and *elm* are cognates could point to the correct translation (although this is far from foolproof — in general, cognate words are not synonymous); but in the case of Greek we have no such clue as to which of the two words 'οξύα, πτελέα means *elm* and which *beech*; we would just have to find a Greek who could tell elms from beeches (or oxya from ptelea). What this illustrates is that it may not be the *typical* speakers' dispositions to assent and dissent that the linguist must seek to discover; because of the division of linguistic labor, it is frequently necessary for the linguist to assess who are the experts with respect to oxya, or wa'arabi, or gavagai, or whatever, before he can make a guess at the socially determined extension of a word. Then this socially determined extension and the stereotype of the *typical* speaker, inexpert though he is, will *both* function as constraints upon the translation-function. Discovery that the stereotype of oxya is wildly different from the stereotype of *elm* would disqualify the translation of oxya by *elm* in all save the most extensional contexts; but the discovery that the *extension* of oxya is not even approximately the class of elms would wipe out the translation altogether, in all contexts.

It will be noted that we have already enlarged the totality of facts counted as evidence for a translation-function beyond the ascetic base that Quine allows in *Word and Object*. For example, the fact that speakers say such-and-such when the linguist's "confederate" points to the word oxya and asks "what does this mean?" or "what is this?" or whatever is not allowed by Quine (as something the linguist can "know") on the ground that this sort of "knowledge" presupposes already having translated the query "what does this word mean?". However, if Quine is willing to assume that one can *somehow* guess at the words which signify assent and dissent in the alien language, it does not seem at all unreasonable to suppose that one can somehow convey to a native speaker that one does not understand a word. It is not necessary that one discover a locution in the alien language which literally means "what does this word *mean?*" (as opposed to: "I don't understand this word," or "this word is unfamiliar to me" or "I am puzzled by this word," etc.). Perhaps just saying the word *oxya,* or whatever, with a tone of puzzlement would suffice. Why should *puzzlement* be less accessible to the linguist than *assent?*

Also, we are taking advantage of the fact that segmentation into *words* has turned out to be linguistically universal (and there even exist tests for word and morpheme segmentation which are independent of meaning). Clearly, there is no motivated reason for allowing the linguist to utter whole sentences and look for assent and dissent, while refusing to allow him to utter words and morphemes in a tone of puzzlement.

I repeat, the claim is not being advanced that enlarging the evidence base in this way solves the problem of radical translation. What it does is add further constraints on the class of admissible candidates for a correct translation. What I believe is that enlarging the class of constraints can determine a unique translation, or as unique a translation as we are able to get in practice. But constraints that go beyond linguistic theory proper will have to be used, in my opinion; there will also have to be constraints on what sorts of beliefs (and connections between beliefs, and connections of beliefs to the culture and the world) we can reasonably impute to people. Discussion of these matters will be deferred to another paper.

A critique of Davidsonian semantic theory. In a series of publications, Donald Davidson has put forward the interesting suggestion that a semantic theory of a natural language might be modeled on what mathematical logicians call a *truth definition* for a formalized language. Stripped of technicalities, what this suggestion comes down to is that one might have a set of rules specifying (1) for each word, under what conditions that word is true of something (for words for which the concept of an extension makes sense; all other words are to be treated as syncategorematic); (2) for sentences longer than a single word, a rule is given specifying the conditions under which the sentence is true as a function of the way it is built up out of shorter sentences (counting words as if they were one-word sentences, e.g., "snow" as "that's snow"). The choice of one-word sentences as the starting point is my interpretation of what Davidson intends; in any case, he means one to start with a *finite* stock of *short* sentences for which truth conditions are to be laid down *directly.* The intention of (2) is not that there should be a rule for each sentence not handled under (1), since this would require an infinite number of rules, but that there should be a rule for each sentence *type.* For example, in a formalized language one of the rules of kind (2) might be: if S is (S_1 & S_2) for some sentences S_1, S_2, then S is true if and only if S_1, S_2, are both true.

It will be noticed that, in the example just given, the truth condition specified for sentences of the sentence type (S_1 & S_2) performs the job of specifying the meaning of "&." More precisely, it specifies the meaning of the structure (—— & ——). This is the sense in which a truth definition can be a theory of meaning. Davidson's contention is that the *entire* theory of meaning for a natural language can be given in this form.

There is no doubt that rules of the type illustrated can give the meaning of some words and structures. The question is, what reason is there to think that the meaning of most words can be given in this way, let alone all?

The obvious difficulty is this: for many words, an extensionally correct truth definition can be given which is in no sense a theory of the meaning of the word. For example, consider *"Water" is true of x if and only if x is H_2O.* This is an extensionally correct truth definition for "water" (strictly speaking, it is not a truth definition but a "truth of" definition — i.e., a *satisfaction-in-the-sense-of-Tarski* definition, but we will not bother with such niceties here). At least it is extensionally correct if we ignore the problem that water with impurities is also called "water," etc. Now, suppose most speakers don't know that water is H_2O. Then this formula in no way tells us anything about the *meaning* of "water." It might be of interest to a chemist, but it doesn't count as a theory of the meaning of the term "water." Or, it counts as a theory of the *extension* of the term "water," but Davidson is promising us more than just that.

Davidson is quite well aware of this difficulty. His answer (in conversation, anyway) is that we need to develop a theory of *translation.* This he, like Quine, considers to be the real problem. Relativized to such a theory (relativized to what we admittedly don't yet have), the theory comes down to this: we want a system of truth definitions which is simultaneously a system of translations (or approximate translations, if perfect translation is unobtainable). If we had a theory which specified what it is to be a good translation, then we could rule out the above truth definition for "water" as uninteresting on the grounds that x is H_2O is not an acceptable translation or even near-translation of x is water (in a prescientific community), even if water = H_2O happens to be true.

This comes perilously close to saying that a theory of meaning is a truth definition plus a theory of meaning. (If we had ham and eggs we'd have ham and eggs — if we had ham and if we had eggs.) But this story suffers from worse than promissoriness, as we shall see.

A second contention of Davidson's is that the theory of translation that we don't yet have is necessarily a theory whose basic units are *sentences* and not *words* on the grounds that our *evidence* in linguistics necessarily consists of assent and dissent from sentences. Words can be handled, Davidson contends, by treating them as sentences ("water" as "that's water," etc.).

How does this ambitious project of constructing a theory of meaning in the form of a truth definition constrained by a theory of translation tested by "the only evidence we have," speakers' dispositions to use sentences, fare according to the view we are putting forward here?

Our answer is that the theory cannot succeed in principle. In special cases, such as the word "and" in its truth-functional sense, a truth definition (strictly speaking, a clause in what logicians call a "truth definition" — the sum total of all the clauses is the inductive definition of "truth" for the particular language) can give the meaning of the word or structure because the stereotype associated with the word (if one wants to speak of a stereotype in the case of a word like "and") is so strong as to actually constitute a necessary and sufficient condition. If all words were like "and" and "bachelor" the program could succeed. And Davidson certainly made an important contribution in pointing out that linguistics has to deal with inductively specified truth conditions. But for the great majority of words, the requirements of a theory of truth and the requirements of a theory of meaning are mutually incompatible, at least in the English-English case. But the English-English case — the case in which we try to provide a significant theory of the meaning of English words which is itself couched in English — is surely the basic one.

The problem is that in general the only expressions which are both coextensive with X and have roughly the same stereotype as X are expressions containing X itself. If we rule out such truth definitions (strictly speaking, clauses, but I shall continue using "truth definition" both for individual clauses and for the whole system of clauses, for simplicity) as

"X is water" is true if and only if X is water

on the grounds that they don't say anything about the meaning of the word "water," and we rule out such truth definitions as

"X is water" is true if and only if X is H_2O

on the grounds that what they say is wrong as a description of the *meaning* of the word "water," then we shall be left with nothing.

The problem is that we want

W is true of x if and only if ——

to satisfy the conditions that (1) the clause be extensionally correct (where —— is to be thought of as a condition containing "x," e.g. "x is H_2O"); (2) that —— be a *translation* of W — on our theory, this would mean that the stereotype associated with W is approximately the same as the stereotype associated with ——; (3) that —— not contain W itself, or syntactic variants of W. If we take W to be, for example, the word "elm," then there is absolutely no way to fulfill all three conditions simultaneously. Any condition of the above form that does not contain "elm" and that is extensionally correct will contain a —— that is absolutely terrible as a *translation* of "elm."

Even where the language contains two exact synonyms, the situation is little better. Thus

"Heather" is true of x if and only if x is gorse

is true, and so is

"Gorse" is true of x if and only if x is heather

—— this is a *theory of the meaning* of "gorse" and "heather"?

Notice that the condition (3) is precisely what logicians do **not** impose on *their* truth definitions.

"Snow is white" is true if and only if snow is white

is the paradigm of a truth definition in the logician's sense. But logicians are trying to give the extension of "true" with respect to a particular language, not the meaning of "snow is white." Tarski would have gone so far as to claim he was giving the *meaning* (and not just the extension) of "true"; but he would never have claimed he was saying *anything* about the meaning of "snow is white."

It may be that what Davidson really thinks is that theory of meaning, in any serious sense of the term, is impossible, and that all that is possible is to construct translation-functions. If so, he might well think that the only "theory of meaning" possible for English is one that says " 'elm' is true of x if and only if x is an elm," " 'water' is true of x if and only if x is water," etc., and only rarely something enlightening like "S_1 & S_2 is true if and only if S_1,S_2 are both true." But if Davidson's "theory" is just Quinine skepticism under the disguise of a positive contribution to the study of meaning, then it is a bitter pill to swallow.

The contention that the only evidence available to the linguist is speakers' dispositions with respect to whole sentences is, furthermore, vacuous on one interpretation, and plainly false on the interpretation on which it is not vacuous. If dispositions to say certain things *when queried about individual words or morphemes or syntactic structures* are included in the notion of dispositions to use sentences, then the restriction to dispositions to use sentences seems to rule out nothing whatsoever. On the non-vacuous interpretation, what Davidson is saying is that the linguist cannot have access to such data as what informants (including the linguist himself) say when asked the meaning of a word or morpheme or syntactic structure. No reason has ever been given why the linguist cannot have access to such data, and it is plain that actual linguists place heavy reliance on informants' testimony about such matters, in the case of an alien language, and upon their own intuitions as native speakers, when they are studying their native languages. In particular, when we are trying to translate a whole sentence, there is no reason why we should not be guided by our knowledge of the syntactic and semantic properties of the constituents of that sentence, including the deep structure. As we have seen, there are procedures for gaining information about individual constituents. It is noteworthy that the procedure that Quine and Davidson claim is the only *possible* one — going from whole sentences to individual words — is the *opposite* of the procedure upon which every success ever attained in the study of natural language has been based.

Critique of California semantics. I wish now to consider an approach to semantic theory pioneered by the late Rudolf Carnap. Since I do not wish to be embroiled in textual questions, I will not attribute the particular form of the view I am going to describe to any particular philosopher but will simply refer to it as "California semantics."

We assume the notion of a *possible world*. Let f be a function defined on the "space" of all possible worlds whose value f(x) at any possible world x is always a subset of the set of entities in x. Then f is called an *intension*. A term T has meaning for a speaker X if X associates T with an intension f_T. The term T is *true of* an entity e in a possible world x if and only if e belongs to the set f(x). Instead of using the term "associate," Carnap himself tended to speak of "grasping" intensions; but, clearly, what was intended was not just that X "grasp" the intension f, but that he grasp *that* f is the intension of *T* — i.e., that he associate f with T in some way.

Clearly this picture of what it is to understand a term disagrees with the story we tell in this paper. The reply of a California semanticist would be that California semantics is a description of an *ideal* language; that actual language is *vague*. In other words, a term T in actual language does not have a single precise intension; it has a set — possibly a fuzzy set — of intensions. Nevertheless, the first step in the direction of describing natural language is surely to study the idealization in which each term T has exactly one intension.

(In his book *Meaning and Necessity*, Carnap employs a superficially different formulation: an intension is simply a *property*. An entity e belongs to the extension of a term T just in case e has whichever property is the intension of T. The later formulation in terms of functions f as described above avoids taking the notion of *property* as primitive.)

The first difficulty with this position is the use of the totally unexplained notion of *grasping* an intension (or, in our reformulation of the position, *associating* an intension with a term). Identifying intensions with set-theoretic entities f provides a "concrete" realization of the notion of intension in the current mathematical style (relative to the notions of possible world and set), but at the cost of making it very difficult to see how anyone could have an intension in his mind, or what it is to think about one or "grasp" one or "associate" one with anything. It will not do to say that thinking of an intension is using a word or functional substitute for a word (i.e., the analogue of a word in "brain code," if, as seems likely, the brain "computes" in a "code" that has analogies to and possibly borrowings from language; or a thought form such as a picture or a private symbol, in cases where such are employed in thinking) which *refers* to the intension in question, since *reference* (i.e., being in the extension of a term) has just been defined in terms of *intension*. Although the characterization of what it is to think of an abstract entity such as a function or a property is certainly correct, in the present context it is patently circular. But no non-circular characterization of this fundamental notion of the theory has ever been provided.

This difficulty is related to a general difficulty in the philosophy of mathematics pointed out by Paul Benacerraf.[24] Benacerraf has remarked that philosophies of mathematics tend to fall between two stools: either they account for what mathematical objects are and for the necessity of mathematical truth and fail to account for the fact that people can *learn* mathematics, can *refer to* mathematical objects, etc., or else they account for the latter facts and fail to account for the former. California semantics accounts for what intensions *are*, but provides no account that is not completely circular of how it is that we can "grasp" them, associate them with terms, think about them, *refer to* them, etc.

Carnap may not have noticed this difficulty because of his Verificationism. In his early years Carnap thought of understanding a term as possessing the *ability to verify* whether or not any given entity falls in the extension of the term. In terms of intensions: "grasping" an intension would amount, then, to possessing the ability to verify if an entity e in any possible world x belongs to $f(x)$ or not. Later Carnap modified this view, recognizing that, as Quine puts it, sentences face the tribunal of experience collectively and not individually. There is no such thing as the way of verifying that a term T is true of an entity, in general, independent of the context of a particular set of theories, auxiliary hypotheses, etc. Perhaps Carnap would have maintained that something like the earlier theory was correct for a limited class of terms, the so-called "observation terms." Our own view is that the verifiability theory of meaning is false both in its central idea and for observation terms, but we shall not try to discuss this here. At any rate, if one is *not* a verificationist, then it is hard to see California semantics as a theory at all, since the notion of *grasping* an intension has been left totally unexplained.

Second, if we assume that "grasping an intension" (associating an in-

[24] See his "Mathematical Truth," *Journal of Philosophy*, 70 (1973): 661–678.

tension with a term T) is supposed to be a *psychological state* (in the narrow sense), then California semantics is committed to both principles (1) and (2) that we criticized in the first part of this paper. It must hold that the psychological state of the speaker determines the intension of his terms which in turn determines the extension of his terms. It would follow that if two human beings are in the same total psychological state, then they necessarily assign the same extension to every term they employ. As we have seen, this is totally wrong for natural language. The reason this is wrong, as we saw above, is in part that extension is determined socially, not by individual competence alone. Thus California semantics is committed to treating language as something private — to totally ignoring the linguistic division of labor. The extension of each term is viewed by this school as totally determined by something in the head of the individual speaker all by himself. A second reason this is wrong, as we also saw, is that most terms are *rigid*. In California semantics every term is treated as, in effect, a *description*. The *indexical* component in meaning — the fact that our terms refer to things which are similar, in certain ways, to things that we designate *rigidly*, to *these* things, to the stuff we call "water," or whatever, here — is ignored.

But what of the defense that it is not actual language that the California semanticist is concerned with, but an idealization in which we "ignore vagueness," and that terms in natural language may be thought of as associated with a set of intensions rather than with a single well-defined intension?

The answer is that an *indexical* word cannot be represented as a vague family of non-indexical words. The word "I," to take the extreme case, is *indexical* but not vague. "I" is not synonymous with a *description*; neither is it synonymous with a fuzzy set of descriptions. Similarly, if we are right, "water" is synonymous neither with a description nor with a fuzzy set of descriptions (intensions).

Similarly, a word whose extension is fixed socially and not individually is not the same thing as a word whose extension is vaguely fixed individually. The reason my individual "grasp" of "elm tree" does not fix the extension of elm is not that the word is vague — if the problem were simple vagueness, then the fact that my concepts do not distinguish elms from beeches would imply that elms are beeches, as I use the term, or, anyway, borderline cases of beeches, and that beeches are elms, or borderline cases of elms. The reason is rather that the extension of "elm tree" in my dialect is not fixed by what the average speaker "grasps" or doesn't "grasp" at all; it is fixed by the community, including the experts, through a complex cooperative process. A language which exemplifies the division of linguistic labor cannot be approximated successfully by a language which has vague terms and no linguistic division of labor. Cooperation isn't vagueness.

But, one might reply, couldn't one replace our actual language by a language in which (1) terms were replaced by coextensive terms which were *not* indexical (e.g., "water" by "H_2O," assuming "H_2O" is not indexical); and (2) we eliminated the division of linguistic labor by making every speaker an expert on every topic?

We shall answer this question in the negative; but suppose, for a moment, the answer were "yes." What significance would this have? The "ideal" language would in no sense be similar to our actual language; nor would the difference be a matter of "the vagueness of natural language."

In fact, however, one can't carry out the replacement, for the very good reason that *all* natural-kind words and physical-magnitude words are indexical in the way we have described, "hydrogen," and hence "H_2O," just as much as "water." Perhaps "sense data" terms are not indexical (apart from terms for the self), if such there be; but "yellow" as a *thing* predicate is indexical for the same reason as "tiger"; even if something *looks* yellow it may not *be* yellow. And it doesn't help to say that things that look yellow in normal circumstances (to normal perceivers) are yellow; "normal" here has precisely the feature we called indexicality. There is simply no reason to believe that the project of reducing our language to non-indexical language could be carried out in principle.

The elimination of the division of linguistic labor might, I suppose, be carried out "in principle." But, if the division of linguistic labor is, as I conjectured, a linguistic universal, what interest is there in the possible existence of a language which lacks a constitutive feature of *human* language? A world in which every one is an expert on every topic is a world in which social laws are almost unimaginably different from what they now are. What is the *motivation* for taking such a world and such a language as the model for the analysis of *human* language?

Incidentally, philosophers who work in the tradition of California semantics have recently begun to modify the scheme to overcome just these defects. Thus it has been suggested that an intension might be a function whose arguments are not just possible worlds but, perhaps, a possible world, a speaker, and a non-linguistic context of utterance. This would permit the representation of some kinds of indexicality and some kinds of division of linguistic labor in the model. As David Lewis develops these ideas, "water," for example, would have the same *intension* (same function) on Earth and on Twin Earth, but a different extension. (In effect, Lewis retains assumption (1) from the discussion in the first part of this paper and gives up (2); we chose to give up (1) and retain (2).) There is no reason why the formal models developed by Carnap and his followers should not prove valuable when so modified. Our interest here has been not in the utility of the mathematical formalism but in the philosophy of language underlying the earlier versions of the view.

Semantic markers. If the approach suggested here is correct, then there is a great deal of scientific work to be done in (1) finding out what sorts of items can appear in stereotypes; (2) working out a convenient system for representing stereotypes; etc. This work is not work that can be done by philosophical discussion, however. It is rather the province of linguistics and psycholinguistics. One idea that can, I believe, be of value is the idea of a *semantic marker*. The idea comes from the work of J. J. Katz and J. A. Fodor; we shall modify it somewhat here.

Consider the stereotype of "tiger" for a moment. This includes such features as being an animal; being big-cat-like; having black stripes on a yellow ground (yellow stripes on a black ground?); etc. Now, there is something very special about the feature *animal*. In terms of Quine's notion of *centrality* or *unrevisability*, it is qualitatively different from the others listed. It is not impossible to imagine that tigers might not be animals (they might be robots). But spelling this out, they must always have been robots; we don't want to tell a story about the tigers being *replaced* by robots, because then the robots wouldn't be tigers. Or, if they weren't always robots, they must have *become* robots, which is even harder to imagine. If tigers are and always were robots, these robots mustn't be too "intelligent," or else we may not have a case in which tigers aren't animals — we may, rather, have described a case in which some robots are animals. Best make them "other directed" robots — say, have an operator on Mars controlling each motion remotely. Spelling this out, I repeat, is difficult, and it is curiously hard to think of the case to begin with, which is why it is easy to make the mistake of thinking that it is "logically impossible" for a tiger not to be an animal. On the other hand, there is no difficulty in imagining an individual tiger that is not striped; it might be an albino. Nor is it difficult to imagine an individual tiger that doesn't look like a big cat: it might be horribly deformed. We can even imagine the whole species losing its stripes or becoming horribly deformed. But tigers ceasing to be animals? Great difficulty again!

Notice that we are not making the mistake that Quine rightly criticized, of attributing an absolute unrevisability to such statements as "tigers are animals," "tigers couldn't change from animals into something else and still be tigers." Indeed, we can describe farfetched cases in which these statements would be given up. But we maintain that it is qualitatively harder to revise "all tigers are animals" than "all tigers have stripes" — indeed, the latter statement is not even true.

Not only do such features as "animal," "living thing," "artifact," "day of the week," "period of time," attach with enormous centrality to the words "tiger," "clam," "chair," "Tuesday," "hour"; but they also form part of a widely used and important system of classification. The centrality guarantees that items classified under these headings virtually never have to be reclassified; thus these headings are the natural ones to use as category-indicators in a host of contexts. It seems to me reasonable that, just as in syntax we use such markers as "noun," "adjective," and, more narrowly, "concrete noun," "verb taking a person as subject and an abstract object," etc., to classify words, so in semantics these category-indicators should be used as markers.

It is interesting that when Katz and Fodor originally introduced the idea of a semantic marker, they did not propose to exhaust the meaning — what we call the stereotype — by a list of such markers. Rather, the markers were restricted to just the category-indicators of high centrality, which is what we propose. The remaining features were simply listed as a "distinguisher." Their scheme is not easily comparable with ours, because they wanted the semantic markers plus the distinguisher to always give a necessary and sufficient condition for membership in the extension of the term. Since the whole thing — markers and distinguisher — were supposed to represent what every speaker implicitly knows, they were committed to the idea that every speaker implicitly knows of a necessary and sufficient condition for membership in the extension of "gold," "aluminum," "elm" — which, as we have pointed out, is not the case. Later Katz went further and demanded that *all* the features constitute an *analytically* necessary and sufficient condition for membership in the extension. At this point he dropped the distinction between markers and distinguishers; if all the features have, so to speak, the infinite degree of centrality, why call some "markers" and some "distinguishers"? From our point of view, their original distinction between "markers" and "distinguishers" was sound — provided one drop the idea that the distinguisher provides (together with the markers) a necessary and sufficient condition, and the idea that any of this is a theory of *analyticity*. We suggest that the idea of a semantic marker is an important contribution, when taken as suggested here.

The meaning of "meaning." We may now summarize what has been said in the form of a proposal concerning how one might reconstruct the notion of "meaning." Our proposal is not the only one that might be advanced on the basis of these ideas, but it may serve to encapsulate some of the major points. In addition, I feel that it recovers as much of ordinary usage in common sense talk and in linguistics as one is likely to be able to conveniently preserve. Since, on my view something like the assumptions (I) and (II) listed in the first part of this paper are deeply embedded in ordinary meaning talk, and these assumptions are jointly inconsistent with the facts, no reconstruction is going to be without some counter-intuitive consequences.

Briefly, my proposal is to define "meaning" not by picking out an object which will be identified with the meaning (although that might be done in the usual set-theoretic style if one insists), but by specifying a normal form (or, rather, a type of normal form) for the description of meaning. If we know what a "normal form description" of the meaning of a word should be, then, as far as I am concerned, we know what meaning is in any scientifically interesting sense.

My proposal is that the normal form description of the meaning of a word should be a finite sequence, or "vector," whose components should certainly include the following (it might be desirable to have other types of components as well): (1) the syntactic markers that apply to the word, e.g., "noun"; (2) the semantic markers that apply to the word, e.g., "animal," "period of time"; (3) a description of the additional features of the stereotype, if any; (4) a description of the extension.

The following convention is a part of this proposal: the components of the vector all represent a hypothesis about the individual speaker's competence, *except the extension*. Thus the normal form description for "water" might be, in part:

Syntactic Markers	Semantic Markers	Stereotype	Extension
mass noun, concrete	natural kind, liquid	colorless, transparent, tasteless, thirst-quenching, etc.	H_2O (give or take impurities)

— this does *not* mean that knowledge of the fact that water is H_2O is being imputed to the individual speaker or even to the society. It means that (we say) the extension of the term "water" as *they* (the speakers in question) use it is *in fact* H_2O. The objection "who are *we* to say what the extension of *their* term is in fact" has been discussed above. Note that this is fundamentally an objection to the notion of *truth*, and that extension is a relative of truth and inherits the family problems.

Let us call two descriptions *equivalent* if they are the same except for the description of the extension, and the two descriptions are co-extensive. Then, if the set variously described in the two descriptions is, *in fact*, the extension of the word in question, and the other components in the description are correct characterizations of the various aspects of competence they represent, *both* descriptions count as correct. Equivalent descriptions are both correct or both incorrect. This is another way of making the point that, although we have to use a *description* of the extension to give the extension, we think of the component in question as being the *extension* (the *set*), not the description of the extension.

In particular the representations of the words "water" in Earth dialect and "water" in Twin Earth dialect would be the same except that in the last column the normal form description of the Twin Earth word "water" would have XYZ and not H_2O. This means, in view of what has just been said, that we are ascribing the *same* linguistic competence to the typical Earthian/Twin Earthian speaker, but a different extension to the word, nonetheless.

This proposal means that we keep assumption (II) of our early dis-

191

cussion. Meaning determines extension — by construction, so to speak. But (I) is given up; the psychological state of the individual speaker does not determine "what he means."

In most contexts this will agree with the way we speak, I believe. But one paradox: suppose Oscar is a German-English bilingual. On our view, in his total collection of dialects, the words *beech* and *Buche* are exact *synonyms*. The normal form descriptions of their meanings would be identical. But he might very well not know that they are synonyms! A speaker can have two synonyms in his vocabulary and not know that they are synonyms!

It is instructive to see how the failure of the apparently obvious "if S_1 and S_2 are synonyms and Oscar understands both S_1 and S_2 then Oscar knows that S_1 and S_2 are synonyms" is related to the falsity of (I), on our analysis. Notice that if we had chosen to omit the extension as a component of the "meaning-vector," which is David Lewis's proposal as I understand it, then we would have the paradox that "elm" and "beech" have the *same meaning* but different extensions!

On just about any materialist theory, believing a proposition is likely to involve processing some *representation* of that proposition, be it a sentence in a language, a piece of "brain code," a thought form, or whatever. Materialists, and not only materialists, are reluctant to think that one can believe propositions *neat*. But even materialists tend to believe that, if one believes a proposition, *which* representation one employs is (pardon the pun) immaterial. If S_1 and S_2 are both representations that are *available* to me, then if I believe the proposition expressed by S_1 under the representation S_1, I must also believe it under the representation S_2 — at least, I must do this if I have any claim to rationality. But, as we have just seen, this isn't right. Oscar may well believe that *this* is a "beech" (it has a sign on it that says "beech"), but not believe or disbelieve that this is a *"Buche."* It is not just that belief is a process involving representations; he believes the proposition (if one wants to introduce "propositions" at all) under one representation and not under another.

The amazing thing about the theory of meaning is how long the subject has been in the grip of philosophical misconceptions, and how strong these misconceptions are. Meaning has been identified with a necessary and sufficient condition by philosopher after philosopher. In the empiricist tradition, it has been identified with method of verifica-

192

tion, again by philosopher after philosopher. Nor have these misconceptions had the virtue of exclusiveness; not a few philosophers have held that meaning = method of verification = necessary and sufficient condition.

On the other side, it is amazing how weak the grip of the facts has been. After all, what have been pointed out in this essay are little more than home truths about the way we use words and how much (or rather, how little) we actually know when we use them. My own reflection on these matters began after I published a paper in which I confidently maintained that the meaning of a word was "a battery of semantical rules,"[25] and then began to wonder how the meaning of the common word "gold" could be accounted for in this way. And it is not that philosophers had never considered such examples: Locke, for example, uses this word as an example and is not troubled by the idea that its meaning is a necessary and sufficient condition!

If there is a reason for both learned and lay opinion having gone so far astray with respect to a topic which deals, after all, with matters which are in everyone's experience, matters concerning which we all have more data than we know what to do with, matters concerning which we have, if we shed preconceptions, pretty clear intuitions, it must be connected to the fact that the grotesquely mistaken views of language which are and always have been current reflect two specific and very central philosophical tendencies: the tendency to treat cognition as a purely *individual* matter and the tendency to ignore the *world*, insofar as it consists of more than the individual's "observations." Ignoring the division of linguistic labor is ignoring the social dimension of cognition; ignoring what we have called the *indexicality* of most words is ignoring the contribution of the environment. Traditional philosophy of language, like much traditional philosophy, leaves out other people and the world; a better philosophy and a better science of language must encompass both.

[25] "How Not to Talk About Meaning," in R. Cohen and M. Wartofsky, eds., *Boston Studies in the Philosophy of Science*, vol. 2 (New York: Humanities Press, 1965).

Reference and Context

"I proceed. 'Edwin and Morcar, the earls of
Mercia and Northumbria, declared for him;
and even Stigand, the patriotic archbishop
of Canterbury, found it advisable — ' "
"Found *what?*" said the Duck.
"Found *it,*" the Mouse replied rather
crossly: of course you know what 'it' means."
"I know what 'it' means well enough,
when *I* find a thing," said the Duck: "it's
generally a frog or a worm. The question is,
what did the archbishop find?"

— Lewis Carroll, *Alice's
Adventures in Wonderland,*
chap. 3

593. A main cause of philosophical disease —
a one-sided diet: one nourishes one's thinking
with only one kind of example.

— Ludwig Wittgenstein, *Philosophical Investigations*

1

Singular reference, in the narrow sense, is a certain kind of connection
between a token of an expression of a language — a "singular term" —
produced by a speaker on a given occasion, and a particular thing — the
"referent" of the term. A speaker "makes a reference" or "refers" when
he establishes such a connection by the act of producing the singular
term (uttering it, inscribing it, etc.). A singular term itself "refers" if
the speaker has made a reference by producing it.

In the broad sense, singular reference is any such connection

AUTHOR'S NOTE: Earlier versions of this approach to the theory of reference were
presented in papers read to philosophy colloquia at the University of Minnesota and
Rockefeller University in 1968 and at Duke University, the University of Toronto,
and Macalester College in 1970, and to a conference on the philosophy of language
held at the Minnesota Center for the Philosophy of Science in August 1968. I owe
a number of useful suggestions to participants in those events, especially Saul Kripke

holding between a particular thing and something which is either a singular term in a language or else is an element in some system which is language-like in the relevant ways. Such "quasi-linguistic" entities include: symbols on maps, figures in paintings, statues, scale models, memory images, the components of one's visual field, thoughts, and so on. Each of these is an element in what I will call a *context*, and it is by virtue of referential connections between their elements and particular things in the world that contexts are anchored down to the things and situations which they are about. I use 'context' as a technical term to cover a class of things for which there is no convenient label,[1] some examples of which I will now discuss briefly, postponing further clarification of the notion until later.

A linguistic discourse is a context whose elements are words, phrases, sentences, paragraphs, etc., produced by one or more speakers. A map is a context whose elements are cartographic symbols signifying buildings, bridges, forests, rivers, and so on. A picture is a context consisting of such elements as blotches of paint on canvas, ink marks on paper, silver deposits on photographic printing paper, patterns of light projected on a screen, etc. My visual field is a context consisting of elements commonly called "visual sensations." In general anything which has *content* is a context, as I use the term. Anything that has *meaning* or *sense* is a context. Anything which *expresses* something or *represents* something is a context.

Some elements of contexts are or can be referentially connected with particular things outside the contexts: the word 'Fred' which I utter in a conversation refers to a friend of mine, this blue line on the map

and Hilary Putnam. Since I first started thinking about these problems I have benefited greatly from discussions with Keith Gunderson, and also from discussions at various times with Paul Benacerraf, Keith Donnellan, David Kaplan, and David Lewis. I also want to thank John Olney of System Development Corporation, Santa Monica, California, under whose auspices — in 1966, as a consultant on a project on discourse analysis and anaphora — I began to consider the relation between discourse structure and singular reference.

[1] Nelson Goodman's notion of a "symbol" comes close, but includes some things I exclude (e.g., most "serious" music) and excludes some things I include (e.g., photographs). (See his *Languages of Art* (Indianapolis and New York: Bobbs-Merrill, 1968), passim.) For further explanation, see section 11. The theory I am going to present concerns the reference of concrete linguistic and quasi-linguistic tokens, not the corresponding types; singular terms, linguistic contexts (= discourses), other singular elements, and the contexts of which they are part are all concrete individuals, actual or possible. For brevity, I will often not make this explicit in what I say.

refers to the Mississippi River, that variegated blotch of paint refers to the Duchess of Alba, this sensation of a white disk on a dark background refers to the moon, and so on. These are, of course, uncommon uses of 'refer' (except for the first one); I will apply this expression not merely to singular terms in a language but to anything which is like a singular term in the relevant ways. I will call such entities "singular elements" of contexts.

When a singular element is not referentially connected with anything outside the context there is a *failure of reference*: singular terms which name nothing, maps with imaginary rivers and mountain ranges, paintings of fictitious scenes, hallucinations, dreams, and so on. It is the business of a theory of reference to explain both failure of reference and successful reference, to give a systematic general account of how singular elements in contexts hook up with items in the world. Such a theory must deal first with linguistic reference, since we understand the linguistic cases best; many important facts about them are open to public observation, and we already have a good deal of knowledge about the syntax, and some knowledge about the semantics, of natural languages. Previous theories have generally tried to explain the connection between a singular term and its referent as a function of the meaning of the term and the properties of the referent, paying little or no attention to the circumstances in which the term is uttered. The most influential theory of this sort has, of course, been Russell's Theory of Descriptions. Strawson's revision of Russell[2] paid conspicuous lip service to the importance of "context" but didn't include a systematic theory of what a "context" is or how contexts affect reference.[3] The first significant advance beyond Russell was made by Keith Donnellan, whose paper "Reference and Definite Descriptions"[4] contained several striking examples of singular reference which could not in principle be explained in Russellian or Strawsonian terms. The theory I will present here is the result of an attempt to do justice to Donnellan's cases while retaining what still seems to be true in the Russell-Strawson tradition. I think that Don-

---

[2] First presented in his "On Referring" (*Mind*, 1950; reprinted in Antony Flew, ed., *Essays in Conceptual Analysis* (London: Macmillan, 1956), pp. 21–52; my references will be to the latter.)

[3] Except for such very general remarks as the following: ". . . by 'context' I mean, at least, the time, the place, the situation, the identity of the speaker, the subjects which form the immediate focus of interest, and the personal histories of both the speaker and those he is addressing." ("On Referring," p. 42.)

[4] *Philosophical Review*, 75 (1966):281–304.

nellan's work shows that a "pure" or "autonomous" theory of linguistic reference is impossible, and that the central concepts ('singular term', 'refer') will have to be generalized far beyond their original application to spoken and written language. I believe that a theory of linguistic reference will have to be combined with a systematic account of certain internal states of the speaker — his thoughts, beliefs, perceptions, memories, and so on — which are, so to speak, the intermediate links connecting the singular terms he utters with their referents out in the world. These intermediaries can themselves be understood only if we treat them as being quasi-linguistic in structure and content — as contexts, in my sense of the term — and as containing elements analogous to singular terms which can be referentially connected with things in the world outside the speaker's skin.

2

Since linguistic reference is a kind of connection between singular terms and things in the world, the first thing a theory of reference has to do is define the class of singular terms. Quine offers this definition: "A term is singular if it purports to name an object (one and only one), and otherwise general." General terms don't purport to name anything at all: "The general term may indeed 'be true of' each of many things, viz., each red thing, or each man [he is discussing 'red' and 'man'], but this kind of reference is not called naming: 'naming', at least as I shall use the word, is limited to the case where the named object purports to be unique."[5] What is "purporting"? It is not a matter of whether the term is in fact true of just one object:

For 'Pegasus' counts as a singular term though true of nothing, and 'natural satellite of the earth' counts as a general term though true of just one object. As one vaguely says, 'Pegasus' is singular in that it purports to refer to just one object, and 'natural satellite of the earth' is general in that its singularity of reference is not something *purported* in the term. Such talk of purport is only a picturesque way of alluding to distinctive grammatical roles that singular and general terms play in sentences. It is by grammatical role that general and singular terms are properly to be distinguished.

The basic combination in which general and singular terms find their contrasting roles is that of *predication* . . . Predication joins a general term and a singular term to form a sentence that is true or false accord-

[5] W. V. Quine, *Methods of Logic* (New York: Holt, 1959), p. 205.

ing as the general term is true or false of the object, if any, to which the singular term refers.[6]

To paraphrase this in my terminology: singular terms are the ones whose role is to be referentially connected with objects, and general terms are the ones whose role is to describe or characterize the objects with which the singular terms are connected. Singular terms can purport to refer without actually referring, for purporting is a purely intra-linguistic affair — this is what Quine is getting at when he says that singular terms are identified in terms of their "grammatical role."

I think that Quine is essentially right[7] — his remarks are, indeed, no more than a sophisticated gloss on sound linguistic common sense — and I will shortly take up the question of exactly what the grammatical role of singular terms is and how it enables them to purport singularity of reference. But first I want to consider an objection put by Strawson, who thinks Quine's approach is inadequate and advocates a different one, based on the notion of "identification." He points out that we can distinguish singular terms from general terms by the consequences of their failing to apply to anything. ('Apply to' means 'either refer to or be true of'.) If I say 'The captain is angry' and 'the captain' applies to somebody, then the failure of 'angry' to apply to him results in the sentence being false. If 'the captain' does not apply to anybody the result is not a false sentence but one that has no truth-value at all — there is a so-called "truth-value gap." In general: "Whether the sentence is true or false depends on the success or failure of the general term; but the failure of the singular term appears to deprive the general term of the chance of either success or failure."[8] So Strawson's version of Quine's distinction is this:

Singular terms are what yield truth-value gaps when they fail in their role. General terms are what yield truth or falsity, when singular terms succeed in their role, by themselves applying, or failing to apply, to what the singular terms apply to. This is more or less what we have. It scarcely seems enough. We want to ask 'Why?'[9]

---

[6] W. V. Quine, *Word and Object* (Cambridge, Mass., and New York: Technology Press of M.I.T. and Wiley, 1960), pp. 95–96.

[7] Though his notion of a singular term is broader than mine, since it includes the variables of quantification.

[8] P. F. Strawson, "Singular Terms and Predication," in P. F. Strawson, ed., *Philosophical Logic* (London: Oxford University Press, 1967), p. 72. See also Quine's reply to this in *Synthese*, 19 (1968):292–297.

[9] Strawson, "Singular Terms and Predication," p. 73.

He tries to answer this question by starting with what he regards as the central cases of singular reference, namely those predications "in which singular and general term alike may fairly be said to be applied to a single concrete and spatio-temporally continuous object." In these cases, at least, the "characteristic difference" between the two kinds of terms

is that the singular term is used for the purpose of identifying the object, of bringing it about that the hearer (or, generally, the audience) knows which or what object is in question; while the general term is not. It is enough if the general term in fact applies to the object; it does not also have to identify it.

But what exactly is this task of identifying an object for a hearer? Well, let us consider that in any communication situation a hearer (an audience) is antecedently equipped with a certain amount of knowledge, with certain presumptions, with a certain range of possible current perception. There are within the scope of his knowledge or present perception objects which he is able in one way or another to distinguish for himself. The identificatory task of one of the terms . . . is to bring it about that the hearer knows which object it is, of all the objects within the hearer's scope of knowledge or presumption, that the other term is being applied to. This identificatory task is characteristically the task of the definite singular term.[10]

In general, the identificatory task of the singular term

is successfully performed if and only if the singular term used establishes for the hearer an identity, and the right identity, between the thought of what-is-being-spoken-of-by-the-speaker and the thought of some object already within the reach of the hearer's own knowledge, experience, or perception, some object, that is, which the hearer could, in one way or another, pick out or identify for himself, from his own resources. To succeed in its task, the singular term, together with the circumstances of its utterance, must draw on the appropriate stretch of those resources.[11]

If the identificatory task is botched and nothing correctly identified, the singular term has failed in its mission and there is a truth-value gap, since nothing has been correctly or incorrectly described or characterized by the speaker for the hearer. Thus we can understand what singular terms really are: they are the terms which are used to identify things for hearers. And we can understand what singular reference is: it is the way singular terms apply to objects.

But there are complications. Not every failure of a singular term to

[10] Ibid., pp. 74–75.
[11] Ibid., p. 78; emphasis in original.

perform the identificatory task results in a truth-value gap. Only the "radical" failures do so, and these occur when there is nothing there to be identified, when "there just is no such particular item at all as the speaker takes himself to be referring to."[12] And this is obviously correct. For example, if I want to tell you that my cat died, and I say 'The cat died', the singular term 'the cat' might fail to identify my cat because you didn't hear what I said, or because you didn't understand it, or because you thought I was talking about the neighbors' cat, and so on. Still, I was talking about her and 'the cat' *did* refer to her — the failure of the singular term was not a *radical* failure, and my assertion has a truth-value. If this is so — if reference can occur without identification — why does Strawson insist that identification is *the* characteristic task of singular terms, that identification is what they are really *for*, and that understanding this enables us to distinguish singular terms from general terms? Why not just say that singular terms are the ones whose characteristic task is singular reference and be done with it? Strawson rejects this because he thinks that 'refer to' is just as ill-understood as 'singular term' and just as much in need of explanation.[13] The same goes for such cognate notions as 'specify'; he says, for example, that a remark by Quine that a singular term "is used purely to specify its object for the rest of the sentence to say something about" is "unsatisfactory, since 'specify' by itself remains vague. To remove the vagueness we need the concept of 'identifying for an audience' which I have just introduced."[14]

I don't believe that the introduction of this notion removes the vagueness. According to Strawson, I am in a position to identify an object for someone only if that object is in fact being "spoken of" by me; that is, only if it is the one "in question"; that is, only if the general term I use is "being applied to" that object rather than some other. My hearer must be able to "distinguish for himself" or "pick out" the object on the basis of some "thought of" it.[15] What is it to "speak of" an object? For it to be the one "in question"? To "think of" it? "Pick it out"?

[12] P. F. Strawson, "Identifying Reference and Truth-Values," *Theoria*, 1964, p. 103. (This paper is an amplification of the account of identification given in "Singular Terms and Predication.")
[13] See "Singular Terms and Predication," p. 74.
[14] *Ibid.*, p. 76.
[15] All the quoted words and phrases come from the passages from Strawson quoted above.

These are at least as vague as the notion of "specifying" which Strawson criticizes Quine for relying on, if not more so. They are themselves instances of what I earlier called singular reference in the broad sense. The connection that holds between a person and an object when he is thinking of the object, or when he distinguishes it or picks it out, or when he intends to speak of it or apply a general term to it, or when it is the one in question, is similar in the relevant respects to the connection that holds when the person has specified or referred to the object by using language. In each case we must take one thing — a spoken remark or an unspoken state of mind — and relate it uniquely to something else: the referent, the thing specified, the object of thought, the thing distinguished or picked out, the thing involved in the speaker's intentions, the object in question. I hope to show that a theory which accounts for such referential connections will in the last analysis have to use the non-linguistic ones to explain the linguistic ones (to this extent I agree with Strawson), but as things presently stand the non-linguistic cases are considerably more obscure and of little use in clarifying linguistic reference. They are part of the problem, not part of the solution.

3

I said earlier that whether and what a singular term "purports" is a purely intra-linguistic affair, and that this was what Quine was getting at when he said that singular terms are distinguished by their grammatical role. In English, for example, which expressions are the singular terms and what is grammatically distinctive about them? Suppose we say tentatively that at least the following kinds of expressions are singular terms:

(a) proper names ('George', 'Leon Trotsky', 'Brazil', etc.);
(b) so-called "definite descriptions" ('the man', 'the man wearing a plumed hat', 'the cat', 'my cat', etc.);
(c) personal pronouns ('he', 'it', 'they', etc.);
(d) so-called "demonstratives" ('this', 'this man', 'that', etc.).

This is more or less the standard roster of expressions used in making singular references. We are already perfectly familiar with these expressions, not just in the sense that we've seen them all before, but, more important, in the sense that we would be able to recognize wholly novel expressions falling under the same categories. New proper names appear continually, and previously unuttered definite descriptions get uttered all the time; it would even be easy, if we wished, to introduce new pronouns (say, 'ger' for old people) or new demonstratives (say, 'thot' for things heard but not seen). Our ability to produce and comprehend new singular terms could no doubt be explicitly represented by a grammar of English, which would enumerate all the definite descriptions, tell us what sentence-frames accept proper names, and so on, but present purposes will be served well enough by exercising the inexplicit ability which such a grammar would represent.

Consider one of the expressions on the list, say 'the man'; since it is a singular term, it purports to refer to exactly one thing; thus it purports to refer to exactly one thing in the following context:[16]

D1: #A man was sitting underneath a tree eating peanuts. A squirrel came by, and the man fed it some peanuts.#

What point are we trying to make when we say that 'the man' in D1 "purports to refer" to exactly one thing? We are, I think, trying to say something about how that remark is to be taken — specifically, we are trying to say that it is appropriate to look for something (a man, presumably) which 'the man' refers to (= which the speaker refers to by uttering 'the man') and which is truly or falsely described by the rest of the sentence. It has often been pointed out that some expressions are grammatically "singular" (at least on a superficial analysis) but couldn't conceivably refer to anything about which something is being truly or falsely said. For example, 'Nothing is under the bed' — 'nothing' doesn't count as a singular term, despite the syntactic similarity of 'Nothing is under the bed' and 'The cat is under the bed.' Singular terms are those which at least *could* be referentially connected with something, and 'nothing' couldn't.[17]

Is it mere membership on the list given above that makes it appropriate to look for something which an expression refers to? No, for 'the man' is on the list and yet it could have no referent in this context, in which D1 is embedded:

D2: #Say, let me tell you a funny story I just made up. A man was

[16] '#' marks the boundary of a discourse.
[17] Unless it is used as an ad hoc substitute for a singular term, say as part of a code. That sort of thing is always possible, and I won't bother to point it out henceforth.

sitting underneath a tree eating peanuts. A squirrel came by, and the man fed it some peanuts. Then the squirrel said . . .#

The possibility of a referential connection between 'the man' and anything outside the context is explicitly canceled by the speaker's announcement that what he is about to say is only a story he made up. That suspends the normal purport which 'the man' has in isolation, which it has in the sentence 'The man fed it some peanuts', and which it retains in some supra-sentential contexts, such as D1. It is a matter of context whether 'the man' purports to refer to anything, and consequently it is a matter of context whether 'the man' is a singular term. The same goes for all the other things on the list, which I will call "singular expressions." An expression is a singular term *in* or *relative to* a given context. Now the question is: When is a singular expression a singular term? In other words: In which contexts does a singular expression purport to refer to exactly one thing? Or: When is it possible for there to be a referential connection between a singular expression in a context and some object outside the context?

The smallest context which a singular expression can have is just the expression itself, in isolation. A singular expression standing alone purports to refer to exactly one thing, by virtue of its "grammatical role," that is, because of the way it enters into predications — so far this just recapitulates Quine's view. Proper names, definite descriptions, pronouns, and demonstratives are therefore all singular terms when uttered in isolation. When such an expression is embedded in a wider context, it may still purport to refer or it may cease to do so. In the latter case, I will say that its purported reference has been *canceled* — in other words, it cannot be referentially connected with anything beyond the boundaries of the context. Purported reference can be canceled by a variety of devices, one of which we have already seen (I will discuss the others later, in section 6). In D2 the announcement at the beginning of the discourse that what followed was fiction canceled the purported reference of the singular expression 'the man', even though it was not canceled by anything in D1. 'The man' is a singular term in D1 but not in D2, and perforce not in any wider context in which D2 might be embedded. (Of course, a fictional context doesn't have to be explicitly marked as such, as long as it is understood in the same way as overtly fictional contexts like D2; and the device which marks a context as fictional doesn't have to be part of the text, properly speaking, but may instead be attached to it in some conventionally understood way, which is what happens when the disclaimer is built into the title (e.g., 'Bertrand and Ludwig: A Novel'), or when the text is preceded by some such warning as: "All the characters in this book are fictitious and any resemblance to actual persons, living or dead, is purely coincidental.")

However, it is not the mere presence of a singular expression within the scope of a remark like 'Say, let me tell you a funny story I just made up' that cancels its purported reference. For example:

> D3: #Say, let me tell you a funny story I just made up. Richard Nixon was playing ping-pong with Mao Tse-tung and . . .#

(As with D2, the reader may complete the story however he wishes.) In this case, a piece of fiction is wrapped around two singular references to actual people, and 'Richard Nixon' and 'Mao Tse-tung' still purport to refer as usual. It is only when a singular expression is introduced within the scope of a fiction-indicating device that its purported singular reference is canceled. 'The man' in D2 is understood on the basis of the earlier expression 'a man', and the latter is, in a familiar sense, its antecedent. If we ask who fed the squirrel, which man did it, one perfectly correct answer would be that it was the man previously mentioned; it is on the basis of that previous predication involving 'a man' that we identify the purported referent of 'the man' — he is not just any man, but, specifically, one sitting underneath a tree eating peanuts. But that man is merely a figment of our tale: he, the tree, the squirrel, and the peanuts were all first mentioned in such a way as to cancel the purported reference of all subsequent mentions of them. We can make what Strawson calls "story-relative identifications"[18] within the context, but we cannot go outside the context and identify that man, that tree, that squirrel, those peanuts as the ones the story was about.

Before going any further I want to introduce some new terminology. Grammarians sometimes talk about "anaphora," by which they mean the kind of relationship that holds between, for example, a pronoun and its antecedent.[19] Let an anaphoric chain be a sequence of singular

---

[18] See his *Individuals: An Essay in Descriptive Metaphysics* (London: Methuen, 1959), p. 18.

[19] The notion seems to be rather loosely employed, but the basic idea is that expressions in different sentences (or, sometimes, in different parts of the same sentence) are related anaphorically if one of them somehow helps us understand or interpret the other.

expressions occurring in a context, such that if one of them refers to something then all of the others also refer to it.[20] 'A man — the man' in D1 and D2 is an anaphoric chain. So are 'A squirrel — it' in D1 and D2, and 'A squirrel — it — the squirrel' in D2. To consider a real-life example:

> D4: #At eleven o'clock that morning, an ARVN officer stood a young prisoner, bound and blindfolded, up against a wall. He asked the prisoner$_1$ several questions, and, when the prisoner$_2$ failed to answer, beat him$_1$ repeatedly. An American observer who saw the beating$_1$ reported that the officer "really worked him$_2$ over." After the beating$_2$, the prisoner$_3$ was forced to remain standing against the wall for several hours.#[21]

This passage, which occurs in the middle of a book, contains the following anaphoric chains:

   I. 'that morning';
  II. 'an ARVN officer — he — the officer';
 III. 'a young prisoner — the prisoner$_1$ — the prisoner$_2$ — him$_1$ — him$_2$ — the prisoner$_3$';
  IV. 'a wall — the wall';
   V. 'an American observer who saw the beating$_1$';
  VI. 'the beating$_1$ — the beating$_2$'.

Unlike the others, (I) is part of a longer chain which begins outside the quoted text in an earlier part of the book. ('That morning' refers, in fact, to the morning of January 8, 1967.) (II), (III), and (IV) all begin with so-called indefinite descriptions and continue with definite descriptions and pronouns. (V) begins the same way and clearly could continue much as they do ('an American observer who saw the beating$_1$ — the American observer — he — the American', for example). (VI) begins with a definite description, 'the beating', but it is clear that the beating in question is the one first mentioned in the second sentence of the text; we can therefore think of the first link of (VI) as being an indefinite description — perhaps 'a beating of the young prisoner by

<hr/>

[20] Zeno Vendler has an interesting discussion of anaphoric relations between singular terms (though he doesn't call them "anaphoric") in his "Singular Terms," which is chap. 2 of his *Linguistics in Philosophy* (Ithaca, N.Y.: Cornell University Press, 1967). His "chains of identification" (see p. 63) resemble but are not the same as my "anaphoric chains."

[21] From Jonathan Schell's *The Village of Ben Suc* (New York: Random, Vintage Books, 1968), p. 54. I have added subscripts to keep track of expressions that occur more than once; they should not be considered as part of the expressions.

the ARVN officer' — which would appear after a deeper analysis of the second sentence.

The difference between an indefinite description and a definite description is, in a sense, merely stylistic. An expression like 'a young prisoner' can occur only at the beginning of an anaphoric chain, not at any later point in it. For example:

> D5: #At eleven o'clock that morning, an ARVN officer stood a young prisoner$_1$, bound and blindfolded, up against a wall. He asked a young prisoner$_2$ several questions, and, when a young prisoner$_3$ failed to answer, beat him repeatedly.#

(This is just the first two sentences of D4 with 'the prisoner' replaced by 'a young prisoner'.) If we want to understand this as normal English we will have to conclude that 'a young prisoner$_1$', 'a young prisoner$_2$', and 'a young prisoner$_3$' do not purport to refer to the same thing and do not together form an anaphoric chain. The reason for this is not that indefinite descriptions cannot refer at all and for that reason cannot be links in anaphoric chains, but rather that whenever an indefinite description enters an anaphoric chain it can enter only as the first link. The difference between indefinite descriptions and definite descriptions is that the former can only be used to initiate anaphoric chains and the latter only to continue them. If we find a definite description, say 'the prisoner', occurring in a text and we want to know what is being talked about, we search the preceding text for an expression which will serve as the *anaphoric antecedent* of 'the prisoner' — 'a prisoner', 'a young prisoner', or what have you. If we don't find an antecedent we say that one is "presupposed" or "understood" or "implicit" in the use of the definite description.[22]

Indefinite descriptions should, therefore, be added to the list of singular expressions, since in many contexts they purport to refer and thus count as singular terms. There is, however, a popular view, first stated by Russell,[23] which denies that indefinite descriptions are on a par with definite descriptions and holds that they cannot refer or pur-

[22] If an anaphoric chain begins with a definite description which is a singular term in that context, then the context is in a sense incomplete or elliptical. Most linguistic contexts that actually get uttered are of this sort.

[23] Perhaps the clearest exposition of it is in his *Introduction to Mathematical Philosophy* (London: Allen and Unwin, 1919), chap. 16. Part of what I say about indefinite descriptions is adapted from Strawson's *Introduction to Logical Theory* (London: Methuen, 1952), pp. 186–187.

port to refer. David Kaplan argues for this view in the following way:
For example:

A senator from New York is supporting Rockefeller. (5)

Now (5) certainly has subject-predicate grammatical form in English, but if you feel that its logical form is the same as

Jacob Javits is supporting Rockefeller. (6)

you can quickly disabuse yourself by comparing:

A senator from New York is supporting Rockefeller, and a senator from New York is not supporting Rockefeller. (7)

with

Jacob Javits is supporting Rockefeller, and Jacob Javits is not supporting Rockefeller. (8)

Sentence (8) is a contradiction, but (7) is true. In fact, isn't it obvious that indefinite descriptions do not even purport to denote a unique object as names do?[24]

No, it isn't. All that (7) shows is that each occurrence of an indefinite description starts a new and different anaphoric chain, like the different occurrences of 'a young prisoner' in D5. It does not show that an indefinite description occurring in its customary place at the beginning of an anaphoric chain does not refer or purport to refer on a par with the definite descriptions occurring later on in the same chain. Since the two occurrences of 'a senator from New York' manifestly belong to different anaphoric chains, they do not purport to refer to the same thing, and it is for that reason that (7) is not a contradiction. (8) is a contradiction precisely because the two occurrences of the proper name 'Jacob Javits' do count as links in the same anaphoric chain, according to our rules for the use of proper names, and thus they purport to refer to the same thing.

Quine argues along similar lines:

The difference between . . . indefinite singular terms [by 'term' he means roughly what I mean by 'expression'] and the ordinary or definite ones is accentuated when repetitions occur. In 'I saw the lion and you saw the lion', we are said to have seen the same lion; indeed 'it' or 'him' could just as well have been used in place of the second occurrence of 'the lion'. But in 'I saw a lion and you saw a lion' there is no such suggestion of identity. . . . There is no one thing named by the indefinite singular term 'a lion'; no one thing even temporarily for the space of the single sentence.[25]

[24] David Kaplan, "What is Russell's Theory of Descriptions?", in D. F. Pears, ed., *Bertrand Russell: A Collection of Critical Essays* (Garden City, N.Y.: Doubleday, Anchor Books, 1972), pp. 230–231.
[25] *Word and Object*, p. 113.

Again, this only shows that an indefinite description can't be repeated without starting a new anaphoric chain; it doesn't show that an indefinite description at the beginning of an anaphoric chain cannot refer or purport to refer. The tacit assumption underlying Quine's and Kaplan's arguments is that all singular terms in English behave like proper names or like individual constants in formalized languages, which can be repeated, at least within a single sentence, without changing their purported reference. This assumption is false. The difference between 'a lion' and 'the lion' is like the difference between 'The lion' and 'the lion' in written English; the latter never occurs at the beginning of a sentence and the former never occurs anywhere else, but that proves nothing about their status as singular terms.

Quine has another argument for the same conclusion:

In 'I saw the lion', the singular term 'the lion' is presumed to refer to some one lion, distinguished from its fellows for speaker and hearer by previous sentences or attendant circumstances. In 'I saw a lion', the singular term 'a lion' carries no such presumption; . . . 'I saw a lion' counts as true if at least one lion, no matter which, was seen by me on the occasion in question.[26]

He apparently assumes that there are only two alternatives: either 'a lion' refers to some one lion, distinguished from its fellows for speaker and hearer by previous sentences or attendant circumstances, or it doesn't refer at all. But there is a third possibility: that 'a lion' refers to some one lion which is not distinguished from its fellows *for the hearer* by previous sentences or attendant circumstances, but which is distinguished from them *for the speaker*. In D4, for example, 'a young prisoner' does not refer to a young prisoner distinguished from his fellows for the hearer (or reader) by attendant circumstances or by previous sentences in that discourse; but it does refer to a young prisoner whom the author of the book was able to distinguish for himself and whom he introduces into the book at that point by using the expression 'a young prisoner'.

Russell presented this argument:

. . . no one could suppose that "a man" was a definite object, which could be defined by itself. . . . when we have enumerated all the men in the world, there is nothing left of which we could say, "This is a man, and not only so, but it is *the* 'a man,' the quintessential entity that is just an indefinite man without being anybody in particular." It

[26] *Ibid.*, p. 112.

is of course quite clear that whatever there is in the world is definite: if it is a man it is one definite man and not any other. Thus there cannot be such an entity as "a man" to be found in the world, as opposed to specific men. And accordingly it is natural that we do not define "a man" itself, but only the propositions in which it occurs.[27]

We would define "a man" itself only if it were a name (by which Russell meant something that "directly designates" an individual, i.e., refers to it). If it were a name it would have to name an indefinite man; since there are no indefinite men, it is not a name.

But what is it for a singular term to name or refer to (or purport to name or refer to) some definite thing? A "definite" thing is not, after all, a *kind* of thing, like a red thing or a spherical thing. 'Definite' and 'indefinite' have to do with the *way* an expression purports to refer, and that is a matter of context. What is it for an expression to purport to refer "definitely" in a given context? Consider the anaphoric chain (III) in D4. Its second link is 'the prisoner', which in that context is a singular term that purports to refer to some prisoner. Which prisoner? If we ask this question *the context will supply us with a definite answer* — for example, "The young prisoner just mentioned, the one that the ARVN officer stood up against a wall." In general, the sort of reference which is purported by a singular term occurring in an anaphoric chain can be made more definite (more specific, more exact) by an appeal to earlier links in the chain and what is predicated of them; but we cannot in that way make the purported reference of the *first* link more definite. To do that we have to go outside the context altogether. Since indefinite descriptions like 'a young prisoner' and 'a lion' and 'a man' can occur only as the initial links in anaphoric chains, we can see why their purported reference cannot be made more definite within the context and why they have the quality of "indefiniteness" which has led some people to deny that they purport to refer.

I am not claiming that indefinite descriptions are *always* singular terms, purporting to refer; like the other singular expressions, including definite descriptions, they qualify as singular terms in some contexts but not in others. When an indefinite description does not purport to refer, then the sentence containing it can be taken as equivalent to an existentially quantified sentence.[28] 'A senator from New York is sup-

---

[27] *Introduction to Mathematical Philosophy*, pp. 172–173. See also Strawson's criticisms of Russell's view about indefinite descriptions in "On Referring," p. 49.

[28] Sometimes but not always. See below, section 6.

porting Rockefeller' sometimes *is* to be read as equivalent to '$(\exists x)(x$ is a senator from New York & x is supporting Rockefeller)', but it doesn't always have to be paraphrased this way. This is perhaps easier to see in the case of 'A senator from New York is supporting Rocke-feller' than it is when we consider some of its stylistic variants, such as 'There is a senator from New York who is supporting Rockefeller', since the latter is closer syntactically to the standard English way of reading the corresponding existential quantification ('There is an x such that x is a senator from New York and x is supporting Rockefeller'). Con-cerning such sentences, Davidson argues:

We recognize that there is no singular term referring to a mosquito in 'There is a mosquito in here' when we realize that the truth of this sentence is not impugned if there are two mosquitos in the room. . . . We learned some time ago, and it is a very important lesson, that phrases like 'a mosquito' are not singular terms, and hence do not refer as names or descriptions do.[29]

If what Davidson says holds for 'There is a mosquito in here' uttered in isolation, it should also hold for the same sentence when embedded in a larger discourse, say this one:

D6: #There is a mosquito in here. You can hear it buzzing. See, it just landed on my left arm. Now it's biting me. [*The speaker swats the mosquito.*] Not much left of it now, is there!#

How do we understand the occurrences of 'it' in such a context? It is clear from the structure of the discourse that all the occurrences of 'it' refer to the same thing if they refer to anything at all.[30] It is also clear that sometimes they do refer, as in this situation: the speaker sees a mosquito and hears it buzzing (there is another mosquito present but he doesn't notice it), and he says: 'There is a mosquito in here. You can hear it buzzing.' Then he sees it land on his left arm and says: 'See, it just landed on my left arm.' Then he feels the bite and says: 'Now it's biting me,' then he swats it, looks at the remains, and says: 'Not much left of it now, is here!' It would be absurd to maintain that on the four occasions he uttered the word 'it' the speaker was not refer-ring to that mosquito, talking about it, making remarks about it, com-menting on its activities, describing what it was doing, and so on. There

[29] Donald Davidson, "The Individuation of Events," in Nicholas Rescher, ed., *Essays in Honor of Carl G. Hempel* (Dordrecht: Reidel, 1969), p. 220.
[30] Unless the setting is such as to force an ad hoc interpretation on what is said.

is (in my jargon) a referential connection between that mosquito and each of those tokens of 'it'. Does such a connection also hold between 'a mosquito' and that mosquito? Davidson denies that it does, on the grounds that the first sentence of D6 is true even though there is a plurality of mosquitoes present, from which it follows that 'a mosquito' is not a singular term. Why does the extra mosquito matter? There are three readings of the sentence to consider. First, we can take it as an existential quantification, to be paraphrased as '$(\exists x)(x$ is a mosquito & x is in here)'. In this case the sentence is true and 'a mosquito' is not a singular term.[31] Second, we can read it as initiating an anaphoric chain ('a mosquito — it — it — it — it'), in which case it is a singular term referring to the mosquito seen, heard, and swatted by the speaker, and the sentence is true if and only if *that* mosquito was in the room. The presence of an additional mosquito not noticed or referred to by the speaker does not falsify the sentence, nor does it prevent him from referring to that one. If it had turned out that the mosquito heard and seen by the speaker was not inside the room but, say, just outside it, the presence of the unnoticed mosquito inside the room would not have saved the sentence from falsity.[32] Third, we can understand the sentence as saying that *exactly* one mosquito is in the room. Now this *is* falsified by the presence of the second mosquito, and it seems to be this case that Davidson had in mind when he argued that 'a mosquito' was not a singular term. But the claim that it is a singular term does not depend on taking the sentence that way.[33]

[31] And neither are the four occurrences of 'it', which are syntactically linked with 'a mosquito' and stand or fall with it. If 'a mosquito' is no more closely connected with the perceived mosquito than with the other one, the occurrences of 'it' are in the same position and do not refer uniquely to the one the speaker perceived. The only way to avoid this obviously false conclusion is to hold that the syntactic connection with 'a mosquito' is of no significance. But that won't do either. The occurrences of 'it' purported to refer to a *mosquito* only because of the anaphoric chain linking them with 'a mosquito'; if we deny this linkage, we can't account for their purported reference, and the first sentence might as well not be in the discourse.

[32] It is easy to overlook this because of the fact that if 'There is a mosquito in here' is true on the reading that takes 'a mosquito' as a singular term, it will also be true on the reading that takes the sentence as an existential quantification, provided that the referent of 'a mosquito' is a mosquito, as it will be in normal cases. (The abnormal cases are the ones pointed out by Donnellan; see below, section 9.)

[33] You are likely to take it that way only if you make both of the following assumptions: (i) if 'a mosquito' is a singular term, it must be understood the same way as 'the mosquito'; (ii) 'the mosquito' must be understood according to Russell's

4

Sentences containing indefinite descriptions are ambiguous. Sometimes 'A mosquito is in here' and its stylistic variant 'There is a mosquito in here' must be taken as asserting merely that the place is not wholly mosquito-less, but sometimes they involve an intended reference to one particular mosquito. Their disambiguation depends on how the speaker intends the contexts containing them to be related to other contexts.

For example, suppose that I am reading the morning newspaper and I come across the following story:

> D7: #Houston, Texas, March 10 (UPI) — Dr. Michael DeBakey stated at a press conference today that an artificial heart could be developed within five years. The famed Baylor University heart surgeon said that such a development would make heart transplants unnecessary.#

I then report this fact to you by saying:

> D8: #A doctor in Texas claims that artificial hearts will be developed within five years.#

Is 'a doctor' in that token of D8 a singular term? Is it possible to trace a referential connection between that expression and a particular person, such that what I said is true if and only if *that* person claimed that artificial hearts will be developed within five years? Or am I merely asserting that the class of Texas doctors claiming that artificial hearts will be developed within five years is non-empty, as the existential quantification reading of D8 would have it? In that case what I said would be true even if the news report about DeBakey were wholly erroneous and DeBakey had never made any such claim but some other doctor in Texas had, say in a private conversation, unknown to the reporter who wrote the story. Which reading is the correct one in this case? Imagine how the conversation might continue: you ask 'Who said that?' and I answer 'Dr. Michael DeBakey.' Or perhaps: you say 'I can't quite believe that' and I say 'Well, it was DeBakey who said it and he ought to know. He's a famous surgeon.' Or perhaps: you say 'What's his name?' and I say 'Michael DeBakey.' Such continuations would be unintelligible on the existential quantification reading, for they pre-

Theory of Descriptions (or something very much like it). (ii) is false, as I will argue below, in section 9.

suppose that one and only one person is being said to have claimed that artificial hearts will be developed within five years; they presuppose that there is a unique referent of 'a doctor' whose name can be requested by asking 'Who?' or 'What's his name?' and who can be identified by saying 'Michael DeBakey'.

Compare that with:

D9: #A doctor normally makes at least $75,000 a year.#

Here the referential purport of 'a doctor' is canceled, and it would be a misunderstanding of the context to ask 'Which one?' or 'What's his name?' There is and could be no particular one. (But notice that 'There is a doctor who normally makes at least $75,000 a year' could be taken either as an assertion about DeBakey, say, who normally makes at least $75,000 a year, or as an assertion that there is at least one such doctor, based, perhaps, on a statistical analysis of medical incomes. D9 would not be naturally read in either of these ways, but rather as an assertion about the normal doctor's income.) The fact that 'a doctor' in D8 was intended to be taken as purporting a singular reference could have been signaled explicitly, for I could have said any of the following instead:

D10: #A doctor in Texas — DeBakey, the heart surgeon — claims that artificial hearts will be developed within five years.#

D11: #A doctor in Texas — namely, Michael DeBakey — claims that . . . #

D12: #A doctor in Texas, Michael DeBakey in fact, claims that . . . #

D13: #A certain doctor in Texas claims that . . . #

And so on. In each of these contexts it is made clear by the speaker that he intends 'a doctor' to be taken as a singular term. In D10, D11, and D12 the appearance of the proper name shows that the discourse is to be understood as being linked with another discourse not uttered by the speaker, which contains the proper name (in this case the newspaper story D7). D13 also implies such a linkage but does not present a name taken from the other context.

When a singular term in one context is connected with another the way 'Dr. Michael DeBakey' in D7 is connected with 'a doctor' in D8,[34] I will say that the two expressions are *referentially linked* in a *referen-*

---

[34] And with 'a doctor' and 'DeBakey' in D10, with 'a doctor' and 'Michael DeBakey' in D11 and D12, and with 'a certain doctor' in D13.

tial chain. If a singular term in a given context is not referentially linked to a singular term in another context, I will call it *referentially isolated*.[35] If the structure and content of the context in which a singular expression appears cancels its referential purport, thereby insuring that the expression will not be referentially linked with another expression in another context, the expression is not only referentially isolated but *referentially segregated* — its isolation is not merely an accidental result of the circumstances in which that token is produced, but is, so to speak, intrinsic to it. 'A doctor' in D9 is an example of this. The fact that the expression is segregated need not be indicated on the surface.

So far, then, we have anaphoric chains within contexts, such that if one expression in the chain refers to a given thing then so do all the others, and we also have referential chains between contexts, for which the same condition holds: if one expression in the chain refers to a given thing then so do all the others. Since the links in referential chains connecting different contexts are also links in anaphoric chains within those contexts,[36] we can see how very lengthy referential chains can be constructed: a singular term T_1 in a context C_1 is anaphorically linked with another singular term T_2 in C_1, which is in turn linked with T_3 in C_1, which is referentially linked with T_4 in another context C_2, which is anaphorically linked with T_5 in C_2, which is referentially linked with T_6 in C_3, and so on, until we come at last to some singular term T_n which refers to some object, which thus counts also as the referent of all the singular terms along the whole chain from T_1 to T_n.[37]

Thus the theory of singular reference is concerned with three broad topics:

(i) *Contexts*. What kinds of contexts are there, linguistic and non-linguistic? When does a singular expression in a linguistic context, or a singular element in a non-linguistic context, purport to refer? How does the content of a context affect the purported reference of a sin-

---

[35] If D8 had been uttered "out of thin air," idly, with no such background as D7 provides, the occurrence of 'a doctor' which it contains would have been referentially isolated. This is not to say that it could have had no referent, but merely that it would not have been connected with its referent via any referential chains. See below, section 8.

[36] Since any singular expression is a link in at least one anaphoric chain, namely the one consisting of that expression itself.

[37] Thus the whole referential chain includes all the anaphoric chains as segments; every anaphoric chain is a referential chain, as I am using the phrase, but not vice versa.

gular term or other singular element? When do singular elements form anaphoric chains? When is referential purport canceled, segregating the context?

(ii) *Connections Between Contexts.* Under what conditions are referential chains formed? When is a context referentially isolated, and when is it linked to another context?

(iii) *Connections Between Contexts and Referents.* When do the links in a referential chain actually refer to something? Which terms are connected with objects in the world "directly" rather than by way of other links in the chain? Under what conditions does this take place? What sorts of objects must the referents be if such connections are to be established?

Before sketching a theory of how such questions should be answered for contexts in general, I will deal first with the problem of linguistic reference, examining some popular theories and attempting to demonstrate their inadequacy. For the time being, then, 'context' will mean 'linguistic context' unless otherwise indicated.

5

So far my discussion has been mostly concerned with anaphoric chains beginning with indefinite descriptions, and referential chains in which a singular term T_1 in a context C_1 is referentially linked with an indefinite description T_2 in a context C_2. Now I want to consider other varieties of anaphoric chains and other modes of referential linkage.

First, proper names. An obvious characteristic of proper names is that generally all occurrences of them in a given discourse produced by a single speaker belong to the same anaphoric chain, unless there is something in the context which explicitly indicates the contrary. In a novel, say, or a biography, there will be a number of occurrences of names of characters — 'Humbert Humbert', 'Lolita', 'Winston Churchill' — and unless the reader is told otherwise he is entitled to assume that, e.g., in *Lolita* all the occurrences of 'Humbert Humbert' belong to one anaphoric chain (including other expressions as well, such as, in this case, a number of occurrences of 'I', since the novel is narrated in the first person) and all the occurrences of 'Lolita' belong to another (which would also include a number of occurrences of 'Dolores Haze', 'Dolly',

'Lo', some but not all occurrences of 'she', and so on). The ability to comprehend a novel, a biography, or even a single paragraph presupposes the ability to keep track of who's who by assigning a given singular expression to the right anaphoric chain, where the latter will normally be a mixture of proper names and other singular expressions. This means, for example, knowing when an occurrence of 'she' in *Lolita* belongs with the occurrences of 'Lolita' and when it belongs with the occurrences of the names of the other female characters in the book. The ability to identify anaphoric chains of this sort is obviously very complex: we employ our knowledge of the syntax and semantics of the language, plus our knowledge of how discourses are constructed, plus our knowledge of whatever special literary or scholarly or other conventions pertain to the genre in question, plus our knowledge of what the writer is likely to have meant, and so on.[38] I think it quite unlikely that anaphoric chains in any moderately lengthy discourse are formed according to rules in the way that grammatical sentences are formed according to rules — that is, in such a way that, for a given discourse, the rules generate all and only the anaphoric chains in it — but whether or not this is so is irrelevant to the present discussion. It is more to the point to consider how proper names get introduced into a discourse, that is, how the first occurrence of a proper name comes about. Various devices are in current use in spoken and written English, including the following (this list doesn't pretend to be exhaustive):

D14: #A man named Fred Schultz runs a liquor store around the corner from here. He was held up yesterday, and . . .#

D15: #A man known to the underworld as Joe the Snake was found dead in a vacant lot this morning. Police spokesmen stated that . . .#

D16: #George Frisbee, a friend of mine from college, just inherited a million dollars. He always used to say that . . .#

D17: #A certain bookie I know whom I'll just call "Harry" told me that . . .#

D18: #Stella Houston (a fictitious name) underwent a sex-change operation six months ago. "She" told this reporter that . . .#

---

[38] The difficulties involved in the disambiguation of even a single sentence taken by itself are great enough; see the discussion of this point by Katz and Fodor in their "The Structure of a Semantic Theory," in J. A. Fodor and J. J. Katz, eds., *The Structure of Language: Readings in the Philosophy of Language* (Englewood Cliffs, N.J.: Prentice-Hall, 1964), especially pp. 486–491.

D19: #It turns out that a man I saw at the beach is a spy. He's been working for the Russians for years. Who is it? Bernard J. Ortcutt.#

D20: #Stately, plump Buck Mulligan came from the stairhead, bearing a bowl of lather on which a mirror and a razor lay crossed. . . .#

The simplest way to introduce a proper name into a discourse is to just start using it, as in D18 and D20. Another way is to make a statement of identity, in one guise or another, utilizing a singular term that has already appeared in the context (D19 is of this sort), or one that is itself used there for the first time (D16 and D17).[39] Sometimes it appears that a proper name is introduced *de novo* into a context, in such a way that it is not referentially linked to another occurrence of the same name in an antecedent context. This is indicated explicitly in D17 by the phrase 'whom I'll just call "Harry"'; the pseudonym is introduced simply in order to have a name *in that context* for the person whom the speaker is referring to by another singular term ('a certain bookie I know'), and previous occurrences of the name in other contexts are irrelevant. It doesn't matter whether anyone has ever referred to that person by the name 'Harry' before, and even if someone did it would have no bearing on the present case. (This is the natural way to take D18 also.) Sometimes the situation is just the opposite: the first occurrence of the name in the context harks back to previous occurrences in other contexts, as with D14, D15, and D19. In D19 the point of what the speaker is saying is lost unless the token of 'Bernard J. Ortcutt' which he utters is referentially linked with another token of 'Bernard J. Ortcutt' in some previous context known to him. Otherwise the speaker would not be *identifying* the man seen at the beach as Ortcutt, for identification requires that the terms in the identity statement be independently connected with the referent.[40] The same goes

[39] In effect, D16 asserts 'George Frisbee = a friend of mine from college' and D17 asserts 'Harry = a certain bookie I know'. D14 and D15 are more complicated; D14 asserts that the man who runs the liquor store is actually *named* 'Fred Schultz' and presumably known by this name to persons other than the speaker; and D15 asserts that the dead man was known to certain people by the nickname 'Joe the Snake'. It is not always clear how the falsity of such assertions would affect the reference of the name. See below, section 14.

[40] Some identity statements don't identify anything; they serve instead to introduce new singular terms, such as 'Harry' in D17. Notice also that in D19 it is plausible, contrary to Strawson's view, to say that the speaker has identified the

for D11 and D12: 'namely, Michael DeBakey' and 'Michael DeBakey, in fact' indicate linkages with antecedent contexts containing 'Michael DeBakey'. Sometimes it isn't clear whether such a referential linkage is present. In D20 the speaker might be initiating an entirely new referential chain; he might be continuing one whose last link was 'Buck Mulligan'; and he might be continuing one whose last link was some other singular term (possibly even another proper name).

Proper names pose no special problems when they hark back to a previous occurrence of the same name in an antecedent linguistic context. Here the relationship is like that between proper names forming an anaphoric chain within a single context. Which other singular terms in antecedent contexts can be linked to proper names? As far as I can see, any category of singular term will do. I said in section 3 that indefinite descriptions appear only as the initial links of anaphoric chains. Chains so initiated can continue in various ways, as D4 showed: 'an ARVN officer — he', 'a young prisoner — the prisoner', and so on. Like definite descriptions, demonstrative phrases can have indefinite descriptions as antecedents. For example:

> D21: #I know a guy who likes to eat grasshoppers. Really! Sometimes this guy will eat five or six at a time. He puts vinegar on them.#

'This guy' has whatever reference 'a guy' has, since it belongs to the anaphoric chain 'a guy — this guy — he'. 'This' — which is sometimes erroneously believed to be a "pure" demonstrative which indicates its referents "directly" or "ostensively" — can function the same way:

> D22: #I've just spotted a large boulder. It seems to contain mica. This is the most interesting rock I've seen so far.#

D22 might be a running report by an explorer communicating by radio with his base camp, for example; in such a situation the token of 'this' which he utters cannot pick out anything "ostensively" for his absent audience, but they would know nevertheless that it had whatever referent 'it' and 'a large boulder' had, by virtue of their common membership in the anaphoric chain 'a large boulder — it — this'.[41] Pronouns with

spy as Ortcutt even though the hearer will not be helped by this if 'Bernard J. Ortcutt' doesn't appear in any antecedent context familiar to him.

[41] Sometimes, of course, 'this' would not continue the anaphoric chain; for example, the speaker could suddenly change the topic: '. . . seems to contain mica. [*He suddenly shifts his attention to another rock.*] This is the most interesting rock . . .' But when this doesn't happen — that is, when 'a large boulder' is an anaphoric ante-

the appropriate syntactic characteristics may be added to any anaphoric chain: 'a man — he', 'Bella — she', 'this guy — he', 'this — it', 'the table — it', and so on. Each of the singular expressions considered so far can initiate an anaphoric chain in a given context, and each can be referentially linked with other contexts.

Obviously there is a great deal more to be said about how anaphoric chains are formed in English discourses, but I don't have a theory that would do justice to this vast subject, and even if I did it would not advance my purpose to present it here, since I am concerned with singular reference in general, not just singular reference in English. My examples are from English because that is the only natural language I am competent to theorize about, but I am concerned not with the details of English discourse structure but with the broad outlines which it shares with other languages: anaphoric chains linking singular terms in a context, however they may be constructed by the speaker or comprehended by the hearer.

6

The referential purport of a singular expression can be canceled in a variety of ways, corresponding to the variety of purposes for which discourses can be produced. Referentially segregated contexts may be involved in all of the following.

(i) *Fiction.* If a discourse is produced as fiction *with respect to a singular expression E* contained in it, then: (a) the speaker produces it with the intention that no referential linkage shall hold between E and any singular term in an antecedent context,[42] and (b) he intends that

cedent of 'this' in D22 — and the context is a running report in which 'a large boulder' is uttered earlier than 'this', so that the former has a reference before the latter is even uttered, it seems clear that 'this' gets whatever reference it has from the earlier phrase, not "ostensively." (Later on I will present an analysis of ostension, in terms of perceptual contexts; see section 13.)

[42] This condition insures that E will not be connected with an extra-linguistic referent by way of an expression in another context; you might object that it does not rule out a "direct" connection with a referent. Later on I will argue that in most cases of linguistic reference it must always be possible to establish a referential linkage with a singular element in an antecedent linguistic or non-linguistic context. See section 9.

You might also think that this formulation ignores cases in which the speaker intends a linkage with another *fictional* context, as when I now say 'Superman is faster than a speeding bullet'. But the referential purport of 'Superman' would be canceled in such antecedent contexts — since they are fiction — and it would not be a singular *term*. If I had been unaware that the Superman stories are fiction,

219

this should be evident to his audience. If a discourse is produced as pure fiction, it is produced as fiction with respect to each of the singular expressions in it.[43] These are necessary, not sufficient, conditions. They don't distinguish between fictional uses of contexts and the other non-factual uses (suppositional, modal, etc.) discussed below; all such contexts meet conditions (a) and (b), and further differences among them must be explained in terms of the purposes for which they are produced rather than the status of the singular expressions they contain. Any such distinctions — e.g., that fiction is used to entertain or edify, that discussions of hypothetical cases are used to illustrate general principles, and so on — are likely to be rather vague, as these are, and we can get along well enough without them. As I pointed out in section 3, the fact that a discourse is fiction can be signaled by putting 'a novel' in the title and by stating that all of its characters are fictitious. The same effect can be achieved by 'Once upon a time . . .', 'Let me tell you a story . . .', and 'Have you heard the one about the traveling salesman who . . .', and the like, depending on the prevailing conventions. Often the non-linguistic circumstances will make it evident to the hearer that the speaker intends what he says as fiction even though that is not indicated openly.

(ii) *Supposition.* The following contexts are referentially segregated:

D23: #Assume that a man deposits $5,000 in a savings account at 5¼% interest, compounded semiannually. After three years, how much money will the man have in his account?#

D23': #Assuming that a man deposits $5,000 in a savings account at 5¼% interest, compounded semiannually, how much money will he have in his account after three years?#

D24: #If a man were to deposit $5,000 in a savings account at 5¼% interest, compounded semiannually, how much money would he have in his account after three years?#

D25: #A man deposits $5,000 in a savings account at 5¼% interest, compounded semiannually. After three years, how much money does he have in his account?#

D25': #If a man deposits $5,000 in a savings account at 5¼%

then (a) would not hold, but then also I would not have produced my remarks as fiction.

[43] Most fiction is not pure. Historical novels, jokes about actual people, political satires, etc., are very impure; even novels whose characters are all fictitious may contain references to actual places, say, and be somewhat impure. Even if the characters of a discourse are all actual people, as in D3, some of the *events* may be fictitious (e.g., 'the ping-pong game' might appear later in D3).

interest, compounded semiannually, how much money does he have in his account after three years?#

These could be used to state a problem in a textbook, or as part of an exposition of the laws of compound interest (in which case the questions would be "rhetorical"). They are available for such purposes because in each case it is explicitly indicated that the referential purport of 'a man — the man' or 'a man — he' is canceled, so that these expressions cannot be referentially linked to singular terms in other contexts. In each case it would be incorrect for a hearer to ask 'Who?', 'Which man?', 'Is that him over there?', etc., and it would be incorrect for the speaker to continue the discourse by identifying the man as this one or that one. The segregation of the context is accomplished by different syntactic forms, but the effect is the same in each case: a supposition is made that such-and-such a person exists and does so-and-so, and a question is asked about the consequences.

There are very similar contexts in which the question of referential linkage can only be settled by the speaker's intentions:

D26: #A man deposited $5,000 in a savings account at 5¼% interest, compounded semiannually. Now, after three years, how much money does he have in his account?#

D27: #A man is depositing $5,000 in a savings account at 5¼% interest, compounded semiannually. How much money will he have in his account after three years?#

These could serve as textbook examples or exam questions, but they also could under certain circumstances be used to talk about particular people. (Imagine D26 continuing: # . . . He asked me to work it out for him. He doesn't trust the people down at the bank.#)

In a similar fashion, referentially segregating the context enables one to make suppositions and then state rather than ask about their consequences:

D28: #Assuming that a man deposits $5,000 in a savings account at 5¼% interest, compounded semiannually, after three years he will have $5,432.10 in his account.#

D29: #If a man were to deposit $5,000 in a savings account at 5¼% interest, compounded semiannually, after three years he would have $5,432.10 in his account.#

D30: #If a man deposits $5,000 in a savings account at 5¼% interest, compounded semiannually, after three years he has $5,432.10

in his account.# (D30 is discussed further in (iv) below.)

Proper names can figure in suppositional contexts:

> D31: #If John Smith were to deposit $5,000 in a savings account at 5¼% interest, compounded semiannually, after three years he would have $5,432.10 in his account.#

Depending on the speaker's intentions, 'John Smith' may be referentially linked with another context (and perhaps ultimately with an actual person), or its referential purport may be canceled. In the former case it would still involve a supposition, of course; not a supposition to the effect that there exists such-and-such a person, but a supposition that an actual person does so-and-so.

(iii) *Modality.* Referentially segregated contexts can be used to make points about what might, or could, or ought to be the case. For example:

> D32: #A visitor might trip over that cord. Then he could sue us.#
>
> D33: #A rattlesnake can kill you.#
>
> D34: #You ought to buy a new car. The bank will loan you the money for it.#

The segregation of 'a new car', 'a rattlesnake', and 'a visitor' in D32–D34 is explicit in their structure; normally understood, they cannot be referentially linked to other contexts via those expressions. But this is not due to the syntactic position occupied by 'a new car' and the rest, as we can see by making some substitutions:

> D35: #Uncle Bert might trip over that cord. Then he could sue us.#
>
> D36: #That rattlesnake over there can kill you.#
>
> D37: #You ought to buy this new Mazda. The bank will loan you the money for it.#

'Uncle Bert', 'that rattlesnake over there', and 'this new Mazda' could each be referentially linked with a term outside the context and ultimately with an extra-linguistic referent. To use the indefinite description instead is to indicate an intention to block the linkage. Yet the possibility of such a hook-up can be reinstated if the right sort of indefinite description is used, as is obvious from the following:

> D38: #A nearsighted man who comes in here occasionally might trip over that cord. Then he could sue us. He's very litigious.#
>
> D39: #A rattlesnake I just saw can kill you. He wasn't defanged like the others here in the zoo.#

D40. #You ought to buy a new car I saw today. It's got a little scratch on the top and the dealer is willing to come down on the price.#

Thus it is not the mere presence of an indefinite description within the range of a term like 'might' or 'can' or 'ought' that cancels its purported singular reference. It is much more complicated than that, and I believe that the preceding examples show that any simple account is likely to fail.

(iv) *Universal Quantification.* D30 can be paraphrased as a universal quantification:

(1)   (x) ((x is a man & x deposits $5,000 in a savings account at 5¼% interest compounded semiannually) ⊃ x has $5,432.10 in his account after three years)

Apparently D30 must always be paraphrased as a universally quantified sentence, in which the singular expressions disappear ('a man' in (1) should be considered part of the predicate). If D30 had contained genuine singular terms, they could not have been brushed away by paraphrase; a singular term possesses at least the *possibility* of referential linkage with other terms in other contexts, but no expression in (1) does so. Thus the fact that (1) is evidently a correct paraphrase for any token of D30 shows that D30 itself does not contain 'a man' or 'he' as singular terms. If the latter had counted as singular terms, then the paraphrase would have had to look like this:

(2)   a deposits $5,000 in a savings account at 5¼% interest compounded semiannually ⊃ a has $5,432.10 in his account after three years

with 'a' as an individual constant. (1) could also have been reached from

D41: #The man who deposits $5,000 in a savings account at 5¼% interest, compounded semiannually, has $5,432.10 in his account after three years.#

D42: #He who deposits . . . #

D43: #If John Smith deposits . . . #

D43, at least, is ambiguous between the two readings, and would sometimes have to be paraphrased as (2).

Sorting out when a discourse has the referential purport of some of

223

its singular expressions canceled by universal quantification[44] is not a simple matter, if you want general principles which, given the syntactic and semantic structure of the discourse, will tell you whether a given singular expression is referentially segregated. As in the case of fiction, supposition, and modality, I offer no such principles, nor do I think they will be easy to find.

(v) *Existential Quantification.* Sometimes the existential quantification

(3) $(\exists x)$ (x has been eating my porridge)

is a correct paraphrase of

D44: #Someone has been eating my porridge.#

and sometimes it is not, as we can see by considering

D45: #Someone has been eating my porridge. She says her name is "Goldilocks." Here she is. What are we going to do with her?#

The same goes for

(4)   The window has been forced & $(\exists x)$ (x is a burglar & x has been in here)

D46: #The window has been forced. A burglar has been in here.#

D47: #The window has been forced. A burglar has been in here. You can see his footprints quite clearly. It's Joe the Snake again — I can spot those shoes of his every time.#

and also for

(5) $(\exists x)(\exists y)(\exists z)$ (x is an enchanted evening & y is a stranger & z is a crowded room & you will see y on x across z)

D48: #Some enchanted evening, you will see a stranger, across a crowded room.#

D49: #Some enchanted evening, you will see a stranger, across a crowded room. She will be Agent 99, and she will give you the recognition signal. You will leave the room with her immediately and . . . #

The fact that D44, D46, and D48 can be expanded into D45, D47 and D49, which cannot take the quantificational paraphrase, shows that their

---

[44] A useful but inaccurate phrase to apply to sentences or discourses which would be paraphrased by universal quantifications. No sentence of English could actually be universally quantified, since English does not possess quantifiers or bindable variables.

referential segregation[45] is not something built into the structure of the discourse.

The following, however, are segregated no matter what the circumstances:

D50: #Has a burglar been in the house?#

D51: #Buy me an ice-cream cone!#

D52: #I promise to give you a bicycle for Christmas.# [46]

D53: #I dreamed that a new continent arose out of the Pacific Ocean.#

D54: #Imagine a mountain. Think of it as a place where the cops have wooden legs, and all the dogs have rubber teeth, and . . . #

Each should be thought of as involving an existential quantification along roughly the following lines:

(6)   Has the following been the case? — $(\exists x)$ $(x$ is a burglar & x is in the house)

(7)   I dreamed that the following was the case: $(\exists x)(x$ is a new continent and x arises out of the Pacific Ocean)

and so on. Putting other singular expressions in place of the indefinite descriptions in D50–D53 may restore the possibility of referential linkage: e.g., 'the burglar' or 'Joe the Snake' or 'he' for 'a burglar', 'that ice-cream cone' for 'an ice-cream cone', 'that bicycle' for 'a bicycle', 'Australia' for 'a new continent'.[47]

(vi) "Generic" Reference. Consider:

D55: #A humpback whale is a whalebone whale with a rounded back and long, knobby flippers. It is on the verge of extinction.#

D56: #The dodo lived on Mauritius. It is extinct.#

---

[45] I.e., their referential segregation with respect to the singular expressions in question; for brevity I will often not state this qualification.

[46] This is borderlinish. I think that #I promise to give you a bicycle for Christmas. It's that one over there.# (indicating a particular one, not a brand or model) is incorrect, but others may think otherwise. It isn't essential that everyone agree with all my examples as long as enough of them are accepted to establish the general claims I am making, for these claims concern the range of possibilities inherent in the structure of the discourses we utter, not the details of how these possibilities are realized.

[47] But the place occupied by 'a mountain' seems absolutely closed; if we try to substitute a proper name or a demonstrative phrase, the result seems somehow malformed: #Imagine Mao Tse-tung. . . . # Imagine him doing or being what? I can't imagine just *him*, for he is already a real person.

'A humpback whale — it' and 'the dodo — it' cannot be referentially connected, directly or by way of referential linkages with singular terms in other contexts, with any particular whale or dodo, for no particular whale or dodo could be extinct — 'extinct' applies to the species as a whole or not at all. For the same reason they cannot be read quantificationally, if the variables range over concrete objects only. They must be taken as referring to the species *Megaptera novaeangliae* and *Raphus cucullatus*, respectively, if they refer to anything. The normal referential purport of the singular expressions has been canceled — they can no longer take concrete things as referents — and another involving abstract objects put in its place.

As with fiction, supposition, etc., some contexts are ambiguous between the two kinds of readings; for example:

D57: #The whale is a mammal.#

Strawson would disagree:

It is obvious that anyone who uttered the sentence, 'The whale is a mammal', would be using the expression 'the whale' in a way quite different from the way it would be used by anyone who had occasion seriously to utter the sentence, 'The whale struck the ship'. In the first sentence one is obviously *not* mentioning, and in the second sentence one obviously *is* mentioning, a particular whale.[48]

But consider:

D58: #No, ladies and gentlemen, not every creature you see here today is a fish. The lobster — he's perched on that rock in the middle of the tank — is a crustacean, and the whale is a mammal. But all the others are fish.#

Unsegregated occurrences of 'The whale is a mammal' may occur less frequently than segregated ones, but that fact is of no theoretical interest.

7

When a singular expression is referentially segregated, we can say that it is to be read "internally" or given an "internal" reading, and that the discourse itself is to be read internally with respect to that expression. When an expression is not segregated, we can say that it is to be read "externally" and that the discourse is to be read externally with respect to it. Thus, for example, 'a mountain' in D54 must be read internally,

[48] "On Referring," p. 21.

and D54 must be read internally with respect to 'a mountain'. I use 'internal' and 'external' instead of the current 'de dicto' and 'de re' because expressions read externally need not refer to a thing — be "de re" for some re — at all, or be intended as such; they may only hark back to a singular expression in another context which itself fails to refer and which may itself be referentially segregated. And I use 'internal' and 'external' instead of Quine's 'referentially opaque' and 'referentially transparent' because his distinction rests on the notion of an expression's having "purely referential position," defined as follows:

When a singular term is used in a sentence purely to specify its object, and the sentence is true of the object, then certainly the sentence will stay true when any other singular term is substituted that designates the same object. Here we have a criterion for what may be called *purely referential position*: the position must be subject to the *substitutivity of identity*.[49]

But substitutivity of identity doesn't even hold in general for the expressions in a single anaphoric chain, as I have already argued, since e.g., 'a man' and 'the man' can be linked in an anaphoric chain and will refer to the same thing if to anything but will never be interchangeable (see section 3). Therefore they are not used "purely to specify" their referents and are thus referentially opaque. But among them some will be read externally and some internally. Thus this distinction cuts across his.

The cases of referential segregation discussed in this section suggest that the unit of analysis — if we are considering a *natural language* — is at the very least a whole discourse, not an isolated single sentence. Those examples showed the great variety of ways in which and purposes for which referential purport can be canceled. By expanding the discourse in one way or another, one or another option is opened or closed, and it seems obvious that we can place no limit on the length of expansions which may turn out to be relevant. It is better to consider the single sentence as a nest of possibilities for anaphoric connection within a larger discourse and for referential linkage to other contexts, with the realization of these possibilities depending on subtle and complex structural properties of the discourse plus the speaker's intentions concerning inter-contextual linkage. (The relation of the sentence to the discourse is analogous to the relation of the word to the sentence.) This makes our problem harder if we want to find the "logical form" of single

[49] *Word and Object*, p. 142.

227

sentences by trying to show how they should be paraphrased in a canonical language, since we will typically end up with several disparate logical forms for each sentence. Quine says in *Word and Object* that

it would be folly to burden a logical theory with quirks of usage that we can straighten. It is the part of strategy to keep theory simple where we can, and then, when we want to apply the theory to particular sentences of ordinary language, to transform those sentences into a "canonical form" adapted to the theory. If we were to devise a logic of ordinary language for direct use on sentences as they come, we would have to complicate our rules of inference in sundry unilluminating ways.[50]

It does not conflict with Quine's principles, though it does conflict with his practice, to attempt the paraphrases only at the level of whole discourses, including some very long ones. Simple sentences in isolation such as 'I saw a lion' will usually be ambiguously paraphrased. To find out whether this token uttered by this speaker on this occasion goes over to '$(\exists x)$ (I saw x & x is a lion)' or to 'I saw a' we must find out what discourse it is part of and how it is linked to other contexts. Once we have that information we can paraphrase the whole discourse, if that is what our purposes require.

I suggest that we take as basic the notion of an extended discourse containing one or more anaphoric chains and having a certain "content." Discourses constructed out of different sentences arranged in different ways will normally contain different anaphoric chains, but may nevertheless have the same content, like D23–D27 above. A discourse with a given content may be used for various purposes: to assert straightforwardly what the speaker takes to be ordinary fact about individual things, or to propound fiction, or to make general statements, or modally, or suppositionally, or what have you. I am not concerned with these other uses here, since they all involve cutting off the possibility of singular reference.

8

If a discourse is not referentially segregated, when and how do its singular terms refer? If T_1 in C_1 refers because it is linked to T_2 in C_2, that only postpones the question. How is it that T_2 refers? How does a referential connection with an extra-linguistic object make its first entry into a linguistic context?

[50] *Ibid.*, p. 158.

If T_2 in C_2 uttered by Y is referentially linked with T_1 in C_1 uttered by X, and X and Y are different people, and T_1 refers to O, and Y's utterance of T_2 is preceded by X's utterance of T_1 and is thus dependent on it for its referential connection with O, then Y in uttering T_2 has made a *secondary reference* to O. For example, taking the reporter who wrote D7 as X and the speaker who uttered D8 as Y, and assuming that X referred to DeBakey in writing 'Dr. Michael DeBakey', we can see that in uttering 'a doctor' Y made a secondary reference to DeBakey. If someone refers but does not make a secondary reference, then he has made a *primary reference*.[51]

What I just said about secondary reference doesn't cover cases in which what X says now is linked "externally" with what he said earlier, as would happen if he developed amnesia and then read things he had written before, or if his memory was bad but not wholly destroyed and he relied heavily on what he himself had previously written or otherwise recorded, and so on. Such cases shade over into ordinary ones familiar to everyone, since nobody's memory is perfect. A complete definition of 'secondary reference' would have to clear up such cases,[52] but that can be ignored here. Secondary reference is explained by primary reference. What accounts for primary reference? I will begin by discussing theories of primary reference for definite descriptions.

If a definite description in a given discourse is a genuine singular term, possessing the possibility of referential linkage outside the discourse, then it cannot be paraphrased away a la Russell, since the paraphrase will not contain any expression capable of being linked with another context; it will only contain quantifiers, bound variables, predicates, and so on, which cannot sustain the linkage. Strawson's account of definite descriptions does not brush them away; they remain singular terms (at least in most cases: it isn't clear to me whether he believes that Russell's analysis *never* holds). Given that T is a definite description which is not referentially segregated and which does not get its reference secondarily, when does it refer and what does it refer to?

[51] I found it necessary to make this distinction because of an important class of cases pointed out to me by Saul Kripke, commenting on an earlier statement of this theory that covered primary reference only. See below, section 14.

[52] By expanding 'X and Y are different people' into 'either X and Y are different people or else they are the same person and X at t_1 stands at Y at t_2, with respect to the utterance of T_1 and T_2, as *if* they are different people', or something of the sort. The 'if . . . then' could then be replaced by 'if and only if'.

I now want to introduce the notion of the "descriptive content" of a singular term in (or relative to) a given discourse.[53] Again:

> D1: #A man was sitting underneath a tree eating peanuts. A squirrel came by, and the man fed it some peanuts.#

The descriptive content of 'the man' in D1 is: man who ate peanuts while sitting underneath some tree. The descriptive content of 'it' is: squirrel which came by some tree underneath which some man was sitting and eating peanuts. I will say that a singular term *denotes* something if that thing satisfies its descriptive content. Thus 'the man' in D1 denotes any man who ate peanuts while sitting underneath some tree, and 'it' denotes any squirrel which came by some tree underneath which some man was sitting and eating peanuts. A singular term *uniquely* denotes something if it denotes it and nothing else.

Thus the descriptive content of a singular term in a discourse is, roughly speaking, what sort of thing the discourse says the referent is supposed to be, what properties it is supposed to have, what sorts of other things it is supposed to be related to and in what ways, and so on. The term not only purports to refer but purports to refer to a thing of a specified kind. Do other links in the same anaphoric chain have the same descriptive content? Not if 'a man' and 'a squirrel' are assigned descriptive content in the way that 'the man' and 'it' were, for the assignment of content to 'the man' and 'it' did not include what was predicated of them at the point at which they appeared in the context. We did not include 'and who fed peanuts to some squirrel which came by' in the content of 'the man' or 'and which was fed peanuts by the man' in the content of 'it', for these were being *asserted* to hold for the referents of 'the man' and 'it', and such assertions are intended to say something new, not merely unfold what is already contained in the terms. Parallel to this, the descriptive content of 'a man' would just be 'man' and that of 'a squirrel' merely 'squirrel', and these terms would denote anything which was a man or a squirrel, respectively.

There is an issue which I have put off considering up to now. Any discourse can appear part by part, one sentence being uttered and then another and so on, possibly separated by periods of silence of various lengths. D6 (the mosquito story) and D22 (the explorer's report) would normally occur that way. On the other hand, a discourse can be pro-

---

[53] This is an adaptation of Paul Ziff's notion of "information-content"; see his *Semantic Analysis* (Ithaca, N.Y.: Cornell University Press, 1960), pp. 97–101.

duced and presented "all at once," as a single unit, like a book or a newspaper story. The audience must take it in part by part, but from the point of view of the speaker it is all of a piece: the order of the sentences and the arrangement of the anaphoric chains partly depend on contingencies of exposition which select one from among a number of different discourses having the same content, such as:

> D59: #A squirrel came by a tree. A man fed it some peanuts. He had been eating them. The man was sitting underneath the tree.#
>
> D60: #A man was eating peanuts. He was sitting underneath a tree. He fed some peanuts to a squirrel which came by.#

The same information about this situation is introduced in a different order in D1, D59, and D60, but what we make of it is the same in the end: man-sitting-underneath-tree-eating-peanuts-and-squirrel-coming-by-and-man-feeding-peanuts-to-squirrel. As far as content goes, D1, D59, and D60 are equivalent. They do not contain the same anaphoric chains — D1 has 'a man — the man', D59 'a man — he — the man', and D60 'a man — the man — he' — but they do contain the same "characters": a man, a tree, a squirrel.[54]

The singular terms in a temporally extended discourse like D6 or D22 must, I think, be assigned descriptive content in the way I just described, since "what sort of thing the referent is supposed to be" depends on what the speaker meant by what he said, how he was putting sentences together in order to construct a story with a certain content. If in the mosquito story the descriptive content of the first occurrence of 'it' is taken to include not only 'mosquito in here' but what was predicated at later links in the chain ('lands on left arm', 'bites', etc.), then the descriptive content has nothing to do with what the speaker meant and what sort of thing he was purporting to refer to, since he did not foresee the events which led him to make those remarks later in the discourse.

---

[54] But we might also have had this, which clearly has the same content as the others: #A feeding of peanuts took place. It was done by a man who had been sitting underneath a tree eating them, to a squirrel which came by.# 'A feeding of peanuts — it' appears here but not in D1, D59, or D60, which do not on the surface contain singular terms purporting to refer to events. Are we to say that such terms are somehow there anyway, that the event is also part of the story, or what? I don't believe that my account of reference depends on resolving this issue one way or the other, so for simplicity I will ignore this kind of problem and assume that all the relevant singular terms are there on the surface and that each discourse has an anaphoric chain for each character in the story. For a discussion of this problem, see Donald Davidson's "The Individuation of Events."

It will, I think, introduce avoidable complications to make allowances for discourses in which the descriptive content of the anaphorically connected singular terms accumulates over time, so henceforth I will assume that every discourse is of the simultaneous variety, in which each term in an anaphoric chain has the same descriptive content as well as the same reference as all the others. I am also going to ignore discourses uttered by more than one speaker, discourses uttered by vacillating speakers who introduce discrepancies into the content of what they are saying by retracting their previous remarks or contradicting themselves, discourses involving questions or commands or other non-assertive speech acts, and so on[55] — everything but the most humdrum factual narration.

How is descriptive content comprehended? How does the hearer manage to grasp the story and identify the characters, assigning each one to the singular terms purporting to refer to him? Like the ability to recognize anaphoric chains (which is, in fact, one of its components), this ability is subtle and complex and difficult to theorize about. Nevertheless, discourses do have content, the stories they tell are identifiable, the descriptive content of their singular terms can be discerned, and we can safely proceed by dealing with clear examples even though we lack an analysis of what makes them examples.

9

The orthodox view of primary reference is a theory which I will call *denotationism*. First I will discuss the denotationist analysis of definite descriptions and later on consider denotationist accounts of other singular terms. Take first the case in which a speaker utters a short discourse of the following sort:

#The king of France is wise.#

The denotationist view — as expounded by Strawson, who discusses this example in "On Referring" — is that in uttering 'the king of France' the speaker is referring to whatever thing happens to be uniquely denoted by 'the king of France'. Thus if he is referring to anything at all he is referring to something which uniquely satisfies the descriptive content of the term, i.e., some person who is sole king of whatever country 'France' refers to. If nothing is denoted by the term, or if more than

---

[55] It is the existence of such discourses that makes it necessary to regard multi-sentence discourses in general as sequences of sentences rather than conjunctions.

one thing is, he has failed to make a reference. Unique denotation is held to be both a necessary and a sufficient condition for primary reference in all cases involving unsegregated definite descriptions.

Denotationism is false, as Keith Donnellan has shown. He introduces a distinction between the "attributive" and the "referential" uses of definite descriptions, as follows:

A speaker who uses a definite description attributively in an assertion states something about whoever or whatever is the so-and-so. A speaker who uses a definite description referentially in an assertion, on the other hand, uses the description to enable his audience to pick out whom or what he is talking about and states something about that person or thing. In the first case the definite description might be said to occur essentially, for the speaker wishes to assert something about whatever or whoever fits that description; but in the referential use the definite description is merely one tool for doing a certain job — calling attention to a person or thing — and in general any other device for doing the same job, another description or a name, would do as well. In the attributive use, the attribute of being the so-and-so is all important, while it is not in the referential use.

To illustrate this distinction, in the case of a single sentence, consider the sentence, "Smith's murderer is insane." Suppose first that we come upon poor Smith foully murdered. From the brutal manner of the killing and the fact that Smith was the most lovable person in the world, we might exclaim, "Smith's murderer is insane." I will assume, to make it a simpler case, that in a quite ordinary sense we do not know who murdered Smith (though this is not in the end essential to the case). This, I shall say, is an attributive use of the definite description.

The contrast with such a use of the sentence is one of those situations in which we expect and intend our audience to realize whom we have in mind when we speak of Smith's murderer and, more importantly, to know that it is this person about whom we are going to say something.

For example, suppose that Jones has been charged with Smith's murder and has been placed on trial. Imagine that there is a discussion of Jones's odd behavior at his trial. We might sum up our impression of his behavior by saying, "Smith's murderer is insane." If someone asks to whom we are referring, by using this description, the answer here is "Jones." This, I shall say, is a referential use of the definite description.

That these two uses of the definite description in the same sentence are really quite different can perhaps best be brought out by considering the consequences of the assumption that Smith had no murderer (for example, he in fact committed suicide). In both situations, in using the definite description "Smith's murderer," the speaker in some sense presupposes or implies that there is a murderer. But when we hypothe-

233

size that the presupposition or implication is false, there are different results for the two uses. In both cases we have used the predicate "is insane," but in the first case, if there is no murderer, there is no person of whom it could be correctly said that we attributed insanity to him. Such a person could be identified (correctly) only in case someone fitted the description used. But in the second case, where the definite description is simply a means of identifying the person we want to talk about, it is quite possible for the correct identification to be made even though no one fits the description we used. We were speaking about Jones even though he is not in fact Smith's murderer and, in the circumstances imagined, it was his behavior we were commenting upon. Jones might, for example, accuse us of saying false things of him in calling him insane and it would be no defense, I should think, that our description, "the murderer of Smith," failed to fit him.[56]

Thus 'Smith's murderer' used referentially in a situation where Jones did not murder Smith does not denote Jones but refers to him, and does not refer to the person who murdered Smith (if there is such a person) even though it denotes him. Unique denotation is neither necessary nor sufficient for reference.

Donnellan's notion of referential use involves a fairly complex situation in which a speaker is attempting to get a message across to a hearer by any means necessary, including non-denoting descriptions. One might think that his point holds only for such situations, but that would be a mistake, as we can see by considering what might happen in a similar case in which the speaker asserts 'Smith's murderer is insane' purely for his own benefit. He might be writing or dictating notes on the trial, or merely expressing aloud his opinion about Jones's mental condition.[57] In uttering 'Smith's murderer' he refers to Jones even though the definite description does not denote Jones. How is this possible? If the referential connection between 'Smith's murderer' and Jones is not established as a result of Jones being denoted by 'Smith's murderer', how is it established?

So far we have been considering only those linguistic contexts which are written or uttered aloud or otherwise publicly produced. What about those which are not? Many if not most people think many if not most of their thoughts in the words of the language they speak, and these words form sentences which form discourses which are in the relevant

---

[56] "Reference and Definite Descriptions," pp. 285–286.

[57] It is not essential that the speaker is making an assertion. He might instead be asking himself a question: 'Is Smith's murderer insane?'

respects the same as the ones we have been discussing. Call these *covert discourses*. Most of the discourses I have discussed so far could appear covertly as well as overtly, could be thoughts as well as express them (that is, some tokens of them could).[58] Like overt discourses, covert discourses consist of expressions some of which will be singular expressions connected together in anaphoric chains. Some of these will count as singular terms which could be referentially linked with singular terms in other contexts. In particular, they could be linked with singular terms in overt discourses produced by the same person. Thus we could have the following:

D62 (covert): #Jones is on trial here for the murder of Smith. He sure looks like a criminal. He's got beady eyes. He must be guilty. He murdered Smith. He's behaving very strangely. He's staring like a madman. Smith's murderer is insane. That's why Smith's body was so horribly mutilated.#

D61 (overt): #Smith's murderer is insane.#

D61 is what the speaker says, D62 is what he thinks (and, in a sense, what he "means" by D61).

In a case in which D61 is not uttered idly but expresses the speaker's thoughts, the relation between D62 and D61 will be like the relation between D7 and D8. D62 contains the anaphoric chain 'Jones — he — he — he — he — he — he — Smith's murderer', and the token of 'Smith's murderer' in D62 is referentially linked with the token of 'Smith's murderer' in D61. The following referential chain extends over the two contexts:

$$\underbrace{\text{Jones — he — } \ldots \text{ — he — Smith's murderer}}_{\text{D62}} \text{ — } \underbrace{\text{Smith's murderer}}_{\text{D61}}$$

'Smith's murderer' in D61 refers to whatever 'Smith's murderer' in D62 refers to, and the latter refers to whatever 'Jones' refers to. In this case 'Jones' refers to Jones the defendant, and so does 'Smith's murderer',

---

[58] This is hardly a novel suggestion. As Plato put it:

Socrates. . . . And do you accept my description of the process of thinking?
Theaetetus. How do you describe it?
Socrates. As a discourse that the mind carries on with itself about any subject it is considering. You must take this explanation as coming from an ignoramus; but I have a notion that, when the mind is thinking, it is simply talking to itself, asking questions and answering them, and saying Yes or No. . . . So I should describe thinking as discourse, and judgment as a statement pronounced, not aloud to someone else, but silently to oneself. (*Theaetetus*, 189E–190)

even though Jones did not murder Smith. The actual murderer is *denoted* by 'Smith's murderer' — whose descriptive content in D61 is 'murderer of Smith' — but is not its referent, since he is not the referent of 'Jones'.

This kind of example makes it possible to understand how a definite description can have a referent even when it obviously would never denote anything uniquely.

D63 (overt) : #The cat is on the mat.#

'The cat' and 'the mat' denote all the cats and all the mats and hence (in this world) denote no cats or mats uniquely. Instead, they get their reference via linkages with singular terms in covert discourses.

I don't claim that a definite description *never* refers by virtue of uniquely denoting something, for we have such examples as these:

D64: #Shoot the first man who comes through that door!#

D65: #Our one-millionth customer will receive a month's free groceries and a check for $100.#

D66: #The last word uttered by the last human being will be "rosebud."#

D64 might be uttered by someone leading a defense against a lynch mob about to invade a building. Although the speaker doesn't know who the first one through the door will be — that is, at the time of utterance he can refer to that person *only* by uttering 'the first man who comes through that door' — it is *that* man, the one uniquely denoted by the definite description, who is to be shot when the time comes. He, whoever he is, is the subject of the command, and whether or not the command is obeyed is a matter of whether or not he is shot. Therefore he is the referent. D65 is similar. Whoever turns out to be uniquely denoted by 'our one-millionth customer' will get the groceries and the money if the pledge is kept, and whether or not the pledge has been kept is a matter of whether that person gets them. D66 is idle speculation, but it is true or false nevertheless, depending on whether 'the last word uttered by the last human being' uniquely denotes something which is as described. That thing will be the referent because it is the thing whose properties determine whether the statement is true or false.

I believe that Donnellan would disagree with this, for he says:

denoting and referring should not be confused. If one tried to maintain that they are the same notion, one result would be that a speaker might be referring to something without knowing it. If someone said, for example, in 1960 before he had any idea that Mr. Goldwater would be the Republican nominee in 1964, "The Republican candidate for president in 1964 will be a conservative," (perhaps on the basis of an analysis of the views of party leaders) the definite description here would *denote* Mr. Goldwater. But would we wish to say that the speaker had referred to, mentioned, or talked about Mr. Goldwater? I feel these terms would be out of place. Yet if we identify referring and denoting, it ought to be possible for it to turn out (after the Republican convention) that the speaker had, unknown to himself, referred in 1960 to Mr. Goldwater. On my view, however, while the definite description used did *denote* Mr. Goldwater . . . the speaker used it *attributively* and did not *refer* to Mr. Goldwater.[59]

I agree, of course, that referring and denoting are distinct notions and I agree that they are not coextensive in any cases except the kind just discussed, but I think that it is possible to explain why one feels reluctant to say that the speaker is "referring" in those cases. For singular terms in general — with the sole exception of terms in referentially isolated linguistic contexts such as D64–D66 and the one Donnellan cites — primary reference can occur only if the speaker has *knowledge* of the referent (I will explain this notion later), which he clearly does not in those examples. I think there is a tendency to use 'refer' when applied to speakers in such a way as to rule out the cases where knowledge of the referent is lacking. For the sake of a uniform terminology I will use 'refer' there too, though it sounds somewhat unnatural, since I want to draw attention to what I think is the most important thing which they all have in common, namely the fact that with both kinds of singular terms there is one and only one thing the facts about which have a decisive bearing on the success or failure of the speech act, on whether the statement is true or false, the command obeyed or disobeyed, the promise kept or broken, and so on. If such a thing is related via a singular term to the speech act performed in producing the context, I will call it the *referent* of the term.

10

Unique denotation doesn't determine reference in the case of a definite description in an overt discourse which forms a referential chain with a singular term in an antecedent covert discourse. The reason for

[59] "Reference and Definite Descriptions," p. 293.

this is that the connection with the term in the other discourse provides a way for the description to be connected with a referent independent of whether its descriptive content happens to fit that referent. Denotation can fail and reference succeed because there is an alternative route to the thing referred to. If this alternative route is removed — that is, if the definite description happens not to be linked with a term in a previous context — then there is nothing left but denotation to fix the identity of the referent, and since the reference is supposed to be to one thing only the denotation must be unique. This explains why the denotationist analysis applies to some cases and also why reference is possible in the other cases. This approach can easily be extended to cover indefinite descriptions.

> D67 (covert): #An insect is crawling up my arm. It looks as if it's going to bite me. It's a mosquito. There, it bit me. That really hurt.#

> D68 (overt): #A mosquito just bit me.#

The referential chain is:

$$\underbrace{\text{an insect} - \text{it} - \text{it} - \text{it} - \text{it}}_{\text{D67}} - \underbrace{\text{a mosquito}}_{\text{D68}}$$

'A mosquito' refers to whatever 'an insect' refers to, and if the speaker is mistaken in identifying it as a *mosquito* then 'a mosquito' — and also any occurrences of 'the mosquito' anaphorically connected with it in expansions of D68 — would denote whatever mosquitoes there are in the world but would refer to whatever non-mosquito 'an insect' referred to: a horsefly glimpsed by a badly astigmatic speaker, a Disney-designed mosquito robot, or what have you. When a term is referentially linked with a previous context, denotation is neither necessary nor sufficient for reference.

This takes care of definite and indefinite descriptions in overt discourses. It obviously applies to pronouns as well:

> D69 (overt): #It bit me.#

'It' could be linked to the occurrences of 'it' in D67; if so, it would refer to whatever they refer to. The same pattern of analysis applies to demonstratives and proper names which are referentially linked to antecedent contexts.

But this doesn't solve the basic problem: when and how do referentially unsegregated singular terms in linguistic contexts overt or covert make primary references? Singular terms like 'Jones' in D62 and 'an insect' in D67 could be referentially linked with other covert discourses, conscious or unconscious, which could themselves be linked with still others, and so on, but tracing the referential chains into the past in this way will, if we have a case of primary reference, lead us eventually to a covert discourse which contains a singular term which has a referent but is not referentially linked to any earlier singular term in any other discourse, overt or covert. When does such a term refer?

A covert discourse like D67 could occur as a kind of internal running commentary on what the speaker was (or thought he was) perceiving: something that looked like an insect, was seen and felt crawling up his arm, looked as if it were going to bite him, and bit him. In such a case there is no mystery about what object he was referring to: it is that object which he perceived as an insect, saw and felt crawling up his arm, and so on. Perhaps that object is really an insect, perhaps not; if it is an insect, perhaps it is a mosquito, perhaps not. Whatever it is, it is the thing he was perceiving, well or poorly, as it really was or as it really wasn't, and *it* is the referent of 'an insect' in the covert discourse and 'a mosquito' (or 'the mosquito') in the overt discourse. It is the referent because a perception of it was connected with the singular terms. If the speaker only *seemed* to perceive something but in fact perceived nothing at all, then there was no referent.

11

In section 1, I said that a context was something which had content, which could represent some thing or situation, and which could contain singular terms or analogous elements which could be referentially connected with objects in the world. Linguistic contexts are discourses, overt or covert. Other overt contexts belong to non-linguistic representational systems — maps, pictures, and so on — and these can appear covertly as well. Everything I regard as a context or an element of a context is a concrete individual object which is spatio-temporally or at least temporally located and which is capable of causal interaction with other objects.[60] As a materialist I hold that every context is *in fact* a physical

---

[60] Borderline case: an eternal God located outside space and time might think in words, i.e., possess covert linguistic contexts. (In which language? Hebrew?) Any-

239

thing of some sort,[61] but a non-materialist, a dualist for example, could still accept my analysis of reference, though we hold different opinions about what certain contexts are made of. He would regard covert contexts as immaterial, since he holds that all mental entities are immaterial, but I would regard them as states of or events in the speaker's brain. No matter how these issues are to be decided, their resolution cannot affect my theory of reference, since all I require is that contexts be concrete things of some sort.

Sometimes one context has the same content as another and can be used for the same purposes. Examples of this are the contexts D28–D30, all of which could be used to make the same point about compound interest (a false one, as it happens). They have the same content because, even though they are referentially segregated, *within* the context the singular expressions count as singular terms and these singular terms purport to refer to exactly the same kind of situation: some man depositing $5,000 in a savings account at 5¼% interest compounded semiannually and after three years having $5,432.10 in his account. With respect to content they are perfect paraphrases of one another. If someone uttered D28 and I then commanded 'Say it another way!' and he uttered D29, he would have said it another way, obeying my command. Paraphrases can be understood as translations within a given language instead of between languages, and like translations they may convey the content of the original well or badly, may stick as closely as possible to the original forms of expression or be highly non-literal, may carry over the content of the original wholly or only in part, and so on. Like translation, paraphrase can decrease the number of distinct anaphoric chains[62] but it ought not to introduce any new ones (at least, not any new ones that were not already somehow implicit in the original). For example, D1 could be paraphrased by

D70: #A man fed a squirrel some peanuts.#

which leaves the tree out of the story, but not by

D71: #A man fed a squirrel some peanuts while a frog croaked loudly.#

one who finds this intelligible will have to reject my claim that all contexts are concrete objects.

[61] Taking 'thing' in a very broad sense, so that it covers such things as events and discontinuous individuals consisting of scattered parts.

[62] That is, distinct maximal anaphoric chains — ones which could not have additional singular expressions already present in the context added to them.

because this introduces a new character into the story.

Paraphrase and translation are processes of transferring content from one linguistic context to another, and if accurate they do not alter the content substantially, except possibly to reduce it.[63] I want to generalize this notion to cover all cases in which, starting from an antecedent context with a given content, there is produced a second context which has all or part of the content of the previous context and which introduces no new content.[64] "Translation" in this extended sense is thus a process involving two temporally related concrete objects such that beginning with the first one a second one with the same or, within limits, partly the same, content is produced.

Examples:

(a) The overt discourse D68 is a (very) partial translation of the covert discourse D67, and similarly for D61 and D62.

(b) Starting from a map of the Northern Hemisphere with Mercator Projection, I draw a map of the Northern Hemisphere with Simple Conic Projection.[65]

(c) Starting from a map I describe its content in words:

D72: #In the center there is a river running roughly north-south.

[63] 'Substantially' is vague. Minute inaccuracies are frequent, e.g., in translations of books, but there are no general criteria setting off bad translations from bad attempts so inaccurate as to be nontranslations. We can live with this vagueness, and assume that in each case where accuracy is at issue some appropriate criteria are in force.

[64] I call this process *translation* in order to underline the essential similarity to ordinary translation, which is itself a consequence of the essential similarity between linguistic and non-linguistic contexts. This is the sort of thing Wittgenstein described in the *Tractatus* (4.0141): "There is a general rule by means of which the musician can obtain the symphony from the score, and which makes it possible to derive the symphony from the groove on the gramophone record, and, using the first rule, to derive the score again. That is what constitutes the inner similarity between these things which seem to be constructed in such entirely different ways. And that rule is the law of projection which projects the symphony into the language of musical notation. It is the rule for translating this language into the language of gramophone records."

[65] The same job could be done by a machine. Human abilities like conscious deliberation are not essential to inter-contextual translation; all that is required is some law-governed process which, operating on things with content, consistently yields other things with content, within prescribed limits of accuracy and reliability. Furthermore, contexts can originate mechanically. Content-bearing objects containing elements capable of singular reference (in the broad sense), and capable of referential linkage with elements in other contexts, are, for example, produced by cameras. Once produced, they can be translated mechanically as well; e.g., a camera in a satellite photographs part of the earth's surface, the negative is processed automatically, the picture is transmitted to earth by radio, and ultimately an image appears on your television screen: hurricane in the Caribbean.

A large swamp lies to the west. It's shaped like an inverted iota. A heavily forested area lies east of the river.#

Most of the information in the map is lost in D72, so unlike (b) it is only a partial translation.

(d) Looking at a painting I describe the scene it depicts by saying:

D73: (overt): #A firing squad is about to execute several prisoners. The soldiers are pointing rifles with fixed bayonets at them, and they have terrified expressions on their faces.#

(e) In the same situation, D73 occurs covertly.

(f) I picture again to myself something I once saw and translate into words:

D74: #The apparition of these faces in the crowd; petals on a wet, black bough.#

This could be either overt or covert.

In all these cases we have antecedent contexts containing singular elements which are referentially linked to singular elements in contexts derived from them by translation. In (c), for example, the map contains, say, a blue line, and this is referentially linked with 'a river' and 'the river' in D72; a portion of the map shaded green is linked with 'a heavily forested area'; and so on. The map could (and typically would) contain singular elements not linked to anything in D72, such as symbols for buildings, roads, etc. With a different selection from the content of the map, different singular terms would have been used in the second context ('a building', 'a road'). In (d) the painting contains a blotch of paint which is linked to 'a firing squad' and 'the soldiers',[66] and in (f) there is some component of the memory image which is linked to 'these faces in the crowd'.

A singular element in the second context refers to whatever the corresponding element in the first context refers to. It is obvious that the denotationist analysis doesn't apply to such non-linguistic elements as maps and pictures.[67] Denotationism holds that they refer to — i.e., are

---

[66] "Plural terms" like 'the soldiers' can be referentially linked to a plurality of elements in an antecedent context. If C_1 has, e.g., # . . . a soldier . . . another soldier . . . another soldier . . . #, C_2 might contain 'several soldiers' or 'the soldiers' or 'the group of soldiers' or 'the group' or 'the squad' and so on. But a detailed analysis of plural terms would involve complications not worth pursuing here. (It should be noted that they count as singular terms in my sense, for they can be referentially linked with other terms in other contexts and ultimately with referents.)

[67] Of course, they can sometimes be referentially isolated and used in much the same way as the definite descriptions in D64–D66.

maps of or pictures of — whatever they uniquely denote, and not to anything else. They uniquely denote something if it and it alone is accurately represented by them (satisfies their descriptive content). Thus a picture of a face[68] uniquely denotes someone if and only if it shows what he looks like ("resembles" him, we usually say), and a map uniquely denotes a tract of land if and only if the latter is as represented by the map: river here, forest there, and so on. Yet clearly we could have a picture of someone painted by an incompetent (or playful, or nearsighted, or avant-garde) portrait artist which failed to denote its subject, much less denote him uniquely; despite the poor likeness, it would still be a portrait of him — would "refer" to him, in my sense. Even if someone else — say an inhabitant of another galaxy who won't be born for a million years — happens to be denoted by the painting, he will not be its referent and it will not be a portrait of *him*.[69] The same goes for maps: an incompetent cartographer might produce a highly inaccurate map of *this* territory, and it would only be a coincidence that it fits *that* one perfectly.

Case (f) involves an antecedent context of the kind I now want to discuss, namely non-linguistic covert contexts involved in memory, imagination, and perception. So far I have stressed the similarity between overt non-linguistic contexts, like pictures and maps, and linguistic contexts which have the same or partly the same content; we have already accepted covert linguistic contexts, the existence of which few would seriously dispute,[70] but covert pictures and maps — "mental images," "mental pictures" — are generally thought to be unfit for regular philosophical service. They are rejected for various reasons: because they are supposed to be essentially "private" and unknowable by anyone other than the person whose mind they are in; because there are no clear criteria for telling whether somebody else has one; because a brain surgeon rummaging through your skull will see nerve tissue but no pictures or

---

[68] Taking 'picture of a face' as a characterization of its content, with no implication that it is referentially connected with this face or that. 'Of' is ambiguous in this respect, and where necessary I will indicate which reading I intend. Goodman would signal the content-characterization reading by the expression 'face-picture'; for a discussion see *Languages of Art*, pp. 21–31.

[69] The distinction is especially apparent for photographs; see David Kaplan, "Quantifying In," *Synthese*, 19(1968): p. 198.

[70] Even behaviorists, abandoning their principles, have been known to accept them. Gilbert Ryle, for example, speaks of "your silent colloquies with yourself" which "I cannot overhear" (*The Concept of Mind* (London: Hutchinson, 1949), p. 184).

maps or images. I will take up these points one by one. First, the privacy objection. As a materialist I hold that covert contexts are physical objects — brain states or events — which are no more private or unknowable than any other physical objects, even though they cannot be casually detected by other people. If a dualist accepts everything else I have said but claims that covert contexts are immaterial entities made of "spook-stuff,"[71] that they are private and unknowable by others, that they are somehow indelibly subjective, then *he* will have problems, but they won't be problems for me.[72]

Second, the "criteria" objection. Suppose you are looking at a painting and I ask you to tell me what you see. It is a picture of a man wearing a plumed hat, and you say 'It's a man wearing a plumed hat.' If there is an epistemological gap between what you saw and what you said — if it was pure coincidence that you uttered words whose content matched that of the painting — then certainly you didn't *report* to me what its content was, and what you said was no indication that you knew what its content was. Conversely, if we concede that you were telling me what you *knew* about its content, that you did make a genuine report, then it would be irrational not to conclude that, between the stimulation of your retina by light from the painting and the articulation of the sounds you uttered, there were inner states having the same content as what you saw and what you said. Such inner states I call "covert contexts." We can at present identify them only in terms of their content, not in terms of their physical structure, but they are not unique in that respect: if a television camera scanned the same painting and if on the screen of a television set hooked up to it by cable there appeared a picture of that painting, then — barring miracles — we would have to conclude that between the lens of the camera and the image on the screen there were inner states of the system which we who know no electronics can describe only as states having the same content as the painting and the picture of it on the screen.[73] If we were to look inside the camera, or dissect the cable, we wouldn't see any pic-

---

[71] As J. J. C. Smart calls it.

[72] What still is a problem is the alleged certainty or incorrigibility of first-person reports on the content of experiences, but the solution, whatever it turns out to be, has no bearing on my claims about reference.

[73] When I perceive my own mental states, I identify them only in terms of their content, with no knowledge of the physical structure which "embodies" or "encodes" it. It is as if I could grasp the content of a painting without seeing the paint.

tures, but we would see things — wires, transistors — which in fact had the same content (at that moment) as the pictures we saw, just as a surgeon who looks at your brain sees no pictures but does see things — masses of nerve cells — which at that moment have the same content as certain pictures.[74] Such objects are what I am calling "covert pictures," "covert maps," "memory images," and so on — all perfectly respectable physical things, no more mysterious than inscriptions on paper or sound waves in the air.

Mental states such as memory, imagination, and perception have in each case a certain content.[75] If I remember, or seem to remember, the appearance of the house I lived in as a small child, then whether or not the house really looked like that, my present "memory image" (to use the everyday expression for it) gives the content of the memory: a one-story house, painted white, with a green roof, and so on. In general a necessary condition for remembering something is possessing a covert context with appropriate content, which can be translated, partially at least, into an overt context, either linguistic ('It was a one-story house . . .') or non-linguistic (e.g., by drawing a picture). I want to stress that I am concerned with memory contexts — whether they are linguistic or non-linguistic, visual, auditory, tactile, or whatever — only as objects having a certain content, and aside from that I make no claims about what their structure might or must be.[76] The same goes for imagination: any analysis will have to include the fact that a person who is imagining is imagining something of a particular sort, some possible thing or event or situation; that is, what he is doing involves a covert context with such-and-such content. If I imagine a golden mountain, then whatever else is going on in me, at least I possess a context whose content includes: golden mountain.

[74] An undeveloped negative has the same content as the print which will eventually be made from it, but you can't see what its content is; that is something you can find out only indirectly.

[75] That is, they have "intentionality." For a useful discussion, see the Chisholm-Sellars correspondence in Herbert Feigl, Michael Scriven, and Grover Maxwell, eds., *Concepts, Theories, and the Mind-Body Problem*, Minnesota Studies in the Philosophy of Science, vol. 2 (Minneapolis: University of Minnesota Press, 1958), pp. 507–539. For an interesting account of the views of Brentano and Meinong, see J. N. Findlay, *Meinong's Theory of Objects and Values* (Oxford: Oxford University Press, 1968), chap. 1, "The Doctrine of Content and Object," pp. 1–41.

[76] C. B. Martin and Max Deutscher call them "structural analogues"; their paper "Remembering" (*Philosophical Review*, 75(1966):161–196) presents a very interesting discussion of the problems involved in a causal analysis of memory.

Charles Chastain

Typically, when someone consciously perceives a physical thing or event, at least two things are happening: first, he is having an experience with a particular content; this determines what sort of thing he "seems to see" (hear, etc.). For example, if I perceive the full moon on a dark night in a cloudless sky, then normally I will possess a context whose content includes at least: white disk on a dark background. (It may, of course, include more, depending on how much detail I see, what other things are visible, what I make of the experience — e.g., not merely 'white disk' but 'moon' or 'goddess' — and so on.) Such contexts I will call *perceptual contexts* (subspecies: visual contexts, auditory contexts, etc.). 'Perceptual' is used noncommittally here: the person who possesses such a context may not be perceiving anything, but at least it appears to him that he is. Secondly, when someone perceives a physical thing or event there is a causal connection between his perceptual context and the thing he perceives: roughly, the occurrence of the former must be explained in terms of the latter. Grice has stated the case for a causal theory of perception,[77] and I won't repeat his arguments here, except to point out that a causal analysis is the only one likely to appeal to materialists. (Non-materialists may accept whatever theory appeals to them and still hold that perception involves having perceptual contexts, but they will have to find some other way of making intelligible the connection between the perceiver and what he perceives.) Again, I want to emphasize that the claims I make about perceptual contexts have to do only with their content and their relations with other contexts via translation and referential linkage, *not* with the ways they happen to be encoded in the brain (or, if you prefer, in some morsel of spook-stuff.) I don't have a materialist analysis of what it is for a perceptual context to have content, but perceptual contexts are no

[77] See H. P. Grice, "The Causal Theory of Perception," *Aristotelian Society Supplementary Volume XXXV* (1961); reprinted in Robert J. Swartz, ed., *Perceiving, Sensing, and Knowing* (Garden City, N.Y.: Doubleday, Anchor Books, 1965). For causal analyses of knowing in general, see Alvin I. Goldman, "A Causal Theory of Knowing," *Journal of Philosophy*, 64(1967):357–372; Peter Unger, "Experience and Factual Knowledge," *Journal of Philosophy*, 64(1967):152–173; and Peter Unger, "An Analysis of Factual Knowledge," *Journal of Philosophy*, 65(1968):157–170. For a non-technical survey of issues in the psychology of perception, see R. L. Gregory, *The Intelligent Eye* (London: Weidenfeld and Nicolson, 1970). Advanced discussions, from rather different points of view, are presented in James J. Gibson, *The Senses Considered As Perceptual Systems* (London: Allen and Unwin, 1968), and Ulric Neisser, *Cognitive Psychology* (New York: Appleton-Century-Crofts, 1967).

worse off in that respect than linguistic contexts; no one has a plausible materialist analysis of what it is for words and sentences to have content (sense, meaning, intension), but they still do.

Like other contexts, the non-linguistic covert contexts involved in memory, imagination, and perception[78] can contain singular elements which, like singular terms in discourses, can be referentially linked with other singular elements in other contexts. Suppose that I am now consciously remembering (or at least it seems to me that I am remembering) the house I lived in as a small child; there is a context which can be translated (fully or partially) into discourses like the following, which could appear either covertly or overtly:

D75: #It is a one-story house, painted white. It has a green roof. . . .#

D76: #The house has one story. It is painted white. It has a green roof. . . .#

D77: #There is a one-story house, painted white. It has a green roof. . . .#

'It' in D75, 'the house' in D76, and 'a one-story house' in D77 are all referentially linked to some singular element in the memory context which has the descriptive content: one-story house, painted white, with a green roof . . . This is parallel to cases (c), (d), and (e); a singular term in a linguistic context is linked with an element of an antecedent non-linguistic context and refers to whatever the latter refers to (taking 'refer' in the broad sense, of course). If in the memory context the element in question refers to object O, then O is the thing which is being remembered. It may or may not have had a green roof, may or may not have been painted white, may or may not have been a house: memories can be inaccurate, even highly inaccurate.[79] Denotation of O by a singular element in a memory context is not a necessary condition for that context's being (or containing, or being contained in)[80] a memory of O. And denotation, even unique denotation, is not a sufficient condi-

---

[78] The present discussion doesn't require any finer distinctions than these. A more detailed analysis would also have to take into account dreams, hallucinations, eidetic imagery, and so on.

[79] How inaccurate can they get without ceasing to be memories? A similar problem arises with perception; see below.

[80] Is the context identical with the memory or is it only one component of it? Can part of a memory context be a memory in its own right? It is not clear how these questions ought to be answered, so I will continue to use these terms rather loosely.

tion for memory: for the thing I remembered might in fact have had a brown not a green roof, or it might have been painted yellow not white, or it might have been a garage not a house, and there might have existed at the same time in another place one and only one house which I had never seen but which had just those properties, but nevertheless the latter would not have been the thing which I remembered. A memory must be causally connected with its object, which must, except in cases of time travel or precognition, exist prior to the memory.[81]

Although contexts involved in episodes of imagining, both linguistic and non-linguistic, can be referentially linked with later contexts — for I can remember and describe what I imagine — they cannot be linked with earlier ones (as long as it is an object which we are imagining, not just some fact about it).[82] Even apart from such linkages their singular elements cannot refer to any actual things.[83] Like fictional contexts, they are referentially segregated, and if a singular element happens to denote something uniquely, the resemblance is "purely coincidental."

This brings us again to perceptual contexts. Like all the others, they can be referentially linked to later contexts, overt and covert, linguistic and non-linguistic, including memory contexts. A person perceives something if and only if he possesses a perceptual context with a singular element which refers to that thing. You may take this as a definition: either of 'perceive', if you think you already understand 'refer', or of 'refer' (for perceptual contexts) if you think you already understand 'perceive'. In the example discussed earlier, I perceive the moon if and only if the white disk in my visual context — strictly speaking, the element in my visual context whose content is 'white disk'[84] — refers to the moon. Either side of this biconditional will be true if and only if there is a certain sort of causal connection between the thing perceived (= the referent) and the perceptual context. *Exactly* what sort of connection? As things stand at present with the theory of perception, one cannot say.[85]

Denotationism is, of course, false for perceptual contexts. What

---

[81] This doesn't cover every use of the words 'memory' and 'remember'; e.g., there is a sense in which I remember the meeting tomorrow if I remember that there will be a meeting tomorrow.

[82] See above, footnote 47.

[83] Leaving open the question of whether they refer to unactualized possibles.

[84] There would literally be a white disk in my visual context only if there were a white disk in my brain (or mind).

[85] See Grice, "The Causal Theory of Perception," V–VI.

counts is the causal pathway along which information passes from the object perceived to the perceptual context; it is this which determines the identity of the thing which is seen, heard, touched, smelled, etc. The information may be degraded or contaminated in transit or distorted by the perceiver, but still it is *that* object which is perceived and not some other one which, quite accidentally, happens uniquely to fit the content of the perceptual context. So, for example, if my visual context has the content white-disk-on-dark-background because I am looking through experimental lenses at a dark disk drawn on a piece of white paper, the latter is the thing I see, even though the way it seems to me is not the way it is; and even if the world happens to contain exactly one white disk on a dark background, which is not causally connected in the right way with my visual context, that thing, though it uniquely satisfies the content of my perceptual context, is not the thing I see. If in this situation I utter the singular term 'the white disk' referentially linked to the visual context, then what I am referring to is the dark disk. If a person utters a singular term which is referentially linked, directly or indirectly, to a singular element in an antecedent visual context, the questions 'What did he refer to?' and 'What did he see?' will have the same answer. In the case of hallucinations, dreams, and some kinds of optical illusion, the answer will be 'Nothing.' In other cases it won't be clear what the answer is. Here, for example, is a hard case described by G. E. M. Anscombe:

An interesting case is that of *muscae volitantes*, as they are called. You go to the doctor and you say: "I wonder if there is something wrong with my eyes or my brain? I see" — or perhaps you say "I *seem* to see" — "floating specks before my eyes." The doctor says: "That's not very serious. They're there all right" (or: "You see them all right") — "they are just the floating debris in the fluids of the eye. You are a bit tired and so your brain doesn't knock them out, that's all." The things he says you see are not *out there* where you say you see them — *that* part of your intentional description is not true of anything relevant; but he does not say that what you are seeing is that debris *only* because the debris is the cause. There really are floating specks. If they caused you to see a little red devil or figure of eight, we should not say you saw them. It may be possible to think of cases where there is nothing in the intentional object that suggests a description of what is materially being seen. I doubt whether this could be so except in cases of very confused perception — how could a very definite intentional description be connected with a quite different material object of seeing? In such cases, if we are

249

in doubt, we resort to moving the supposed material object to see if the blurred, not colour-true, and misplaced image of it moves.[86]

An even thornier case is provided by the "fortification figures" experienced by people who have migraine headaches:

The visual disturbance usually precedes the headache and can occur without any headache. It generally begins near the center of the visual field as a small, gray area with indefinite boundaries. If this area first appears during reading, as it often does, then the migraine is first noticed when words are lost in a region of "shaded darkness." During the next few minutes the gray area slowly expands into a horseshoe, with bright zigzag lines appearing at the expanding outer edge. These lines are small at first and grow as the blind area expands and moves outward toward the periphery of the visual field. The rate of expansion of both the arc formed by the zigzag lines and its associated band of blindness is quite slow: some 20 minutes can elapse between their initial appearance near the center of the visual field and their expansion beyond its limit. It is then that the headache usually begins, behind or above the eyes. It is the only unpleasant aftereffect of a spectacular visual display.[87]

The figures look about like this:

It is claimed that they are caused by a wave front of electrical activity moving across the visual cortex and interacting with neurons packed in a hexagonal lattice: "the advancing waves of disturbance draw continuous traces across the cortex and in less than half an hour reveal part of the secret of its neuronal organization."[88] I would count this as visual perception, since an element in a visual context is produced by an event in such a way as to convey information about it; the shape of the fortification figure and the speed at which it expands can be explained by the

[86] In her paper "The Intentionality of Sensation: A Grammatical Feature," in R. J. Butler, ed., *Analytical Philosophy, Second Series* (Oxford: Blackwell, 1965), pp. 178–179. An "intentional description" is one which would express the content of the perceptual context.

[87] Whitman Richards, "The Fortification Illusions of Migraines," *Scientific American*, May 1971, p. 89.

[88] *Ibid.*, p. 94.

properties of the waves and the way the neurons are packed together. Assuming that the migraine victim visually perceives events occurring in his visual cortex, does he literally *see* them? I am inclined to say that he doesn't. Seeing ought to involve light and eyes, or things functionally equivalent to them.

12

If the first (i.e., earliest) link in a referential chain is a singular element E in a perceptual context, and the last is a singular term T in an overt linguistic context, then the following are true under all the same circumstances:

(i)   the speaker perceives, via E, an object O;
(ii) E refers to O;
(iii) T refers to O;
(iv) the speaker in uttering T refers to O.

Sometimes the perceptual context occurs immediately before the utterance of the linguistic context, and the latter serves as a kind of "running commentary" on it; this is how D6, D22, and D62 can be accounted for. Sometimes the perceptual context occurred at some time in the past and is linked by memory to the present utterance.

However, this does not account for all cases of primary reference,[89] for I can refer to objects which I have never perceived. In the remainder of this section I will present an analysis which I believe is adequate and show how it applies to the most important cases.

Excluding cases of secondary reference, the following condition holds: a singular element E in a context C possessed or produced by a person P refers to an object O if and only if either (i) E in C is referentially linked with an element E' in an antecedent context C' and E' in C' refers to O, or (ii) P has *knowledge of* O via E in C. If C is an overt linguistic context, then P in uttering E is referring to or making a reference to O. Clauses (i) and (ii) are not mutually exclusive; e.g., to remember something is to have knowledge of it, but the singular element in the memory context is linked to an element in an antecedent context (or else it wouldn't be memory). C can be either overt or covert.

What does (ii) mean? What is it to have "knowledge of" something?

---

[89] I.e., primary reference by singular terms that are not referentially isolated, excluding cases like D64–D66 in which the term refers to whatever it uniquely denotes. I will not state this qualification explicitly in the following discussion.

If someone perceives something, he has knowledge of it. If he detects or observes it, he has knowledge of it. If he introspects it, he has knowledge of it. If he becomes aware of it by clairvoyance or telepathy, he has knowledge of it. If he precognizes it, he has knowledge of it. If he is in mystical rapport with it, he has knowledge of it. And so on. The notion of someone possessing "knowledge of" something has been neglected in recent epistemological theorizing, which has tended to focus exclusively on "knowing how" and "knowing that"; but it has appeared before in the theory of reference. Russell distinguished between "knowledge by acquaintance" and "knowledge by description," which he characterized as follows:

I say that I am *acquainted* with an object when I have a direct cognitive relation to that object, i.e. when I am directly aware of the object itself. . . . it is natural to say that I am acquainted with an object even at moments when it is not actually before my mind, provided it has been before my mind, and will be again whenever occasion arises. . . . When we ask what are the kinds of objects with which we are acquainted, the first and most obvious example is *sense-data*. When I see a colour or hear a noise, I have direct acquaintance with the colour or the noise. . . . we have also (though not quite in the same sense) what may be called awareness of *universals*. . . . It will be seen that among the objects with which we are acquainted are not included physical objects (as opposed to sense-data), nor other people's minds. These things are known to us by what I call "knowledge by description," which we must now consider. . . . What I wish to discuss is the nature of our knowledge concerning objects in cases where we know that there is an object answering to a definite description, though we are not *acquainted* with any such object. This is a matter which is concerned exclusively with *definite* descriptions. . . . I shall say that an object is "known by description" when we know that it is "*the* so-and-so," i.e. when we know that there is one object, and no more, having a certain property; and it will generally be implied that we do not have knowledge of the same object by acquaintance.[90]

This epistemological distinction runs parallel to an equally basic distinction in the theory of reference: *logically proper names* can be used simply and directly to refer to the objects of acquaintance, but everything else must be referred to by means of definite descriptions. A speaker can refer to something by uttering a logically proper name only if he has knowledge of it by acquaintance. Knowledge of the refer-

[90] Bertrand Russell, *Mysticism and Logic* (Garden City, N.Y.: Doubleday, Anchor Books, 1957), pp. 202–207 *passim*.

ent is a necessary condition for referring by means of logically proper names. What I am advocating is a kind of generalization of this principle which takes into account the facts about referential linkage and which applies to things which we cannot be acquainted with (in Russell's sense). What I call "singular terms" survive paraphrase as well as Russellian logically proper names do, and for the same basic reason: they refer, not because they uniquely denote their referent, but because they are connected with it in a more direct way, and this connection can hold even when denotation fails or when (as in the case of logically proper names) they have no descriptive content and hence no denotation. Any paraphrase which reveals their logical form will have to contain some expression which can sustain this kind of connection.

"Perception" is a loose notion. We can perceive physical objects and events external to us by seeing, hearing, smelling, tasting, and touching them; if I detect the presence of a burglar by seeing the marks he left on the window frame and smelling the smoke of his cigar, do I perceive *him*? Perhaps, perhaps not — but I do have knowledge of him, even if we don't know whether to call it perception. Seeing the marks is not seeing the burglar; smelling the smoke is not smelling the burglar; but if I see the marks and smell the smoke and say 'A burglar has just been here', then 'a burglar' and other terms anaphorically connected with it later on, such as 'the burglar' and 'he', refer to whoever left those marks and that smoke. If it turns out that they were not left by a burglar but by a policeman planting a hidden microphone in my house, then he is the referent; and he remains the referent even if there was also a burglar there who left no spoor. I *detected* the burglar — or whoever it was who was there — and even if this isn't perception in the ordinary sense it is close to it. (You might object that it differs from ordinary perception because there has been an inference from evidence; but ordinary seeing and hearing involve inferences, often highly complicated ones, though they are normally unconscious.[91])

Even if perception is taken broadly enough to cover such examples, there are others where it seems clear that nothing is literally perceived but there is — or at least there is alleged to be — knowledge of various objects. Russell claimed that we know universals by acquaintance ("though not quite in the same sense") and can on this basis refer to

[91] See the books by Gregory and Neisser in footnote 77. Analogous things can be said about 'observe', especially in scientific usage.

them by using logically proper names; but only metaphorically do we "perceive" them. I can't perceive something that doesn't exist yet, but sometimes I can have knowledge of an event which is about to happen. If I see a stick of dynamite attached to a burning fuse, I can by uttering 'the explosion which is about to occur here' refer to the explosion which occurs shortly thereafter, for it is within the scope of my knowledge. If the dynamite turns out to be a dud but a leaking gas main which I was completely unaware of explodes anyway, then even though that explosion is uniquely denoted by the definite description I was not referring to it: I had no knowledge of it.

The theory of reference merges on its periphery with the theory of knowledge. The denotationist analysis of reference is inadequate precisely to the extent that reference depends ultimately on knowledge of the referent by the speaker at the time of utterance, or at some earlier time, or (in cases of secondary reference) by someone else. One can have knowledge of something, by perception for example, even though one's conception of it — as given by the descriptive content of the singular element in the perceptual context — is highly inaccurate; a singular term which has that content will fail to denote the object in question, but it will still refer to it if it is referentially linked to the singular element by which the object is perceived.

There is some truth in denotationism — any popular philosophical theory has some truth in it — but for the most important uses of singular terms it fails to identify the principles by which the referent is selected. This inadequacy originated in Russell's epistemology: if the only concrete things which can be the referents of genuine singular terms are the objects of acquaintance, and acquaintance extends only to sense-data, then any non-sense-datum which is referred to at all will have to be referred to in some way that does not involve singular terms which survive paraphrase into a "logically perfect" language. The solution is the Theory of Descriptions: any sentence containing a definite description is understood as asserting that exactly one thing satisfies its descriptive content, that is, exactly one thing is uniquely denoted by it. We get at physical objects only by a semantic shot in the dark: we specify properties or relations and hope that they are uniquely exemplified. The limitations of this way of setting up referential connections were well understood by Russell:

When we, who did not know Bismarck, make a judgment about him, the description in our minds will probably be some more or less vague mass of historical knowledge — far more, in most cases, than is required to identify him. But, for the sake of illustration, let us assume that we think of him as "the first Chancellor of the German Empire." Here all the words are abstract except "German." The word "German" will again have different meanings for different people. To some it will recall travels in Germany, to some the look of Germany on the map, and so on. But if we are to obtain a description which we know to be applicable, we shall be compelled, at some point, to bring in a reference to a particular with which we are acquainted. Such reference is involved in any mention of past, present, and future (as opposed to definite dates), or of here and there, or of what others have told us. Thus it would seem that, in some way or other, a description known to be applicable to a particular must involve some reference to a particular with which we are acquainted, if our knowledge about the thing described is not to be merely what follows logically from the description. For example, "the most long-lived of men" is a description which must apply to some man, but we can make no judgments concerning this man which involve knowledge about him beyond what the description gives. If, however, we say, "the first Chancellor of the German Empire was an astute diplomatist," we can only be assured of the truth of our judgment in virtue of something with which we are acquainted — usually a testimony heard or read. . . . All names of places — London, England, Europe, the earth, the Solar System — similarly involve, when used, descriptions which start from some one or more particulars with which we are acquainted.

. . . knowledge concerning what is known by description is ultimately reducible to knowledge concerning what is known by acquaintance.

The fundamental epistemological principle in the analysis of propositions containing descriptions is this: *Every proposition which we can understand must be composed wholly of constituents with which we are acquainted.*[92]

If we accept this principle we will have to find ways of translating each sentence used in talking about concrete objects into one containing no singular terms except logically proper names referring to sense-data. If this is impossible — and it is now generally agreed that it is — then the principle will have to be abandoned, and with it Russell's view that only sense-data can be referred to by paraphrase-immune ("logically proper") singular terms. If singular terms refer to things other than the objects of immediate acquaintance, we have to explain how this is possible. Strawson and many others believe that some version of denota-

[92] *Mysticism and Logic*, pp. 209–212.

tionism will do; I have argued that it will not, and the present approach is proposed as an alternative.

It does not contain, nor is it committed to, any particular theory of how we have knowledge of objects or any theory of what objects are there for us to know. Any theory which correctly accounts for ordinary perception and everything analogous to it, up to and including such things as clairvoyance and mystical insight, if there are such things, will be compatible with this theory of reference. Anyone who accepts a non-causal theory of perception (as I do not) can accept what I say about the connection between perception and reference, though he will reject some of my examples. Anyone who believes (as I do not) that we have knowledge of abstract entities can argue that that puts us in a position to make singular references to them, and what he says will be fully compatible with this analysis. That is why it would be incorrect to call this a "causal" theory of reference, as has been suggested to me. If it must be pigeonholed it would be more accurate to call it an "epistemic" theory; what gives it a "causal" flavor is the incidental fact that the processes of inter-contextual translation and the formation of referential chains are intelligible to most people only on a causal analysis, as are perception and memory; but that is not essential to the theory.

13

What about "demonstratives" like 'this' and 'that'? What about other indexical expressions like 'I' and 'here' and 'now'? We have already seen that 'this' sometimes has whatever reference it has because of an anaphoric connection with an antecedent singular term (see D21 and D22 in section 5). 'This' and 'that'[93] can also be referentially linked to previous contexts, linguistic or non-linguistic, overt or covert. The most important cases are those in which they are linked to singular elements in perceptual contexts. If I see a tree in front of me, my perceptual context containing some singular element whose descriptive content is, say, 'tree with white blossoms along its branches', then I can say or think 'This is the most beautiful thing I've seen today', where 'this' is referentially linked to that element. Since the latter refers to that tree (= I perceive the tree by way of it), 'this' refers to the tree also; it is the thing whose beauty or lack of it fixes the truth-value of my remark. The connection

[93] And their plurals 'these' and 'those'; see above, footnote 66.

between the word and the tree is mediated by that element in the perceptual context; if it were not, we would be unable to explain why the word hooked up with the tree rather than any of the other things present: birds, people, shrubbery, houses, or what have you.

'This' accompanied by pointing is more complex. I point only for your your benefit, in order to establish a referential link between the word 'this' which I utter and some element E in your visual context; if I am successful, my remark that "this is a so-and-so" will be understood as predicating so-and-so-ness of the thing you see via E. There will be a number of such elements at a given time, for at a given time you see, or seem to see, a number of things; to zero in on the right one I introduce into your visual context a new element, one by whose presence there you see my finger (or a stick, or whatever); this new element (think of it as the "finger-percept") is geometrically related to E in such a way as to get you to single out E as the recipient of the linkage with 'this'. There are more involved cases to which the same sort of analysis applies (e.g., I point not at the thing but at a picture of it and say 'this', referring to the thing *through* the picture of it, not to the picture itself), but I won't go into those, nor will I discuss the nonvisual analogs of pointing.

'This' without pointing can produce the right linkage if the circumstances are right. I say 'This is Fred Schultz'; no other person is in sight; you link your Schultz-percept with the token of 'this'. The less prominent the objects are, the chancier such linkages become. The theory of reference need not concern itself very much with these cases, nor with other problems of reference *to* or *for* another. I cannot refer to O for you unless I can refer to it for myself, and all the interesting problems are already posed by the latter.[94]

It is almost correct to say that 'I' refers to whoever utters or inscribes it. "Almost" because we have to rule out some obvious exceptions: numerous occurrences of 'I' in *Lolita* were inscribed by Vladimir Nabokov but none of them refer to him; a translation of a speech in a foreign language might contain occurrences of 'I', and they would refer not to the translator who uttered them but to the original speaker, who uttered not 'I' but 'ich' (or whatever); and so on. The first case can be disposed

---

[94] This is why I don't think that Strawson's notion of "identification for an audience" can play the leading role he has cast it in; to understand identification we must already understand reference. (See above, section 2.)

of by ruling out occurrences of 'I' which are referentially segregated, as fictional occurrences are; in the second case the word uttered by the translator is linked to the word uttered by the original speaker and for that reason refers to him, not the translator.

When 'I' is uttered unsegregated and without linkage to an antecedent context, then it does refer to the person who utters it. But who is that? Compare:

D78: #After I die, I will be buried.#

D79: #After I die, I will no longer exist.#

(These could be either overt or covert.) The statements made by someone who utters D78 and D79 can both be true only if the reference of 'I' is not the same in both discourses. Taking an individual to be an aggregate of temporal stages, the referent of 'I' in D78 must, and the referent of 'I' in D79 must not, include some corpse-stages. Descartes considered excluding everything which is not a thinking-being-stage: ". . . it might indeed be that if I entirely ceased to think, I should thereupon altogether cease to exist."[95] The scope of 'I' can vary from one context to another, depending on what descriptive content it has: 'I [body]', 'I [person]', 'I [thinking being]', and so on. Thus 'I' is similar to some definite descriptions used denotatively, namely the ones which get at their referents by building on another singular term. In D64, starting with the referent of 'that door', we get the referent of 'the first man who comes through that door' denotatively: it is whoever bears to that door the relation of coming through it first. Starting with the present temporal stage of the speaker, we get the whole referent of 'I' denotatively, according to the descriptive content of the term in that context, e.g., all body-stages causally related to the present one and satisfying certain criteria of continuity.

'Here' behaves in a similar way. If it is referentially segregated, as in a work of fiction, it refers to no place at all. If unsegregated, it can be referentially linked with a singular term in some antecedent context, perhaps a previous occurrence of 'here', perhaps some other term which can refer to a spatial location ('the Amazon basin', '50°N, 40°W', 'this room', 'the place where my body is presently located', etc.). The latter can, but need not, refer to the place where the speaker is presently located, or to something in his vicinity; e.g., I point to a map of Yosemite

---

[95] *Second Meditation.*

Valley and say 'Here is where I would like to be.' What if 'here' occurs neither segregated nor linked? A popular answer is to say that in such cases 'here' refers to "the place of utterance." But what place is that? Strictly speaking, the utterance is a process occurring in the speaker's mouth and throat, but that region obviously doesn't comprise the whole of the referent of 'here' in very many cases. What else is to be included? This will vary from context to context, depending on what further content attaches tacitly to 'here'. Suppose I am talking to you long distance and I say: 'It's very cold here. That's why I've spent the day indoors. But something's wrong with the thermostat — it's very hot here.' The content of the first 'here' includes something like 'this city' or 'this region' and that of the second includes something like 'this room' or 'this building'. The vagueness of the additional content insures that the regions referred to will have equally vague boundaries (just as the vagueness of 'body' and 'person' insure that the referent of 'I' — that is, the speaker himself — will begin and end fuzzily).

As with 'here', so with 'now'. Referential segregation is possible, in which case 'now' will name no time. Referential linkage is possible, in which case 'now' can name some time other than the time of utterance. If unsegregated and unlinked, 'now' names "the time of utterance," but the latter's identity depends on further conditions which vary from case to case and are likely to be vague.

When a token of 'I' or 'here' or 'now' is neither segregated nor referentially linked with an antecedent term, must the speaker have knowledge of the referent, as claimed in section 12? The notion of self-knowledge is certainly obscure, but it seems likely that any plausible analysis of it will hold that a competent speaker who consciously and deliberately utters 'I' at least knows of his own existence. Thus 'I' appears to be no exception to my claim. What about 'here' and 'now'? The difficulty is in deciding what counts as knowledge of places and times. Can I perceive or observe or detect or witness or experience or be aware of a place or a time other than by perceiving, observing, etc., some object or event located in it or at it? If knowledge of a place or time merely amounts to knowledge of something located there, then 'here' and 'now' are not exceptions to my claim; the speaker will always have knowledge of at least one thing which is located in the place and at the time in question, for he himself is located there. If knowledge of a place or time amounts to more than knowledge of something located there, then we will have to wait and see what more is required before deciding whether 'here' and 'now' — in the relatively infrequent cases in which they are not referentially linked even to an antecedent perceptual context — can refer to things of which the speaker has no knowledge.

14

We have already seen that a proper name can be introduced *de novo* into a linguistic context by means of an identity statement:

D17:   #A certain bookie I know whom I'll just call "Harry" . . . #

'Harry' is anaphorically linked to 'a certain bookie I know' and has whatever reference it has. Denotationism cannot account for proper names which get their reference this way, since it cannot explain the reference of the definite descriptions, pronouns, etc., with which such proper names are linked. Proper names can, of course, be linked to definite descriptions used denotatively — e.g., in D64 we could have had something like: 'the first man through that door — call him "Primus" ' — just as they can be referentially segregated, as in fiction, but such uses don't pose any new problems; neither do examples of proper names with secondary reference, although the latter provide some especially striking counter-examples to denotationism. (See below.)

A proper name can also be linked to singular elements in earlier nonlinguistic contexts, for example perceptual contexts. I see the moon and call it 'Luna'; the name is linked to the singular element in my visual context by which I see the moon and which, say, has the descriptive content 'white goddess in the sky' (that is, I see the moon as a white goddess in the sky). 'Luna' refers to the object I see: the moon, not some goddess. The name 'Luna' denotes nothing, but refers to the moon, because it is referentially linked with an element in my perceptual context which is in turn causally related to the moon. (Referential linkages are themselves to be given a causal analysis, of course, but I have nothing useful to say on that topic.)

What about "pure" cases of proper names, ones which don't derive their reference from linkages with other singular terms? Several denotationist theories have been developed to account for them, and they all pursue the same general strategy: (i) find, or construct, a context containing all occurrences of the proper name which according to certain criteria are relevant to determining its reference; (ii) determine the content of the name as it appears in this context; (iii) find something which uniquely satisfies the content (or a certain part of the content) of the name — that unique *denotatum* will be the referent.

In an interesting paper entitled "Substances Without Substrata,"[96] Neil Wilson proposes the following procedure to find the referent of a proper name, as used by a given individual at a given time:

Let us suppose that somebody (whom I am calling "Charles") makes just the following five assertions containing the name "Caesar." Let us suppose in addition that we know the significance which Charles attaches to expressions other than "Caesar" and that, in the beginning at least, we are ignorant of Roman history.
(1) Caesar conquered Gaul.
(2) Caesar crossed the Rubicon.
(3) Caesar was murdered on the Ides of March.
(4) Caesar was addicted to the use of the ablative absolute.
(5) Caesar was married to Boadicea.

. . . We have Charles' five assertions. We now conduct an empirical investigation, examining all the individuals in the universe. We might suppose that Charles intends the word "Caesar" to signify or designate Prasutagus (who, as every schoolboy knows, is the husband of Boadicea). On this supposition (5) could be called true and all the rest would have to be called false. Or we might suppose that "Caesar" signifies the historical Julius Caesar, in which case (1)–(4) could be called true and (5) would have to be called false. There do not seem to be any other candidates since any number of persons must have conquered Gaul and/or crossed the Rubicon and/or used the ablative absolute to excess. And so we act on what might be called the Principle of Charity. We select as designatum that individual which will make the largest possible number of Charles' statements true. In this case it is the individual, Julius Caesar. We might say the designatum is that individual which satisfies more of the asserted matrices containing the word "Caesar" than does any other individual.

In my terminology, this amounts to the following: (i) We construct a discourse consisting of all the assertions containing the name in question which the speaker would be willing to make (he may or may not have actually uttered this discourse), e.g.:

D80: #Caesar conquered Gaul. He crossed the Rubicon. He was murdered on the Ides of March. He was addicted to the use of the ablative absolute. He was married to Boadicea.#

(ii) We decide what is the descriptive content of the name in that con-

[96] *Review of Metaphysics*, 12(1959):521–539; the quotes are from pp. 530–532.

text (the whole story about "Caesar"). (iii) We find out what individual in the universe satisfies more of the descriptive conditions than does any other individual. That individual need not be uniquely denoted by all of the conditions taken together, but he must be uniquely denoted by a greater portion of them than the nearest rival. If no individual satisfies any of the conditions, or if two or more individuals are tied for first place in satisfying them, then the name has no referent. (As Wilson points out, we might wish to weight the conditions for importance, or to count some of them as absolutely essential, but these complications don't affect the main point.)

This sort of approach makes possible a denotationist account of proper names in which they are not merely abbreviated definite descriptions but have an independent status. Proper names could thus be used to initiate referential chains as well as to continue those started by other terms. John Searle proposes a similar procedure:

. . . though proper names do not normally assert or specify any characteristics, their referring uses nonetheless presuppose that the object to which they purport to refer has certain characteristics. But which ones? Suppose we ask the users of the name "Aristotle" to state what they regard as certain essential and established facts about him. Their answers would be a set of uniquely referring descriptive statements. Now what I am arguing is that the descriptive force of "This is Aristotle" is to assert that a sufficient but so far unspecified number of these statements are true of this object. . . . The question of what constitutes the criteria for "Aristotle" is generally left open, indeed it seldom in fact arises, and when it does arise it is we, the users of the name, who decide more or less arbitrarily what these criteria shall be. If, for example, of the characteristics agreed to be true of Aristotle, half should be discovered to be true of one man and half true of another, which would we say was Aristotle? Neither? The question is not decided in advance.[97]

No matter how ties are to be judged, the referent of 'Aristotle' must satisfy at least *some* of the agreed conditions:[98]

. . . if a classical scholar claimed to discover that Aristotle wrote none of the works attributed to him, never had anything to do with Plato or Alexander, never went near Athens, and was not even a philosopher but was in fact an obscure Venetian fishmonger of the late Renaissance,

[97] "Proper Names," in Charles E. Caton, *Philosophy and Ordinary Language* (Urbana: University of Illinois Press, 1963), pp. 158–159.

[98] As Searle asserts in his *Encyclopedia of Philosophy* article on names (vol. 6, p. 490).

then the "discovery" would become a bad joke. The original set of statements about Aristotle constitute the descriptive backing of the name in virtue of which and only in virtue of which we can teach and use the name. It makes sense to deny some of the members of the set of descriptions of the bearer of the name, but to deny them all is to strip away the preconditions for using the name at all.

Presumably, the referent of the name must also satisfy *more* of the conditions than any other individual (for then why wouldn't that other individual be the referent?) — as required by Wilson's Principle of Charity.

Similar views have been developed by P. F. Strawson[99] and Paul Ziff.[100] Can any such view be correct? Remember that such an analysis cannot account for proper names which are referentially linked to antecedent singular terms (or to singular elements in non-linguistic contexts), for those names take whatever referents their antecedents have, whether or not the referent satisfies their descriptive content, or satisfies more of it than any other object. If I christen the moon 'Luna' (e.g., link it to a moon-percept in a visual context) and then proceed to use the name with the content 'white goddess in the sky', it nevertheless names the moon, which is no goddess, and would not name any white goddess who happened to be lurking, unglimpsed by me, in the sky that night. Similar examples can be constructed in which one's use of the name is linked to a memory context which derives from some past encounter with the object but which has with time become very inaccurate. The present memory and the linguistic context based on it determine the content of the name, which because of the inaccuracy doesn't denote the object, but the connection with the past experience of the object makes it *that* object which is named by the name.[101] If we consider cases of secondary reference — which can be understood as a kind of social analog to individual memory — the same kind of examples crop up, such as this case described by Donnellan:

. . . The sort of description generally mentioned as helping to pick out,

[99] See *Individuals*, pp. 190–192.
[100] See his *Philosophic Turnings* (Ithaca, N.Y.: Cornell University Press, 1966), pp. 94–95.
[101] For an example of how a proper name could be based on confused memory, see Keith Donnellan, "Proper Names and Identifying Descriptions," *Synthese*, 21 (1970):343. His discussion of denotationist theories of proper names based on what he calls the "principle of identifying descriptions" applies to proper names the insights about reference contained in "Reference and Definite Descriptions."

say, Thales, is such as 'the Greek philosopher who held that all is water'. Nothing is made of the fact that such descriptions are given by us
derivatively. We might be pardoned if we supposed that the referent
of 'Thales' is whatever ancient Greek happens to fit such descriptions
uniquely, even if he should turn out to have been a hermit living so
remotely that he and his doctrines have no historical connection with
us at all.

But this seems clearly wrong. Suppose that Aristotle and Herodotus
were either making up the story or were referring to someone who neither did the things they said he did nor held the doctrines they attributed to him. Suppose further, however, that fortuitously their descriptions fitted uniquely someone they had never heard about and who
was not referred to by any authors known to us. Such a person, even if
he was the only ancient to hold that all is water, to fall in a well while
contemplating the stars, etc., is not 'our' Thales.

Or . . . suppose no one to have held the ridiculous doctrine that all
is water, but that Aristotle and Herodotus were referring to a real person — a real person who was not a philosopher, but a well-digger with
a reputation for saying wise things and who once exclaimed, "I wish
everything were water so I wouldn't have to dig these damned wells."
What is the situation then regarding our histories of philosophy? Have
they mentioned a non-existent person or have they mentioned someone
who existed but who did not have the properties they attribute to him?
My inclination is to say the latter.[102]

And thus we can see how Aristotle *could* turn out to have been an
obscure Venetian fishmonger of the late Renaissance, if enough docu-
ments we now think are genuine and accurate happen in fact to be
spurious or mistaken. (Recall Russell's example of the world having

---

[102] "Proper Names and Identifying Descriptions," pp. 352–353. This kind of example, in which a remote historical figure turns out to be totally different from
what we thought, was first pointed out to me by Saul Kripke, as a counter-
example to denotationism in which the speaker does *not* have knowledge of the
referent. His example concerned Jonah, who is supposed to have gone to sea, been
thrown overboard, been swallowed by a great fish, etc., but who might in fact have
been a landlubber about whom people told tall tales which were eventually recorded
in the Book of Jonah and about whose life we actually know nothing at all. Kripke's approach to the theory of reference is presented in his "Naming and Necessity," in Donald Davidson and Gilbert Harman, eds., *Semantics of Natural Language*
(New York: Humanities, 1972). (This paper was written before I had an opportunity to read "Naming and Necessity," and what I have said about Kripke's views
derives from a conversation with him; I don't think, however, that there is any
important difference between the former and the latter.) A similar "causal" theory
of secondary reference is developed by David Kaplan in "Quantifying In."
Although examples of secondary reference would refute denotationism even if
nothing else did, I don't believe that it poses any important problems not already
present when the referential chains are confined to a single speaker.

been created five minutes ago, complete with prefabricated dinosaur bones, histories of philosophy, etc.)

As with perception and memory, names which hook up with their referents via secondary reference are counter-examples to any denotationist analysis. (Unlike the former, they do not satisfy the condition that the speaker have knowledge of the referent.) Is anything left for denotationism to be correct about?

First, we have to exclude names of historic figures like Thales and Aristotle, whom we can pick out only by making secondary references. (For the sake of brevity I am ignoring proper names of things other than people.) Second, we have to exclude many names of contemporary persons of whom the speaker has no knowledge but whose names he has heard or seen, often in a context which supplies them with minimal descriptive content (e.g., 'John Smith, San Francisco' at the bottom of a petition). Third, we have to exclude all names which are linked to present or past perceptual contexts containing singular elements referring to the persons named; this eliminates all names whose use is based on personal encounters with their referents. Fourth, we have to exclude all names which are linked to non-linguistic contexts like photographs, TV images, and the like. But we can stop here, for it is obvious that virtually all of the names uttered by speakers of natural languages have already been eliminated. Might there nevertheless be some examples, however remote, of proper names for which the denotationist analysis is correct? Perhaps something like the following: several outstandingly horrible murders are committed under similar conditions, with the same technique, against the same kind of victim. We begin to say that the city is menaced by "Jack the Ripper." We don't claim that Jack the Ripper committed *all* of the murders, for some of them might be due to coincidence, or to imitators of the earlier ones, or to persons with saner motives using the Ripper murders as a cover. We don't claim of any given one of the murders that *it* was committed by Jack the Ripper — not even the first one, for it is perfectly possible that somebody else did that one and thus unwittingly inspired Jack the Ripper to do all the others. But we do claim that Jack the Ripper committed a substantial number, probably most, of them, and that he certainly committed more of them, probably a lot more of them, than anyone else. If we now assert that Jack the Ripper is a raving lunatic, what we say is true if and only if there is someone sat-

265

isfying those conditions and *he* is a raving lunatic. If half the murders were committed by A and half by B, then we might refuse to concede Ripperhood to either of them. If the murders turn out to have been committed by a woman (Jill the Ripper?), we might say that there was no such person as Jack the Ripper, on the grounds that maleness is essential to him. The denotationist analysis of proper names fits this kind of example perfectly, but it is easy to see how rare this kind of example is, how rare it is for the referent of a proper name always to be lurking unobserved just beyond our view, never quite definitely pinned down to this or that alleged fact about him, never identified as someone already known to us. Is this an exception to my claim that primary references generally involve knowledge of the referent? What *do* we know about Jack the Ripper? Do we really know anything about him? I am inclined to say that if he exists at all then we have knowledge of him: we detected his presence among us even though we never actually observed him. (To simplify matters I assume that none of us ever observed at any other time the person who was Jack the Ripper.) It is like the example in section 12 of the burglar who leaves traces but whom I never actually perceive. The distinction between having knowledge of a thing and merely having knowledge of its effects is a vague one, as is the distinction between seeing a thing and merely seeing some of its effects (or some of its parts). The notions of perceiving, observing, detecting, "finding," or in some other way coming to have knowledge of an object may depend for their general utility on the fact that we generally can get a firmer grip on the objects we are interested in than was ever gotten on Jack the Ripper.

Finally, I must correct an oversimplification which I have indulged in for the sake of orderly exposition. All along I have spoken as if a singular term which is introduced by way of an identity statement linking it anaphorically with another term in the same context, or which is referentially linked to an antecedent singular term or other singular element in an earlier context, were linked with only one such antecedent, which would provide the only possible route connecting it with its referent. In other words I have assumed that all referential chains look like this

$$T_1 \longrightarrow T_2 \longrightarrow T_3 \longrightarrow \cdots$$

266

and that none look like this

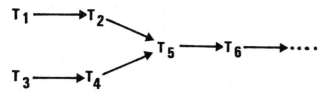

with two or more chains converging at a single point where a new term is introduced. Yet this assumption is obviously false. Suppose that my use of a proper name is linked to independent earlier uses of it by two different people, who I assumed were referring to the same person. If my assumption was correct, then all is well: both paths lead back to the same referent, which is thus the referent of the tokens of the name which I utter. But what if they lead back to different people? To whom am I referring? Both persons? Neither?

Consider an example not involving secondary reference. I see someone enter a building and a minute later I see someone who looks and is dressed exactly like him come out. Later on I utter (or think) the definite description 'the man I saw at City Hall', where this term is linked both to the visual context by which I saw someone entering the building and the visual context by which I saw someone leaving. (They are given equal weight.) If the two percepts hook up with the same person, then all is well: he is the referent. But what if the person I saw entering was the identical twin of the person I saw leaving? Am I referring to both? To neither?

If two or more referential chains converge at the point where a new term is introduced, I will call the latter *multiply linked*. Each of the terms to which the new term is linked could itself have been introduced by a multiple linkage, and multiple linkages obviously can involve many more than two antecedent terms, so that the possibility of there being some discrepancies in the pedigree of a singular term involving memory or secondary reference is very real indeed. Thus it is entirely possible that some name I now use with no qualms is connected with several different objects, as a result of my own or somebody else's past mistakes, but that I will never be in a position to discover this fact: skepticism about one's references is as easy to slide into as skepticism about one's claims to know facts about the things one is referring to. When such a mistake has been made, am I unwittingly

267

talking about two (or more) different people and perhaps uttering truths about one of them and falsehoods about the other? Or am I talking about neither and thus about no one at all?

If one found out about the mistake, then perhaps in most cases common sense would find a way of repairing the damage, e.g., by disconnecting my present token of 'Franklin Roosevelt' from a single maverick referential chain which leads back to Teddy Roosevelt. But what if the mistake is never discovered? It is irrelevant that *if* it were discovered we would know how to correct it and would subsequently be able to make a definite reference to a single object, for the problem is: what are we *now* referring to with the uncorrected version? Furthermore, some mistakes seem to be wholly incorrigible, no matter how much we might find out about the circumstances surrounding them (e.g., my example about 'the man I saw at City Hall'). Such attempted references would have to be abandoned altogether.

Do we live in a world in which such mistakes occur very often? Minor mistakes like the one about 'Franklin Roosevelt' can simply be ignored, even though it seems likely that they occur frequently, just as we ignore the fact that there are no perfectly straight lines or perfectly flat surfaces in the world and continue to employ geometrical concepts which presuppose that there are. Major mistakes, irreparable ones which spoil the reference beyond repair, may not occur very often in our world, at least among the more rational inhabitants of it, but if they do there are no fail-safe devices to insure that the references go through anyway. To suggest that a version of Wilson's Principle of Charity be employed here, that the referent of the term is the one which would make the greatest number of our assertions true, looks like an evasion of the problem, a desperate attempt to insure that despite our cognitive mishaps we are still referring to *something*. I see no reason why the theory of reference should be loaded in favor of our generally saying things which are true (or at least truth-valued), no more than the theory of knowledge should be loaded in favor of our generally remembering things as they were, seeing them as they are, or predicting them as they will be. Multiply linked singular terms are inherently risky, like memory, perception, and induction.

15

Let me conclude this paper by borrowing the words with which Carnap concludes *Meaning and Necessity*. It seems to me that his remarks, although written more than twenty-five years ago, still apply to the present situation (except, perhaps, that instead of 'the best method' I might prefer to say 'the true theory'):

Let me conclude our discussions by borrowing the words with which Russell concludes his paper ["On Denoting"]. It seems to me that his remarks, although written more than forty years ago, still apply to the present situation (except, perhaps, that instead of 'the true theory' I might prefer to say 'the best method'):

'Of the many other consequences of the view I have been advocating, I will say nothing. I will only beg the reader not to make up his mind against the view — as he might be tempted to do, on account of its apparently excessive complication — until he has attempted to construct a theory of his own on the subject of denotation. This attempt, I believe, will convince him that, whatever the true theory may be, it cannot have such a simplicity as one might have expected beforehand.'

Language, Thought, and Communication

I shall discuss two apparently conflicting views about our use of natural language. The first view, that language is used primarily in thought, has rarely been given explicit formulation but may be associated with the theories of W. V. Quine and Wilfrid Sellars. The second view, that language is used primarily in communication, has been explicitly put forward by Noam Chomsky, J. A. Fodor, and J. J. Katz, among others, and may also be associated (I think) with the theories of Paul Ziff and Donald Davidson. I shall describe each view and then try to say where I think the truth lies.

1. The view that language is used primarily in thought. This view is not that all or even most thinking or theorizing is in some natural language. We may reasonably suppose that animals think, that children can think before they learn a natural language, and that speakers of a natural language can have thoughts they cannot express in language. The view is rather that anyone who fully learns a natural language can and does sometimes think in that language. More precisely, it is that some of a speaker's so-called propositional attitudes are to be construed as, at bottom, attitudes toward sentences of his language. A speaker of English may believe that the door is open by believing-true the sentence, "The door is open." Another may fear that the door is open by fearing-true "The door is open." A third may think of the door's being open by adopting the appropriate attitude toward "The door is open."

Strictly speaking, sentential attitudes involve sentences conceived under one or another, more or less detailed grammatical analysis. I shall return to this point near the end of this essay. For now, I shall speak loosely of sentential attitudes as attitudes toward sentences.

The view that the primary use of language is in thought has roughly the following implications for the theory of communication. Linguistic

AUTHOR'S NOTE: The preparation of this paper was supported in part by grants from NEH and NSF.

communication is the communication of thought. The parties involved typically communicate with the language they use in thinking. The words used to communicate a thought are the same as or similar to those one "says to oneself" when one has that thought. Linguistic communication does not typically require any complicated system of coding and decoding. Our usual translation scheme for understanding others is what Quine calls the homophonic one. Words are used to communicate thoughts that would ordinarily be thought in those or similar words. It is true that allowance must sometimes be made for irony and other such devices; but in that case the thought communicated is some simple function of what would be normally communicated by a literal use of those words.

More precisely, linguistic communication typically involves communication of what is sometimes called "propositional content." A speaker says, "The door is shut," 'Shut the door," "Is the door shut?" or some such thing. He does so in part to get his hearer to think of the door's being shut. This first view holds that in such a situation, if communication is successful, the hearer will think of the door's being shut by adopting the appropriate sentential attitude. We might say that the hearer attends to the sentence, "The door is shut," where this is a technical sense of "attends to."

Notice that the claim is not that a person can think of the door's being shut only by attending to the sentence, "The door is shut." He might instead attend to an "equivalent" sentence, where the relevant sort of equivalence is that discussed near the end of this paper; he might even attend to a nonlinguistic representation that is part of a system of representation he uses in thinking, as long as the representation is relevantly equivalent to "The door is shut" in English. The claim is rather that normally, when a speaker successfully communicates in English by saying, "The door is shut," etc., the hearer thinks of the door's being shut by attending to the English sentence, "The door is shut." There will normally be a relatively simple grammatical relationship between the sentence the speaker uses to communicate certain "propositional content" and the sentence to which the hearer must attend if communication is to be successful. The "deep structure" of the latter sentence is the same as or a part of the deep structure of the former. Since sentential attitudes involve sentences conceived under grammatical analyses, it is sufficient that the hearer should attend to the sentence uttered conceived under the appropriate analysis. In this sense, linguistic communication does not typically make use of complicated principles of coding and decoding and our usual translation scheme is the homophonic one. The hearer need only hear the sentence uttered as having the appropriate structural description. He does not need to go on to translate the sentence, under that description, into anything else in order to understand it.

Proponents of the view that language is primarily used in thought can point out that, although one might use a natural language as a code, so that one's listeners would have to use complicated principles of decoding in order to understand what has been said, this would not be an ordinary case of linguistic communication. They would also point out that, when a person learns a second language, he may at first have to treat the new language as a code; but hopefully he soon learns to think directly in the second language and to communicate with other speakers of that language in the ordinary way, which does not involve complex coding and decoding or any sort of translation.

Furthermore, proponents can say, when a person thinks out loud, it is not always true that he has to find a linguistic way to express something that exists apart from language. Without language many thoughts and other propositional attitudes would not even be possible. In learning his first natural language, a child does not simply learn a code which he can use in communicating his thoughts to others and in decoding what they say. He acquires a system of representation in which he may express thoughts made possible by that very system. This is obvious when one acquires for the first time the language of a science or of mathematics. The claim is that it is no less true when one learns his first natural language.

That provides a rough sketch of the view that language is used primarily in thought. I shall say more about that view below. Now I want to describe the apparently conflicting theory that communication provides the primary use of language.

2. *The view that language is used primarily in communication.* J. J. Katz states explicitly a view that is implicit in many things said by other linguists:

Roughly, linguistic communication consists in the production of some external, publicly observable, acoustic phenomenon whose phonetic and syntactic structure encodes a speaker's inner, private thoughts or ideas and the decoding of the phonetic and syntactic structure exhibited in such a physical phenomenon by other speakers in the form of an inner, private experience of the same thoughts or ideas.[1]

Katz takes seriously the notion that linguistic communication ordinarily and typically involves such coding and decoding. He does not agree that typically one thinks in language and that the most usual "code" used in communication is the homophonic one. He does not believe that the words used in communication are usually the same as or similar to those that make up the thought communicated. As a result he sees linguistic communication as a relatively complicated business and he takes the main use of language to be its use in communication:

Natural languages are vehicles for communication in which syntactically structured and acoustically realized objects transmit meaningful messages from one speaker to the other. . . . The basic question that can be asked about natural languages is: what are the principles for relating acoustic objects to meaningful messages that make a natural language so important and flexible a form of communication?[2]

I think that in theory Noam Chomsky rejects Katz's view of communication and natural languages. He takes the primary function of language to be its use in the free expression of thought. He speaks approvingly of Humboldt's emphasis on the connection between language and thought, especially the way in which a particular language brings with it a world view that colors perception, thought, and feeling. According to Chomsky's description of Humboldt's view, to have a language is to have a system of concepts

and it is the place of a concept within this system (which may differ somewhat from speaker to speaker) that, in part, determines the way in which the hearer understands a linguistic expression . . . [T]he concepts so formed are systematically interrelated in an "inner totality," with varying interconnections and structural relations . . . This inner totality, formed by the use of language in thought, conception, and expression of feeling, functions as a conceptual world interposed through the constant activity of the mind between itself and the actual objects, and it is within this system that a word obtains its value . . . Consequently, a language should not be regarded merely, or primarily, as a means of communication . . . and the instrumental use of language

[1] J. J. Katz, The Philosophy of Language (New York: Harper and Row, 1966), p. 98.
[2] Ibid. Cf. J. J. Katz, "Recent Issues in Semantic Theory," Foundations of Language, 3(1967):125.

(its use for achieving concrete aims) is derivative and subsidiary. It is, for Humboldt, typical only of parasitic systems (e.g. . . . the lingua franca along the Mediterranean coast).[3]

Chomsky's approval of Humboldt suggests that he would accept the view that language is used primarily in thought and would reject Katz's remarks about language and communication. But other speculations of Chomsky's seem to make sense only if he does at least unconsciously accept a view like Katz's, only if he does see communication as the central function of language, only if he does think linguistics is primarily concerned with the speaker-hearer rather than the thinker, and only if he does take communication to involve a complicated process of decoding.

Consider the view, held by both Katz and Chomsky, that speakers of a language have (unconscious) knowledge of the grammatical rules of the language. We would not ordinarily say that a typical speaker of English has knowledge of the rules of a transformational grammar of English. We would ordinarily attribute such knowledge to a grammarian if to anyone. It has not been easy to discover why Chomsky and Katz attribute such knowledge to speakers generally. I and others have hypothesized that Chomsky and Katz confuse *knowing how* to use a language that is described by the grammatical rules with *knowing that* the grammar is described by those rules.[4] I have also suggested that Chomsky may be confusing "knowing that certain sentences are grammatically unacceptable, ambiguous, etc., with knowing the rules of grammar by virtue of which sentences are unacceptable, ambiguous, etc."[5] I have also suggested that Chomsky fails "to distinguish principles that account for how a person's mind works from principles a person accepts as true."[6]

Although these suggestions may contain part of the truth, they do not account for the tenacity with which Chomsky (for one) holds on to

[3] Noam Chomsky, "Current Issues in Linguistic Theory," in J. A. Fodor and J. J. Katz, eds., *The Structure of Language* (Englewood Cliffs, N.J.: Prentice-Hall, 1964), pp. 58–59. See also Chomsky's remarks about Humboldt in his book *Cartesian Linguistics* (New York: Harper and Row, 1966).

[4] Three authors endorse this suggestion in a single issue of *Journal of Philosophy*, 64 (1967): N. L. Wilson, "Linguistic Butter and Philosophical Parsnips," pp. 55–67; Henry Hiz, "Methodological Aspects of the Theory of Syntax," pp. 67–84; and Gilbert Harman, "Psychological Aspects of the Theory of Syntax," pp. 75–87.

[5] Harman, "Psychological Aspects of the Theory of Syntax," p. 82.

[6] Harman, review of Chomsky's *Cartesian Linguistics* in *Philosophical Review*, 72 (1968):234.

the view that speakers have (unconscious) knowledge of the rules of grammar. In a paper presented to the 1968 New York University Institute of Philosophy[7] Chomsky surveys criticism of his views and denies they are based on the above confusions. He acknowledges that it would be "absurd" to suppose a typical speaker to have knowledge *that* the language is described by certain rules of grammar and claims he wants to say only that a speaker has knowledge *of* the rules of grammar. On the other hand, there and elsewhere Chomsky describes the speaker's knowledge of the rules as "a system of beliefs" and as a "theory"; and he says that a child who acquires knowledge of the rules "determines *that* the structure of his language has the specific characteristics that empirical investigation of language leads us to postulate . . ."[8] Surely it will be interesting to see what has led Chomsky to such a position.

Simple common sense can account for his denial that a typical speaker of a language has unconscious knowledge *that* the language is described by certain rules of grammar. But it is difficult to say what leads him simultaneously to assert that the speaker has knowledge *of* the rules, knowledge which consists of a system of beliefs or of a theory that the speaker developed in the course of *determining that* those rules describe the language. Chomsky denies that he makes the mistakes suggested above. We can only take him at his word. Therefore we must look elsewhere for an explanation of his view.

I suggest we need look no farther than Katz's book on the philosophy of language. It is easy to see why he attributes to speakers unconscious knowledge of the rules of grammar. He takes linguistic communication to involve complex coding and decoding. Since such communication typically involves novel messages not previously encountered, he thinks speakers and hearers will need to know rules that relate sequences of code to messages. Such rules are provided by transformational generative grammars that define the sound-meaning relationship for the language in question. Therefore he thinks that speakers and hearers must have knowledge of rules that are equivalent to the rules of an adequate transformational generative grammar of the language.

[7] "Linguistics and Philosophy," in *Language and Philosophy*, ed. Sidney Hook (New York: New York University Press, 1969), pp. 51–94.

[8] This last remark is from Chomsky's part in a symposium on "Recent Contributions to the Theory of Innate Ideas" (*Boston Studies in the Philosophy of Science*, vol. 3, p. 86). Most of Chomsky's N.Y.U. paper is devoted to clarifying that earlier paper and responding to remarks in that symposium by Putnam and by Goodman. (The emphasis on *that* is mine.)

Gilbert Harman

To understand the ability of natural languages to serve as instruments for the communication of thoughts and ideas we must understand what it is that permits those who speak them consistently to connect the right sounds with the right meanings.

It is quite clear that, in some sense, one who knows a natural language tacitly knows a system of rules. This is the only assumption by which we can account for a speaker's impressive ability to use language creatively. Fluent speakers both produce and understand sentences that they have never previously encountered, and they can do this for indefinitely many such novel sentences.[9]

Given that fluent speakers are fluent because of their knowledge of the rules of the language and that linguistic communication is a process in which the meaning that a speaker connects with the sounds he utters is the same meaning that the hearer connects with these same sounds, it seems necessary to conclude that speakers of a natural language communicate with each other in their language because each possesses essentially the same system of rules. Communication can take place because a speaker encodes a message using the same linguistic rules that his hearer uses to decode it. This becomes clearer when we think of how we learn a foreign language in the classroom. Our teacher and text present us with a more or less accurate approximation of the rules that any speaker of the foreign language tacitly knows. Our task is to learn them well enough for us to produce utterances that can be decoded by speakers of that foreign language and to understand utterances of those speakers themselves. This sort of example brings out the fact that our competence in a foreign language depends on whether, and to what extent, the rules we have been taught are equivalent to those that speakers of the foreign language acquired naturally. But it also shows that each speaker of the foreign language must use essentially the same system of rules . . .

. . . Roughly, and somewhat metaphorically, we can say that something of the following sort goes on when successful linguistic communication takes place. The speaker, for reasons that are linguistically irrelevant, chooses some message he wants to convey to his listeners: some thought he wants them to receive or some command he wants to give them or some question he wants to ask. This message is encoded in the form of a phonetic representation of an utterance by means of the system of linguistic rules with which the speaker is equipped. This encoding them becomes a signal to the speaker's articulatory organs, and he vocalizes an utterance of the proper phonetic shape. This is, in turn, picked up by the hearer's auditory organs. The speech sounds that stimulate these organs are then converted into a neural signal from

[9] Katz, *The Philosophy of Language*, p. 100.

which a phonetic representation equivalent to the one into which the speaker encoded his message is obtained. This representation is decided into a representation of the same message that the speaker originally chose to convey by the hearer's equivalent system of linguistic rules. Hence, because the hearer employs the same system of rules to decode that the speaker employs to encode, an instance of successful linguistic communication occurs. . . .

. . . A linguistic description . . . describes the knowledge whose possession permits a fluent speaker to communicate with other speakers of his language L and whose absence prevents those who only speak another language from communicating with normal monolingual speakers of L in their language.[10]

Chomsky says nothing quite as explicit; but a similar view of linguistic communication seems to lie behind his version of the thesis that speakers of a language have unconscious knowledge of the grammatical rules of that language. If Chomsky sees communication as requiring the hearer to decode a linguistic message into the appropriate nonverbal thought, then it would be natural for him to believe that a "speaker-hearer" must know the appropriate code, must know the rules for coding and decoding. The speaker has never been explicitly taught the code, so he must have developed a *theory* about it, a *system of beliefs* about it, on the basis of his observation of other speaker-hearers. He must have *determined that* the rules are whatever they are. Since rules linking sound and meaning are what the grammar provides, the speaker-hearer has unconscious knowledge of the rules of grammar.

Some such line of thought is responsible for Chomsky's view of linguistic competence. That he does not go on to say (in fact he explicitly denies) that the speaker has unconscious knowledge *that* the rules are whatever they are must be chalked up to the sudden intrusion of common sense plus the fact that the above picture of communication is not his explicit view but only his tacit background assumption.

Let me try to document this. In Chomsky's explicit remarks about the relations between thought, language, and communication, he endorses the Humboldtian view which is on the whole an elaboration of part of the first sort of view, which I described at the beginning of this essay. Language is primarily an instrument for the free expression of

---

[10] *Ibid.*, pp. 102–105. Cf. J. J. Katz, "Mentalism in Linguistics," *Language*, 40 (1964):124–137; reprinted in Leon A. Jakobovits and Murray S. Miron, eds., *Readings in the Psychology of Language* (Englewood Cliffs, N.J.: Prentice-Hall, 1967).

thought; the instrumental use of language, e.g., in communication, is secondary and based on its primary use; etc.[11] In practice, however, Chomsky appears to take a different line: "The *central* fact to which any significant linguistic theory must address itself is this: a mature speaker can produce a new sentence of his language on the appropriate occasion, and other speakers can understand it immediately, though it is equally new to them."[12] He begins his book, *Aspects of the Theory of Syntax*, with the remark, "Linguistic theory is concerned primarily with an ideal speaker-listener . . ."[13] Similarly, he begins a recent survey article by saying, "At the crudest level of description, we may say that a language associates sound and meaning in a particular way; to have command of a language is to be able, in principle, to understand what is said and to produce a signal with an intended semantic interpretation."[14] In these and many similar passages Humboldt seems to have been forgotten. Language is no longer seen primarily as an instrument for the free expression of thought but as an instrument for the communication of thought.

I think that Chomsky must be taking the thought communicated to be nonverbal. Otherwise he would not be so impressed by what he calls the "creativity of language": "The most striking aspect of linguistic competence is what we may call the 'creativity of language', that is, the speaker's ability to produce new sentences, sentences that are immediately understood by other speakers although they bear no physical resemblance to sentences which are 'familiar'.[15] This aspect of linguistic competence will only seem "striking" to someone who does not think that the relevant thoughts are in words and that the relevant principle for interpreting what others say is to use the homophonic system of translation from words said to words thought. It seems to me that Chomsky's explanation of this "striking aspect of linguistic competence" is like Katz's: The speaker knows what the sound-meaning correspondence for the language is, he knows the rules of grammar that specify this correspondence. Chomsky has put his claim this way:

[11] See above, pp. 273–74 and footnote 3.
[12] Chomsky, "Current Issues in Linguistic Theory," p. 50; my emphasis.
[13] P. 3.
[14] Chomsky, "The Formal Nature of Language," Appendix A to Eric H. Lenneberg, *Biological Foundations of Language* (New York: Wiley, 1967), p. 397.
[15] Chomsky, *Topics in the Theory of Generative Grammar* (The Hague: Mouton, 1966), p. 11. Cf. Chomsky, "Recent Contributions to the Theory of Innate Ideas," pp. 82–83.

The grammar as a whole can thus be regarded, ultimately, as a device for pairing phonetically represented signals with semantic interpretations . . . In performing as a speaker or hearer, [one] puts this [grammar] to use. Thus as a hearer, his problem is to determine the structural description assigned by his grammar to a presented utterance (or, where the sentence is syntactically ambiguous, to determine the correct structural description for this particular token), and using the information in the structural description, to understand the utterance.[16]

It is quite obvious that sentences have an intrinsic meaning determined by linguistic rule and that a person with command of a language has in some way internalized the system of rules that determine both the phonetic shape of the sentence and its intrinsic semantic content — that he has developed what we will refer to as a specific linguistic competence. . . . [T]he technical term 'competence' refers to the ability of the idealized speaker-hearer to associate sounds and meanings strictly in accordance with the rules of his language. The grammar of a language, as a model for idealized competence, establishes a certain relation between sound and meaning — between phonetic and semantic representations . . . Clearly, in understanding a signal, a hearer brings to bear information about the structure of his language.[17]

Chomsky and Halle put the point this way:

The person who has acquired knowledge of a language has internalized a system of rules that determines sound-meaning connections for indefinitely many sentences. Of course, the person who knows a language perfectly has little or no conscious knowledge of the rules that he uses constantly in speaking or hearing, writing or reading, or internal monologue. It is this system of rules that enables him to produce and interpret sentences that he has never before encountered.[18]

What I have been trying to show is that Chomsky's and Katz's talk about unconscious knowledge of the rules of grammar represents more than a careless use of the words "know" and "knowledge." More is involved than a confusion between knowing-that and knowing-how, between knowing that a sentence is ambiguous and knowing the grammatical rules by virtue of which it is ambiguous, or between principles that account for knowledge of a language and knowledge of those principles. Their talk about unconscious knowledge of the rules of grammar

[16] Chomsky, "Current Issues in Linguistic Theory," p. 52.
[17] Chomsky, "The Formal Nature of Language," pp. 397–399.
[18] Noam Chomsky and Morris Halle, *The Sound Pattern of English* (New York: Harper and Row, 1968), p. 3. Cf. Chomsky, "Recent Contributions to the Theory of Innate Ideas," pp. 83–84.

Gilbert Harman

ultimately reflects a basic conception of our use of language and a definite picture of linguistic communication.

Let me note in passing that we can now understand Chomsky's and Katz's otherwise puzzling acceptance of what they call a theory of "innate ideas." They think of the child about to learn a language as faced with a gigantic cryptogram, a code-breaking problem of the toughest sort. They are impressed that children are able to break the code so fast, without any training, all by themselves. Perhaps they think about how long it took a whole group of adults to break the Japanese code. In order to account for a child's performance they infer that he must have had information about the nature of the code ahead of time; and they equate such innate information with what used to be called innate ideas.[19] In this case too, Chomsky and Katz have not simply misused terms like "empiricism," "rationalism," "innate ideas," etc. That's why complaints about their usage fail to move them. Their talk about innate ideas is ultimately based on their conception of the use of language in communication.

3. *Two views of semantics.* What I will call a compositional theory of meaning holds that a hearer determines what the meaning of an utterance is on the basis of his knowledge of the meaning of its parts and his knowledge of its syntactic structure. Such a view follows naturally from the picture of communication that takes it to involve complex coding and decoding. On that picture, to understand (the meaning of) a sentence is to know what (nonverbal) thought or thoughts the sentence encodes. Meanings are identified with the relevant thoughts. Hence the view is that to know the meaning of an expression is to know what meaning, i.e., thought, the code associates with that sentence. On this view, a general theory of meaning of a language is given by the principles of the code that defines the sound-meaning correspondence for the language. Because of the unbounded nature of language, these principles would have to be compositional or recursive.

Recall that for transformational linguistics, the rules of coding and decoding are given by the grammar. This has three components, a *syntactic component* that connects a *phonological component* with a *semantic component*. And as Katz describes it, "whereas the phonological

[19] Chomsky, *Aspects of the Theory of Syntax*, pp. 25–59; *The Philosophy of Language*, pp. 240–282. See also the papers by Chomsky and responses to them in the works cited in footnotes 7 and 8 above.

component provides a phonetic shape for a sentence, the semantic component provides a representation of that message which actual utterances having this phonetic shape convey to speakers of the language in normal speech situations."[20] Katz adds that

The semantic component . . . must contain rules that provide a meaning for every sentence generated by the syntactic component. . . . These rules, therefore, explicate an ability to interpret infinitely many sentences. . . . The hypothesis on which we will base our model of the semantic component is that the process by which a speaker interprets each of the infinitely many sentences is a compositional process in which the meaning of any syntactically compound constituent of a sentence is obtained as a function of the meanings of the parts of the constituent. Hence, for the semantic component to reconstruct the principles underlying the speaker's semantic competence, the rules of the semantic component must simulate the operation of these principles by projecting representations of the meaning of higher level constituents from representations of the meaning of the lower level constituents that comprise them . . .

This means that the semantic component will have two subcomponents: a *dictionary* that provides a representation of the meaning of each of the words in the language, and a system of *projection rules* that provide the combinatorial machinery for projecting the semantic representation for all supraword constituents in a sentence from the representations that are given in the dictionary for the meanings of the words in the sentence.[21]

The projection rules of the semantic component for a language characterize the meaning of all syntactically well-formed constituents of two or more words on the basis of what the dictionary specifies about these words. Thus, these rules provide a reconstruction of the process by which a speaker utilizes his knowledge of the dictionary to obtain the meanings of any syntactically compound constituent, including sentences.[22]

Here meanings are to be identified with "readings."

Projection rules operate on underlying phrase markers that are partially interpreted in the sense of having sets of readings assigned only to the lower level elements in them. They combine readings already assigned to constituents to form derived readings for constituents which, as yet, have had no readings assigned to them . . . Each constituent of

[20] Katz, *The Philosophy of Language*, p. 151.
[21] *Ibid.*, pp. 152–153. Cf. J. A. Fodor, "Some Remarks on the Philosophy of Language," in F. H. Donell, ed., *Aspects of Contemporary American Philosophy* (Wurzburg-Wien: Physica-Verlag, 1965), pp. 82–83.
[22] *Ibid.*, pp. 161–162.

an underlying phrase marker is thus assigned a set of readings, until the highest constituent, the whole sentence, is reached and assigned a set of readings, too.[23]

It is easy to begin to see what is going on here. This theory would certainly be appropriate as an account of the meaning of expressions in one language, e.g., Russian, given in another language taken to be antecedently understood, e.g., English. In that case one would want some general principles for translating Russian into English. Such principles would enable one to know the meaning of Russian expressions because one already knows (in the ordinary sense of this phrase) the meaning of the corresponding English expressions. Katz tries to make the same trick work in order to give an account of the meaning of sentences in English. In effect he envisions a theory that gives principles for translating from English into Mentalese. He thinks these principles are sufficient because one already knows Mentalese.

Proponents of the view that language is primarily used for thought will raise two objections to Katz's theory of communication. They will argue first that the theory is circular, since (as they maintain) Mentalese is simply English used to think in. Second, they will claim that, even apart from that, Katz's maneuver simply shifts the problem back one step. For what would it be to give an account of meaning for Mentalese? One cannot continue forever to give as one's theory of meaning a way to translate one system of representation into another. At some point a different account is needed. Katz's theory only delays the moment of confrontation.

There are two distinct issues here. The first is the difficult quasi-empirical question whether thought is in the relevant sense verbal. I shall postpone discussion of that issue. The second is the methodological question whether a semantic theory may presuppose a theory of the nature of thought. Proponents of the view that language is primarily used for thought take semantics to be part of such a theory of the nature of thought. They argue, as it were, that semantics must be concerned in the first instance with the meaning of thoughts.

Katz seems to assume that no account need be given of the meaning of thoughts, as if Mentalese were intrinsically intelligible. (Cf. Chomsky's remark that "sentences have an *intrinsic meaning* determined by

[23] *Ibid.*, pp. 164–165.

linguistic rule."[24]) That is exactly the sort of view Bloomfield[25] attacked as "mentalistic." In an influential article,[26] Katz attempts to answer Bloomfield but only blurs the issue, since he never considers whether Bloomfield's criticisms of mentalism apply against a theory which assumes that semantics can be unconcerned with the meaning of thoughts because of their supposed intrinsic intelligibility.

Thus, distinguish a theory of the meaning of thoughts from a theory of the meaning of messages. I would argue that a theory of the meaning of messages presupposes a theory of the meaning of thoughts.[27] The former theory might resemble that proposed by Paul Grice.[28] According to Grice, the thought meant is the one the speaker intends the hearer to think the speaker has, by virtue of his recognition of the speaker's intention. However that may be, in normal linguistic communication, speakers and hearers rely on a regular association between messages and thoughts. Those philosophers and linguists who think that language is used primarily in communication suppose that this association is between sentences in, e.g., English (conceived under particular structural descriptions) and, as it were, sentences in Mentalese. Those who believe that language is used primarily in thought suppose that the association is a trivial one, since the language used to communicate with is normally the same as that used to think with.

A theory of the meaning of thought is either a theory of meaning for natural languages (as is claimed by those who take language to be used primarily in thought) or a theory of Mentalese (if those philosophers and linguists are right who take natural language to be used primarily in communication). In either case, it must exploit the Humboldtian insight that the meaning of a linguistic expression is derived from its function in thought as determined by its place in one's total conceptual scheme.[29] One must consider the influence of perception on thought, the role of inference in allowing one to pass from some thoughts to others, and the way thought leads to action. The theory we want will

[24] Chomsky, "The Formal Nature of Language," p. 397.
[25] Leonard Bloomfield, *Language* (New York: Holt, 1933), and "Linguistic Aspects of Science," *International Encyclopedia of Unified Science*, vol. 1 (Chicago: University of Chicago Press, 1955).
[26] "Mentalism in Linguistics."
[27] See my "Three Levels of Meaning," *Journal of Philosophy*, 65 (1968):590–602.
[28] Paul Grice, "Meaning," *Philosophical Review*, 66 (1957):377–388; "Utterer's Meaning and Intentions," *Philosophical Review*, 78 (1969):147–177.
[29] See above, pp. 273–74 and footnote 3.

be like that proposed by Wilfrid Sellars in his paper on "language games."[30] Sellars identifies the meaning of an expression with its (potential) role in the evidence-inference-action language game of thought. Similar theories have been proposed by various philosophers, e.g., Carnap, Ayer, and Hampshire.[31] And Quine[32] has argued that meaning at this level admits of a special sort of indeterminacy.

According to proponents of the view that language is used primarily in thought, transformational semantics, with its compositional theory of meaning, provides neither an account of the meaning of language as used to think with nor an account of the meaning of language as used to communicate with. They claim that it cannot provide a theory of the meaning of thought, since a speaker does not understand the words he uses in thinking by assigning readings to them, and that it cannot provide a theory of the meaning of a message, since it treats a relatively simple problem of interpretation as if it were quite complicated.

4. Composition and communication. Compositional theories of meaning depend on the view that language is used primarily in communication. Thus Davidson argues as follows. He says, first, "we are entitled to consider in advance of empirical study what we shall count as knowing a language, how we shall describe the skill or ability of a person who has learned to speak a language."[33] He wants to argue for the condition "that we must be able to specify, in a way that depends effectively and solely on formal considerations, what every sentence means. With the right psychological trappings, our theory should equip us to say, for an arbitrary sentence, what a speaker of the language means by that sentence (or takes it to mean)."[34] That last reference, to what a speaker *takes* the sentence to mean, sounds suspiciously like Katz's view

---

[30] Wilfrid Sellars, "Some Reflections on Language Games," in *Science, Perception, and Reality* (London: Routledge and Kegan Paul, 1963).

[31] E.g., Rudolf Carnap, "Testability and Meaning," *Philosophy of Science*, 3 (1936) and 4 (1937); "Meaning and Synonymy in Natural Languages," *Philosophical Studies*, 7 (1955):33–47. A. J. Ayer, *Language, Truth, and Logic* (1936), reprinted as a Dover paperback; *The Foundations of Empirical Knowledge* (London: Macmillan, 1940). Stuart Hampshire, *Thought and Action* (London: Chatto and Windus, 1959).

[32] W. V. Quine, *Word and Object* (Cambridge, Mass.: M.I.T. Press, 1960).

[33] Donald Davidson, "Theories of Meaning and Learnable Languages," in Y. Bar-Hillel, ed., *Logic, Methodology, and Philosophy of Science: Proceedings of the 1964 International Congress* (Amsterdam: North-Holland, 1965), p. 387.

[34] *Ibid.*

described above, which assumes that to give an account of meaning for some language is to say how a "speaker-hearer" is able to correlate meanings, qua thoughts, with sentences. Our suspicions are confirmed by Davidson's explicit argument for the compositional theory:

These matters appear to be connected in the following informal way with the possibility of learning a language. When we can regard the meaning of each sentence as a function of a finite number of features of the sentence, we have an insight not only into what there is to be learned; we also understand how an infinite aptitude can be encompassed by finite accomplishments. Suppose on the other hand the language lacks this feature; then no matter how many sentences a would-be speaker learns to produce and understand, there will remain others whose meanings are not given by the rules already mastered. It is natural to say such a language is *unlearnable*. This argument depends, of course, on a number of empirical assumptions: for example, that we do not at some point suddenly acquire an ability to intuit the meanings of sentences on no rule at all; that each new item of vocabulary, or new grammatical rule takes some finite time to be learned; that man is mortal.[35]

This argument makes sense only in the presence of an assumption, which Davidson explicitly acknowledges elsewhere, that "speakers of a language can effectively determine the meaning or meanings of an arbitrary expression (if it has a meaning),"[36] where that is understood to mean that a speaker (hearer) understands a sentence by translating it into its Mentalese counterpart (and where Mentalese is not the language used in communication). If speakers of a language can effectively determine the meaning of messages in ordinary linguistic communication by using the homophonic mapping of verbal message onto verbal thought, the assumption does not support Davidson's argument for a compositional theory of meaning.

In the end Davidson argues for a version of the theory that meaning is given by truth conditions.[37] That is no real improvement over Katz's theory, from the point of view of those who take language to be used primarily in thought. According to Katz, a speaker knows the meaning of sentences of his language because he has mastered the complicated rules (of a transformational generative grammar of his language) that correlate sentences with thoughts. According to Davidson, a speaker knows

---

[35] *Ibid.*, pp. 387–388.
[36] Donald Davidson, "Truth and Meaning," *Synthese*, 17 (1967):320.
[37] *Ibid.*, pp. 304–323.

the meaning of sentences of his language because he has mastered the complicated rules (of a truth definition for his language) that correlate sentences with truth conditions. Katz's theory is in trouble if the relevant thoughts are verbal. The same difficulty faces Davidson in a slightly different form, if the relevant *knowledge* of truth conditions will be verbal. Katz would say that the speaker understands the sentence "Snow is white" by virtue of the fact that he has correlated it with the thought that snow is white. Davidson would (presumably) say that the speaker understands that sentence by virtue of the fact that he knows it is true if and only if snow is white. The difficulty in either case is that the speaker needs some way to represent to himself snow's being white. If the relevant speaker uses the words "snow is white" to represent in the relevant way that snow is white, both Katz's and Davidson's theories would be circular. And, if speakers have available a form of Mentalese in which they can represent that snow is white, so that the two theories avoid circularity, there is still the problem of meaning for Mentalese.

The point is that no reason has been given for a compositional theory of meaning for whatever system of representation we think in, be it Mentalese or English. This point has obvious implications for linguistics, and for philosophy too if only of a negative sort. For example, Davidson uses his theory in order to support objections to certain theories about the logical form of belief sentences.[38] Since his argument for a compositional theory of meaning fails for the language one thinks in, those objections have no force against theories about the logical form of belief sentences used in thinking or theorizing.

Similar remarks apply to the compositional theory of meaning in Paul Ziff's *Semantic Analysis*.[39] He argues that "the semantic analysis of an utterance consists in associating with it some set of conditions [and] that the semantic analysis of a morphological element having meaning in the language consists in associating with it some set of conditions . . ."[40] Very roughly speaking, the relevant conditions are those that must obtain if something is to be uttered without deviance from rele-

[38] See Davidson, "Theories of Meaning and Learnable Languages," and reference therein.

[39] Ithaca, N.Y.: Cornell University Press, 1960.

[40] Paul Ziff, "On Understanding Understanding Utterances," in Fodor and Katz, *The Structure of Language*, and in Ziff, *Philosophic Turnings* (Ithaca, N.Y.: Cornell University Press, 1966), sec. 3.

vant nonsyntactic semantic regularities.

Ziff says, "In formulating the theory presented here [in *Semantic Analysis*] I have had but one objective in mind, viz. that of determining a method and a means of evaluating and choosing between competing analyses of words and utterances."[41] If "analysis" here means "philosophical analysis," Ziff's enterprise must be counted a success, especially in the light of his careful discussion of the analysis of the word "good" in the final chapter. And since analysis is perhaps a kind of translation or decoding, it may be possible to defend a compositional theory of meaning as a compositional theory of analysis. (One must see that a proposed analysis of a word is adequate for various contexts and is consistent with analyses suggested for other words.)

But Ziff does not give quite that argument for a compositional theory of meaning; and the argument he does give indicates that he wants more from his theory than a way of evaluating philosophical analyses. His own argument seems to assume that a *speaker* understands sentences by virtue of being able to give analyses or explications of them. In *Semantic Analysis* the argument goes like this:

In a general form, the principle of composition is absolutely essential to anything that we are prepared to call a natural language, a language that can be spoken and understood in the way any natural language can in fact be spoken or understood.

How is it that one can understand what is said if what is said has not been said before? Any language whatever allows for the utterance of new utterances both by the reiteration of old ones and by the formation of new ones out of combinations of old elements. Hence any natural language whatever allows for the utterance of both novel utterance tokens and novel utterance types. If a new utterance is uttered and if the utterance is not then and there to be given an arbitrary explication, that one is able to understand what is said in or by uttering the utterance must in some way at least be partially owing to one's familiarity with the syntactic structure of the utterance.[42]

In "On Understanding Understanding Utterances" Ziff is more explicit:

part of what is involved in understanding an utterance is understanding what conditions are relevantly associated with the utterance. . . .

Someone says 'Hippopotami are graceful' and we understand what is said. In some cases we understand what is said without attending to the discourse the utterance has occurred in or without attending to the con-

[41] *Semantic Analysis*, sec. 201.
[42] *Semantic Analysis*, sec. 64.

text of utterance. How do we do it?

It seems reasonable to suppose that part of what is involved is this: Such an utterance is understood on the basis of its syntactic structure and morphemic constitution.

Assuming that part of what is involved in understanding an utterance is understanding what conditions are relevantly associated with the utterance, this means that we take a certain set of conditions to be associated with such an utterance on the basis of its syntactic structure and morphemic constitution.[43]

To this the same remarks apply as to the theories of Katz and of Davidson. A speaker can *understand* that certain conditions are associated with an utterance and can *take* certain conditions to be associated with an utterance only if he has some way to represent to himself that the conditions are associated with the utterance. And even if the speaker uses Mentalese to represent utterance-conditions correlations, the problem of meaning is merely pushed back one step to Mentalese. Ziff fares no better than Katz or Davidson in showing that we need a compositional theory of meaning for the system of representation that we think with. If speakers of English think in English and we rely on that fact in communication, Ziff gives us no reason why we need a compositional theory of meaning for English.

5. Language used in thought. According to those who say that speakers of a natural language also think in that language, there are levels of meaning. A theory of the first level must account for the meaning of an expression as a function of its role in thought. A theory of the second level must account for the meaning of an expression used to communicate a thought. A theory of the second level must presuppose the first, since linguistic communication typically communicates a thought that can be expressed (roughly speaking) in the same words used for communication. Such exploitation is even involved in ironic and other nonliteral remarks, since in such cases the words of the relevant thought will be some simple function of the words in the message.

Compare this view with the one expressed in Katz and Fodor's paper, "The Structure of a Semantic Theory."[44] The theory is put forward as a level-two theory of meaning in communication. In particular it is sup-

[43] "On Understanding Understanding Utterances," secs. 3, 4.
[44] J. J. Katz and J. A. Fodor, "The Structure of a Semantic Theory," *Language*, 39(1963):170–210; reprinted in Fodor and Katz, *The Structure of Language*, and in Jakobovits and Miron, *Readings in the Psychology of Language*. Page reference is to this latter source.

posed to account for "the way that speakers understand sentences."[45] That suggests it is the sort of theory Grice has tried to develop. But the authors go on to describe it as an account of the meaning or meanings a sentence has when taken in isolation from possible settings in actual discourse. In other words their theory is restricted to giving an account of meaning for those cases in which the message communicates a thought that (on the other view) can be expressed in (roughly) the same words as those in which the message is expressed. They argue that another theory would have to account for the interpretation or interpretations assigned when a sentence occurs in a particular setting. Furthermore they argue that the latter theory must presuppose the one they present. Thus they come close to the other view's distinction between levels one and two, and in a way they attempt to provide a theory of level one. Or rather, Katz and Fodor see the need for three theories, where on the other view only two are needed. First, there is what we have been calling a theory of the meaning of thought, an account of meaning for whatever system of representation one uses to think with. Katz and Fodor say nothing about this theory, perhaps because they take thoughts to be intrinsically intelligible. On the other hand the other view takes this theory to be the most important part of the theory of meaning. Second, there is the theory that associates sentences (under various structural descriptions) used in communication with meanings they have (in isolation from setting and discourse). This theory associates, e.g., sentences of English (conceived under their structural descriptions) with "readings" in the system of representation used in thinking. Katz and Fodor take this theory to be the central part of the theory of meaning, which amounts to a theory of how sentences in, e.g., English are to be translated into Mentalese. The other view takes this part of the theory to be trivial, on the grounds that speakers of English think in English and can use what amounts to the homophonic scheme of translation. (I shall say more about this in a moment.) Third, there is the theory of the meaning of a message in a particular linguistic and nonlinguistic setting. Both sides argree that this is an important part of a theory of meaning and that it presupposes the other parts.

How are we to decide between a view like Katz and Fodor's, which takes language to be used primarily in communication, and the alterna-

[45] *Ibid.*, p. 399.

tive, which takes language to be used in thought? I tend to favor the latter view, mainly on grounds of simplicity. But on the other side it might be argued that the former theory is needed to account for all the facts. Katz claims that his and Fodor's (and Postal's[46]) semantic theory can explain a great number of different things:

> beside requiring a semantic component to predict semantic anomaly and ambiguity, we also require it to predict such other semantic properties and relations as synonymy, paraphrase, antonymy, semantic distinctness, semantic similarity, inclusion of senses, inconsistency, analyticity, contradiction, syntheticity, entailment, possible answer to a question, and *so forth*.[47]

He may justly complain that we have no right to reject transformational semantic theory unless we have some other method to explain or explain away the phenomena in question. In the remainder of this essay I want to describe and defend one alternative method. In the process I hope to say just what (in my opinion) is and what is not salvageable in the views of Chomsky, Katz, and Fodor discussed above.

I shall be concerned mainly with the level-one theory of meaning, i.e., the theory of the meaning of language used to think in. I shall say something about level two, which is concerned with the meaning of messages, only indirectly and in passing.

What is it for an expression to have a meaning on level one? It is certainly not that the relevant person, the "thinker," can assign it one or more "readings." It is rather that he can use it in thought, i.e., that it has a role in his evidence-inference-action language game. He may be able to use the expression in direct perceptual reports. He must certainly be able to use it in (theoretical and practical) reasoning.

Some reasoning is relatively formal. It depends only on the logical or grammatical form of relevant expressions and is not a function of nonlogical or nongrammatical vocabulary. In order to account for a person's ability to reason formally, it is plausible to suppose that he thinks of the sentences he uses in thought as grammatically structured. He views them or conceives them as having one or another "deep structure" grammatical analysis. I shall return to this point in a moment. First I note that it is not sufficient for full understanding of an expression that

---

[46] J. J. Katz and Paul M. Postal, *An Integrated Theory of Linguistic Description* (Cambridge, Mass.: M.I.T. Press, 1964).

[47] J. J. Katz, "Recent Issues in Semantic Theory," *Foundations of Language*, 3(1967):133.

one be able to make formal moves with that expression, at best this shows merely that one understands the expression as having a particular grammatical form. One must also be able to give paraphrases and make inference that involve changes in nonlogical and nongrammatical vocabulary. One must be able to see what sentences containing the relevant expression imply, what they are equivalent with, etc.

It is here, of course, that Katz and others have imagined that appeal must be made to meaning, to entailment by virtue of meaning, to equivalence by virtue of meaning, etc. But that is a mistake. The relevant notions of equivalence and of implication are the ordinary ones: equivalence or implication with respect to one's background assumptions, where no distinctions need be made between analytic or synthetic background assumptions. One has an understanding of an expression to the extent that one can paraphrase sentences containing it, can make inferences involving such sentences, etc. It adds nothing to one's understanding if one can distinguish "analytic" equivalence and implications from "synthetic" ones. In fact, most people cannot do so. Only those who have been "indoctrinated" can; and they are not the only ones who understand the language they think in.[48]

Here it might be objected that sentences used in thinking are often ambiguous. How can we account for that — and for a person's ability to understand or interpret a sentence one way at one time and another way at another time? Can we account for it without assuming that he assigns an interpretation or "reading" to the sentence in the way in which Katz suggests? Well, we can and we can't. An expression is ambiguous if a person can sometimes treat it as having one sort of role and at other times treat it as having a different role. Treated one way it admits of paraphrases it does not admit when treated the other way. This difference in paraphrasability represents the difference in interpretation; but recall that paraphrasability is relative to background knowledge and need not permit any analytic-synthetic distinction.[49]

To account for the difference in the ways a person can use an ambiguous expression we must suppose that he does not simply view it as a sequence of words. He views it, or hears it, as having a particular syntactic structure and as containing words in one or another of their pos-

[48] I have discussed this issue from a slightly different point of view in "Quine on Meaning and Existence, I," *Review of Metaphysics*, 21 (1967–1968) :124–151.
[49] *Ibid.*, pp. 150–151.

sible senses. Let us consider each of these things in turn.

Syntactically ambiguous sentences may be heard as having either of two (or more) different syntactic structures. They are like the lines on paper which may be seen as a staircase viewed from the back or from the front. Or they are like a group of dots that may be seen as two groups in one way or as two other groups in another way. Or they are like the figure that can be seen either as a duck or as a rabbit. Thus consider, "They are visiting philosophers." We may see or hear this in two different ways, depending on whether we take "are visiting" together or "visiting philosophers" together. We hear the sentence as admitting either of two groups of paraphrases, either "They are philosophers who are visiting," etc. or "They are visiting some philosophers," etc. Similarly, consider "Visiting philosophers can be unpleasant." The difference in interpretation again depends on how we conceive that sentence's grammatical structure, although the difference is not simply a matter of grouping on the surface. It is rather a matter of what transformational theorists refer to as "deep structure." It is a matter of how we conceive the source of "visiting philosophers." It may be heard as coming from "someone visits philosophers" or from "philosophers visit someone." It is true that the average person is quite ignorant of transformational grammar. But that does not mean he fails to hear that sentence as having one or another of the indicated structures. A person can see lines on a page as forming one or another three dimensional structure without knowing any geometry. Therefore I think that in order to account for the way in which a person can deal with ambiguous sentences, we must assume that in some sense he conceives a sentence used in thought as having one or another syntactic structure. As I have already said above, I think that the same conclusion can be reached if we attempt to account for the formal inferences a person can make, even if we ignore ambiguity.

Similarly, in order to account for the way in which a person deals with ambiguous words, we must assume that he distinguishes a word used in one sense from the same word used in another sense. But we can do that without assuming that he makes the distinction by assigning different *readings* to the word. He may mark the distinction in sentences by a device as simple as a subscript. The inferences and paraphrases that are then permissible depend in general on the subscript selected.[50] One reason why no more is needed than a subscript is that the relevant sorts of inference and paraphrase are those possible by virtue of background information. Such background information itself must be "stored" as sentences under certain structural descriptions including subscripted words. But a person can understand the expressions he uses without having divided his background information into a part true by "definition" (the dictionary) and another part not true by definition (the encyclopedia).

We think with sentences conceived under particular structural descriptions, where we may count the subscripts that distinguish word senses as part of the structural descriptions. More precisely, so-called propositional attitudes are sentential attitudes, where the relevant sentences are conceived under particular structural descriptions. In understanding what someone else says to us, we must determine the content of his utterance taken literally. It is sufficient that we should attend to the sentence he utters conceived under the relevant structural description. Therefore, usually we must assign a structural description to his words in the sense that we must hear his words as having a particular syntactic structure. That is not to say that we come (even unconsciously) to know that it has a particular structure; and it is certainly not to say that we have knowledge of the principles that relate phonetic representations to structural descriptions. The situation is strictly analogous to other cases of perceiving something *as* something. We can perceive a series of lines as a particular three-dimensional structure without thinking that it has that structure, indeed without knowing anything about geometrical structure. We can certainly do so without knowing rules that relate two-dimensional figures and three-dimensional structures.

(To say that a speaker of a language must have unconscious knowledge of the grammar of his language is strictly analogous to saying that a person who sees the world in three dimensions must have unconscious knowledge of geometry. To say that a speaker has internalized a generative grammar is like saying a perceiver has internalized a geometry. Perhaps this is sometimes a harmless way of speaking. But one must be careful lest it illegitimately lead one to a compositional theory of meaning or to a rationalist theory of concept formation.)

[50] Cf. Harman, "What an Adequate Grammar Could Do," *Foundations of Language*, 2(1966):134–136.

Sounds reach a person's ears and, after physiological processes we know little about, he perceives or conceives a sentence under a particular structural description. His understanding of that sentence is represented not by his having assigned it a "reading" in Mentalese but rather by his being able to use the sentence under that structural description in his thought. If we like, we may still speak of "decoding" here, in a sense familiar perhaps from information theory. One decodes certain sounds into a sentence plus structural description. Although this point may be partially responsible for Katz's remarks about decoding, it is not exactly what he had in mind. In the present case we have decoding of sound into sentence under structural description. Katz speaks of decoding the phonetic representation of the sentence. He says, you will recall:

The speech sounds that stimulate these organs are then converted into a neural signal from which a phonetic representation equivalent to the one into which the speaker encoded his message is obtained. This *repre-sentation* is decoded into a representation of the same message that the speaker originally chose to convey by the hearer's equivalent system of linguistic rules.[51]

But that is not what usually happens. One does not first perceive the sounds as a sequence of words and then assign a structural description. It is a commonplace of transformation theory that one's understanding of the sentence will partially determine what words he hears. Whatever "decoding" takes place generally translates certain sounds or (perhaps) neural signals directly into a sentence with structural description. Such a "sound-meaning correspondence" is not given simply by the rules of grammar.

I conclude from the above considerations that it is possible to give an account of how people understand sentences that incorporates the insights gained in transformational grammar without leading one to postulate that a speaker knows that sentences have certain structural descriptions, that he knows the rules of grammar, or that he understands sentences by assigning readings to them.

6. Anomaly and Synonymy. Now Katz and Fodor still have one more card up their collective sleeve. They claim that their semantic theory can account for at least two things that it would be difficult to handle in any other way. First they think their theory can show how certain interpretations of a sentence that are grammatically possible are seman-

---

[51] Katz, *The Philosophy of Language*, p. 104; my emphasis.

tically ruled out, so that they can account for a certain amount of disambiguation with their theory.[52] Second, at least Katz thinks that the theory shows how certain sentences are analytic, others contradictory, and still others synthetic, so that the theory can account for native speakers' intuitive judgments about analyticity, etc.[53]

But there is no such thing as semantic anomaly and no such distinction as the analytic-synthetic distinction. To believe otherwise is to suffer from a lack of imagination. For example, Katz and Fodor argue as follows:

> Now let S be the sentence The paint is silent. English speakers will at once recognize that this sentence is anomalous in some way. For example, they will distinguish it from such sentences as The paint is wet and The paint is yellow by applying to it such epithets as 'odd,' 'peculiar,' and 'bizarre.' . . . Hence, another facet of the semantic ability of the speaker is that of detecting semantic anomalies.[54]

But surely, whatever anomaly there is in the phrase "silent paint" is due to the fact that paints do not emit noise. If some paints did and some did not, "silent paint" would not be anomalous. And that is to say that there is nothing peculiarly semantic about the anomaly. Its anomaly is a result of our general nonlinguistic knowledge of the world.

Again, Katz and Fodor argue that a theory that incorporates only the sort of considerations I have sketched above

> will not be able to distinguish the correct sense of seal in One of the oil seals in my car is leaking from such incorrect senses as 'a device bearing a design so made that it can impart an impression' or 'an impression made by such a device' or 'the material upon which the impression is made' or 'an ornamental or commemorative stamp' and so forth, since all of these senses can apply to nominal occurrences of seal.[55]

But it requires only a little imagination to see that "seal" may have any of these senses, although in ordinary discourse it would be more likely to have the sense the authors have in mind.[56]

---

[52] Katz and Fodor, "The Structure of a Semantic Theory," pp. 408–410.

[53] Katz, "Analyticity and Contradiction in Natural Language," in Fodor and Katz, The Structure of Language, pp. 530–541; The Philosophy of Language, pp. 193–220; and "Some Remarks on Quine on Analyticity," Journal of Philosophy, 64 (1967):35–52.

[54] Katz and Fodor, "The Structure of a Semantic Theory," pp. 402–403.

[55] Ibid., p. 409.

[56] For further discussion of the question whether it is possible to distinguish semantic anomaly or disambiguation from anomaly or disambiguation due to our generally accepted beliefs, see Dwight Bolinger, "The Atomization of Meaning,"

Gilbert Harman

McCawley has argued a similar point with respect to purported syntactic anomalies of a certain type:

Moreover, it appears incorrect to regard many so-called "selectional violations" as not corresponding to possible messages, since many of them can turn up in reports of dreams:

(2)   I dreamed that my toothbrush was pregnant.
(3)   I dreamed that I poured my mother into an inkwell.
(4)   I dreamed that I was a proton and fell in love with a shapely green-and-orange-striped electron.

or in reports of the beliefs of other persons:

(5)   John thinks that electrons are green with orange stripes.
(6)   John thinks that his toothbrush is trying to kill him.
(7)   John thinks that ideas are physical objects and are green with orange stripes.

or in the speech of psychotics. While one might suggest that a paranoid who says things like

(8)   My toothbrush is alive and is trying to kill me.

has different selectional restrictions from a normal person, it is pointless to do so, since the difference in "selectional restriction" will correspond exactly to a difference in beliefs as to one's relationship with inanimate objects; a person who utters sentences such as (8) should be referred to a psychiatric clinic, not to a remedial English course.[57]

I agree and think that supports the view that there is no real distinction between semantic anomaly and anomaly due to "extralinguistic factors."

Similar points apply to Katz's claims about analyticity, etc. First, the fact that people have "intuitions" about analyticity shows at best that there is a distinction between "seems analytic to certain people" and "seems synthetic to them." It does not show that there are sentences that are really analytic as opposed to others that are synthetic. The fact that people once had "intuitions" that some women were witches and others not, certainly fails to show that there were women that really were witches as opposed to others that were not.[58]

Language, 41(1965):555–573, reprinted in Jakobovits and Miron, Readings in the Psychology of Language, pp. 432–448.

[57] James D. McCawley, "Where Do Noun Phrases Come From?" in R. A. Jacobs and P. S. Rosenbaum, eds., Readings in English Transformational Grammar (Waltham, Mass.: Ginn, 1970), p. 168.

[58] Cf. N. L. Wilson, "Linguistic Butter and Philosophical Parsnips," p. 65; Harman, "Quine on Meaning and Existence, I," pp. 137–138.

Second, the intuitive distinction Katz and others make between analytic and synthetic truths is easily explained away without appeal to transformational semantic theory: people who have such intuitions suffer from a lack of imagination. The intuitions come from their inability to imagine that certain sentences are false. But as many philosophers have pointed out, after a little practice such things can be imagined.[59]

7. *Final remarks.* Consider the two views. On the first view, a person who speaks a natural language can think in that language and does not need to translate sentences conceived into some other system of representation, Mentalese. On this view, Mentalese incorporates one's natural language. On the second view, a person cannot think in language and must translate sentences of his language into Mentalese. The incorporation view has the following advantages over the translation view: (1) The incorporation view provides a natural explanation of the way in which learning one's first language makes possible thoughts and other propositional attitudes one would not otherwise have. The translation view must invoke some special principle to account for this. (2) Similarly, the incorporation view suggests an explanation of the way in which unconscious thinking makes use of verbal punning, as revealed in psychoanalytic studies of dreaming, slips of the tongue, and free association.[60] It is not clear how the translation view could explain such things. (3) The translation view is unnecessarily complicated because it cannot explain anything that is not explicable on the incorporation view. (4) The translation view suggests things that are false about anomaly, synonymy, etc.

Such considerations seem to be the relevant ones to use in deciding which of the two views is more correct. For example, it is difficult to see how neurophysiological evidence could support one of the views against the other (except by pointing to further relatively behavioral phenomena to be explained). For, as descriptions of the mechanism of the brain, the two views must be taken to be descriptions of a fairly

---

[59] See e.g., Hilary Putnam, "It Ain't Necessarily So," *Journal of Philosophy*, 59(1962):660; J. M. E. Moravcsic, "The Analytic and the Nonempirical," *Journal of Philosophy*, 62(1965):421–423; Harman, "Quine on Meaning and Existence, I." Quine speculates on the mechanism of analyticity "intuitions" in *Word and Object*, pp. 56–57, 66–67.

[60] I owe this point to Lucy Harman.

abstract sort. And it is not easy to see how any neurological mechanism that might account for the relevant behavior and that could be interpreted as an instantiation of one of the descriptions could not also be interpreted as an instantiation of the other. Therefore, I see no reason not to accept the incorporation view.[61]

[61] I discuss these issues further in *Thought* (Princeton, N.J.: Princeton University Press, 1973).

Knowledge of Language

There are a number of different questions that I would like to touch upon in these lectures, questions that arise at various levels of generality and that grow out of different, though not totally unrelated concerns. I want to present a certain framework within which, I believe, the study of language can be undertaken in a very fruitful way — not the only framework, to be sure, and one better suited to certain problems than to other, equally legitimate ones. Within this framework I would like to discuss some technical questions that are at or near the borders of research. At this level of discussion, I will be presenting some material that is internal to the theory of transformational generative grammar. But I would also like to suggest that this rather technical material is potentially of quite general interest, that by studying it we can hope to learn some important things about the nature of human intelligence and the products of human intelligence, and the specific mechanisms that enable us to acquire knowledge from experience, specific mechanisms that, futhermore, provide a certain structure and organization for, and no doubt certain limits and constraints on, human knowledge and systems of belief. I think that the work I will describe at least hints at a concept of man that is rather different, in interesting respects, from others that have been implicit in much modern thinking about these matters, and would like to elaborate on this question as well. I have in mind, then, a large enterprise, of which only a small part can be carried out with a satisfactory degree of clarity and precision. Still, I think it is useful to consider this small part against the background of what might ultimately be achieved.

The study of language, as I will be considering it here, can be regarded as a part of human psychology. It forms a part of the general study

AUTHOR'S NOTE: This essay is the first of six John Locke Lectures delivered at Oxford University in May–June 1969. Parts appeared in the London *Times Literary Supplement*, May 15, 1969. The other lectures have not yet been published.

of human cognitive structures, how they relate to behavior, and how they are acquired. This point of view is not at all characteristic of study of language in the modern period. In fact "psychologism" has been often stigmatized as one of the most grievous sins, a danger that linguists must scrupulously avoid. No doubt one can avoid this sin and still do quite useful work. But I do not see the force of the injunction. I believe, and will try to show, that by disregarding it, by developing the study of language frankly and openly as a branch of theoretical psychology, we can considerably advance our understanding of language and arrive at interesting conclusions with regard to human intelligence and the innate structures of mind that permit us to act as free and creative beings, undetermined, even probabilistically, by stimulus conditions, yet at the same time functioning within a certain system of rules and principles that are in part a product of intelligence, and in part a fixed and immutable framework within which such products of intelligence are constructed. These notions are rather vague. I will try to give them a sharper form in the course of the discussion.

I have tried, a number of times, to relate these considerations to some long-standing issues in the history of Western thought — not too persuasively, to judge by recent commentary. However, I think a strong case can be made that this interpretation, both of history and of current work, is quite legitimate.

There has been a fair amount of controversy and debate over these matters in the past few years, much of it in my opinion beside the point because of faulty and misleading construction of the issues. I don't propose to spend much time reviewing these debates. Let me illustrate with just one example, perhaps extreme in the degree of misunderstanding but not, I am afraid, totally untypical. A recent book contains several essays on the issues that I want to raise here.[1] In one contribution entitled "Innate Knowledge,"[2] Rulon Wells offers a refutation of what he takes to be my basic assumption, namely, that "the difference between automaton and brute, on the one hand, and man on the other, is just the difference between finite and infinite power." He argues at length that "an automaton (and presumably a brute) can be regarded as incorporating an infinite power, and . . . to finish the picture, a hu-

[1] *Language and Philosophy*, ed. Sidney Hook (New York: New York University Press, 1969).
[2] *Ibid.*, pp. 99–119; the quotation is on p. 118.

man being, equally with an automaton and a brute, will have a 'part two' that imposes a finitude limitation. The capacity of man may be ever so much greater than the capacity of automaton or of brute; but the difference will be the difference between one finite magnitude and another . . ."

It is surely true that automata, animals, and man can be regarded as incorporating an infinite power as well as a " 'part two' that imposes a finitude limitation." Wells's mistake is double: first, to suppose that this has ever been an issue, and second, to maintain that "the capacity of man may be ever so much greater than the capacity of automaton or of brute," where capacity is measured in numerical terms, in terms of range of possible responses. In fact suppose that we abstract away from "part two," the finitude limitation imposed by finiteness of life or shortness of breath. This abstraction is surely legitimate and enlightening; it enables us to study the inherent capacity, the "generative capacity," of the mechanisms underlying behavior in the case of automaton, animal, or man. If we do not carry out this abstraction, if we consider generative capacity together with the finitude limitation, then Wells is surely incorrect in his claim that human capacity is greater in scope than that of animal or automaton. But if we make the abstraction, as we must to move to a significant further investigation of capacity, then he is wrong for more interesting reasons. It was obvious even to the Cartesians that "because there may be an innumerable variety in the impressions made by the objects upon the sense, there may also be an innumerable variety in the determination of the Spirits to flow into the Muscles, and by consequence, an infinite variety in the Motions of Animals." (François Bayle, 1669.)

In short an animal can operate on the principle of the speedometer, producing a potentially infinite, in fact in principle continuous, set of signals as output in response to a continuous range of stimuli — the signaling system can be infinite, in fact continuous, in the only sense in which any physical signaling system can be regarded as continuous. The range of behavior, the scope of the signaling system, is thus in principle greater than that of human language — which is a discrete communication system — not by a difference of degree, but by a difference of kind.

Examples can easily be found. Von Frisch's beautiful work on the dance language of the bees is one — but considerable doubt has been cast on it in some recent studies (Adrian Wenner, Patrick Wells, Den-

301

nis Johnson, *Science*, 4 April 1969). Consider therefore a case described by W. H. Thorpe, the case of a bird song in which the rate of fluctuation between high and low pitch signals the intensity with which territory will be defended. This is in principle a continuous system — again, in the only sense in which a physical signaling system can be regarded as continuous. Its scope is therefore greater in kind, not degree, than that of discrete human language. (In passing, I should note that there is a gestural element in human language that is also continuous in scope, but this is not to the point here.) This signaling system differs from human language not only in that it is far greater — not numerically more limited — in scope, but also, more importantly, in that it is directly associated to external stimulus conditions; it is, in short, a signaling system, like a speedometer, and not a language in the human sense, a system that is available for free expression of thought precisely because it is not tied directly to external stimuli. The point was emphasized, again, by the Cartesian linguist-psychologists. Human speech is infinitely varied in scope and appropriate to situations, but is not controlled by external stimuli in the sense of animal signaling systems of the sort just described. This fact poses interesting problems for inquiry; the Cartesians took it as one demonstration that man escapes the inherent limits of mechanical explanation, as they understood the latter notion. Whatever the explanation and analysis may be, the point is that human language differs qualitatively in this respect from animal signaling systems, not in that its scope is numerically greater in finite magnitude, as Wells erroneously believes, nor in that it is infinite as compared to the finiteness of animal behavior, in accordance with the quite ridiculous view that Wells refutes and that no one has ever proposed, but rather in that it is in principle an infinite discrete system rather than a continuous system, and that it is related to external stimuli not by the mechanism of stimulus control, but by the much more obscure relation of appropriateness.

When we consider automata, the matter is still more clear. There is, in fact, an interesting literature on the generative capacity of restricted infinite automata, and there has been much effort — so far inconclusive — to place human languages in a hierarchy of restricted infinite automata, in terms of generative capacity. In the light of this work, Wells's misunderstanding is remarkable. In any event it completely misconceives the issue.

There is, in fact, an interesting issue. Every animal communication system that is known operates on one of two principles: either the principle of the speedometer, as described a moment ago, or else a principle of strict finiteness; that is, the system consists of a finite number of signals, each produced under a fixed range of stimulus conditions. Human gestural systems are not well understood, but it is reasonable to suppose that they too observe these limitations. Human language, however, is entirely different. A person who knows a language has mastered a set of rules and principles that determine an infinite, discrete set of sentences, each of which has a fixed form and a fixed meaning or meaning-potential. Even at the lowest levels of intelligence, the characteristic use of this knowledge is free and creative in the sense that utterances are not controlled by external stimuli but are appropriate to situations, and in that one can instantaneously interpret an indefinitely large range of utterances, with no feeling of unfamiliarity or strangeness — and of course no possibility of introspecting into the processes by which the interpretation of these utterances, or the free and creative use of language, takes place. If this is correct, then it is quite pointless to speculate about the evolution of human language from animal communication systems, as pointless as it would be to speculate about the evolution of language from gesture. It is an interesting question whether properties of human language are shared by other cognitive systems. But no dogmatic assumptions are in order — that much seems clear.

The set of rules and principles that determine the normal use of language I will refer to as the 'generative grammar', or simply the 'grammar' of the language. There is an ambiguity in the usage of this term that should be noted; that is, the term 'grammar' is also used for the explicit theory, constructed by the linguist, which purports to be a theory of the rules and principles, the grammar in the first sense, that has been mastered by the person who knows the language. No confusion should arise if the distinction is kept in mind. The linguist's grammar is a theory, true or false, partial or complete, of the grammar of the speaker-hearer, the person who knows the language. The latter is the object of the linguist's study. He cannot, of course, observe it directly and can only attempt to construct a theory of the speaker-hearer's grammar, making use of whatever evidence he can obtain by observation or introspection, all such evidence, of course, being fallible and subject to correction and revision.

303

Noam Chomsky

The linguist's grammar is, thus construed, a psychological theory. It is an attempt to account for evidence of behavior and introspection by ascribing to the language-user a certain system of rules and principles that he applies in language use, as a speaker and hearer. It is postulated, then, that a person who knows a human language has internalized, has developed a mental representation of a grammar, a set of rules and principles, of which the linguist attempts to construct a precise and explicit model. How this abstract representation is realized is another question, concerning which we have no serious information at the moment. Conceivably such evidence might be forthcoming, but nothing is known today of much significance.

To allay misunderstanding, let me make clear that I am not proposing that this conclusion is one of logical necessity. Obviously not. It is an empirical hypothesis, to be judged in terms of its success in explaining and accounting for certain phenomena, observations that can provide evidence for or against certain explicit assumptions about this grammar which, it is postulated, has been internalized by the language-user. What is postulated is that to know a language is to have a certain mental constitution which is characterized by the linguist's grammar. There is nothing mystical about this approach, contrary to what is sometimes believed. It is precisely the approach that would be taken by a scientist or engineer who is presented with a black box that behaves in a certain fashion, that evidences a certain input-ouput relation, let us say. The scientist will try to construct a theory of the internal structure of this device, using what observations he can as evidence to confirm his theory. If he is unable to investigate the physical structure of the device, he will not hesitate to ascribe to the device a certain abstract structure, perhaps a certain system of rules and principles, if this turns out to be the most successful theoretical approach. There is no reason to adopt some different standpoint when the object under investigation is the human being.

In the case of the black box, the scientist may be mistaken in attributing to it an abstract grammar as its internal structure. He may be wrong in postulating that a certain specific set of rules and principles has been internalized by the device and constitutes its 'mental state', if one wishes to use this terminology. He may be wrong in that he has selected the wrong rules and principles, the wrong grammar. Or he may be wrong, more deeply, in that he has taken entirely the wrong approach to under-

304

standing the device. To make the analogy closer, let us use the terminology that Wells suggests and distinguish, in our theory of the device, two components: part one, a grammar that has a certain generative capacity, and part two, a set of conditions that impose a finitude restriction. The latter can be elaborated in the case of study of mind; we can distinguish certain conditions that have to do with organization of memory and perceptual strategies, perhaps certain belief systems that are extraneous to language, and so on. Abstracting away from these elements of the total theory of mind, we can focus attention on the grammar, the set of rules and principles that determine the form and meaning of an infinite number of sentences. The abstraction is legitimate. It may be misguided, as may any theoretical construction, but I do not think that it is, in this case. My belief that it is the proper abstraction, that the theory of mind is best conceived in these terms, is grounded in two sorts of considerations: First, its success in explaining a variety of phenomena. Second, the difficulty of constructing a coherent alternative. I will discuss the first kind of grounds in subsequent lectures. Let me briefly turn to the second.

One of the very few attempts to construct an alternative to the approach I've just briefly outlined is by Gilbert Harman.[3] He argues that the goal of linguistic description should be not a grammar, abstracted as described above, but rather a performance model, which he outlines (p. 80) as a device with a perceptual and phonetic input, and with an output consisting of bodily movements and sentences. The internal structure of his performance model includes a representation of beliefs, of plans, goals and intentions, and a component labeled "phonetic, syntactic, semantic rules of the grammar." It contains a central "thinker and decider" which makes use of perceptual input, plans, goals and intentions, beliefs, and rules of grammar, in producing its output. Harman proposes this as an alternative to what I described, say, in *Aspects of the Theory of Syntax*. On the contrary, it is virtually identical to what is described in *Aspects*, and — the crucial point here — incorporates an abstract component containing the phonetic, syntactic, and semantic rules of grammar exactly as in the model I described. In attempting to construct an alternative, Harman ends by simply presenting the model

[3] "Psychological Aspects of the Theory of Syntax," *Journal of Philosophy*, 64, no. 2 (1967):75–87.

to which he objects, the model which, he believes (mistakenly, in my opinion), embodies all sorts of confusions and problems.

Harman argues that from the fact that the speaker relates sound and meaning, it does not follow logically that he uses a representation of the rules of grammar to carry out this act. To assume the contrary would lead to an infinite regress, by a familiar argument used, for example, by Ryle in a similar connection. The argument is in part correct. The evidence of language use does not logically imply that the language-user has internalized the rules of a grammar. Rather, the latter conclusion is an instance of what Harman has elsewhere termed induction to the best available explanation. Furthermore, as noted, his alternative is not different from the concepts that he criticizes, in that his model of mind also incorporates a generative grammar as one component. This component of his model of mind contains a representation of the phonetic, syntactic, and semantic rules. It is only a shift of terminology to go on to say that the mind, so modeled, is postulated to contain an internal representation of a grammar. If we want to go on to study acquisition of language, we will have to consider the question how this grammar is constructed on the basis of experience, and what kind of structure must be attributed to the mind to account for this achievement. When misunderstandings and misinterpretations are cleared away, I see no issue raised in Harman's critique. Harman's several papers are important, in this connection, because they represent the only serious attempt, to my knowledge, to construct an alternative to the conception outlined earlier, taking into account the actual empirical problems.

On the surface the behaviorist account of language use proposed by many philosophers, psychologists, and linguists appears to be a genuine alternative approach. However, the behaviorist alternative, as actually formulated, contains so many escape hatches that it ultimately has no empirical content whatsoever, so far as I can see. The matter is worth a few moments' discussion, since the development of the behavioral approach in psychology and the social sciences has been heralded as a major breakthrough. My personal opinion is rather different. I think that it is an intellectual and social calamity. However this may be in general, the behaviorist position to language collapses when the issue is pressed.

I think that Quine's recent writings are quite informative in this regard. Quine has been the leading and certainly the clearest exponent of a behaviorist position with regard to human language, its use and acquisition. He has frequently indicated that he sees himself as developing a view rather like Skinner's. The latter proposition, incidentally, seems to me without content. Skinner, so far as I can see, has no position at all with regard to human language. He has made a variety of terminological proposals; in particular he insists that the words 'stimulus', 'response', 'reinforcement', and several neologisms be used in describing language use, but he proceeds to deprive these terms of any content. For example, his notion of 'reinforcing stimulus' includes as a special case stimuli that do not impinge on the organism at all, but are merely hoped for or imagined. Quine too uses the term 'reinforcement' in a purely ritual fashion. Thus he suggests that in the case of language learning the child's reinforcement may be the "corroborative usage" of the speech community. Anyone would agree that corroborative usage, that is, data, is required in language learning. Quine in fact insists that his behaviorism is virtually empty. In his discussion of this matter in *Language and Philosophy*, he rejects the restriction of "behaviorism" to the theory of conditioning and says this: "When I dismiss a definition of behaviorism that limits it to conditioned response, am I simply extending the term to cover everyone? Well, I do think of it as covering all reasonable men. What matters, as I see it, is just the insistence upon couching all criteria in observation terms." All conjectures, he says, must "*eventually* be made sense of in terms of external observation."[4] This is, to be sure, a sense of "behaviorism" that would cover all reasonable men.

Quine states explicitly that "conditioning is insufficient to explain language-learning." In fact, he states that this was in essence the content of his doctrine of indeterminacy of translation. That is, he interprets this doctrine as asserting that language learning cannot be explained in terms of conditioning. I find it difficult to read this interpretation into his exposition in *Word and Object*, for example, but instead of pursuing the matter of "indeterminacy of translation," let us consider rather the notion 'conditioning'. In *Word and Object* Quine states that a theory, in particular a language, can be characterized as "a fabric of sentences variously associated to one another and to non-verbal stimuli by the mechanism of conditioned response" (p. 11). On the face

---

[4] "Linguistics and Philosophy," p. 97; emphasis added.

of it, this definition seems inconsistent with his assertion, in *Language and Philosophy*, that "conditioning is insufficient to explain language-learning." If the latter is true, then a language will not be a fabric of sentences and stimuli associated by the mechanism of conditioned response.

I think that the solution lies in the fact that Quine has not only virtually abandoned behaviorism and the concept 'reinforcement', but also the notion 'conditioned response'. We can see this by considering his account of language-learning, what he calls "learning of sentences." In *Word and Object* he specifies three mechanisms by which sentences can be learned, three mechanisms for language-learning. The first is association of sentences with sentences; the second, association of sentences with stimuli. These two methods would, it is true, lead to a fabric of associated sentences and stimuli. But there is a third method that is left rather obscure in *Word and Object*, namely, learning of sentences by what he calls "analogic synthesis." He gives only one example, which I quote in full:

"It is evident how new sentences may be built from old materials and volunteered on appropriate occasions simply by virtue of analogies. Having been directly conditioned to the appropriate use of 'Foot' (or 'This is my foot') as a sentence, and 'Hand' likewise, and 'My hand hurts' as a whole, the child might conceivably utter 'My hand hurts' on an appropriate occasion, though unaided by previous experience with that actual sentence." [5]

In Quine's terminology, the sentence "My hand hurts" might be learned by this method, the method of analogic synthesis. Clearly this is a curious use of the word 'learning'. Putting that matter aside, however, consider the consequences for his theory of language. Suppose that the sentence "My hand hurts" is 'learned' in this manner, and consider now the assumption that a language is a fabric of sentences associated by the mechanism of conditioned response. Then the sentence "My hand hurts," in the given example, is associated to the complex containing "This is my foot," "foot," "My foot hurts," and "hand" by the mechanism of conditioned response. To say this is to deprive the notion "conditioned response" of its strict meaning, or anything resembling this meaning. The responses and stimuli entering into the relationship of "conditioning" need not even appear together. Having deprived the no-

[5] W. V. Quine, *Word and Object* (Cambridge, Mass.: M.I.T. Press, 1960), p. 9.

tion of any content, we are free, without fear of contradiction, to describe a language as a fabric of sentences and stimuli associated by the mechanism of conditioned response.

In a response in *Synthese*[6] Quine emphasizes that one cannot regard sentences as associated with one another and "learned" as "unstructured wholes." Rather, they are associated and learned by various modes, such as "analogic synthesis," not merely as unstructured wholes. It must be stressed that this remark is, in its entirety, his positive theory of language learning. Again, we have a sense of "behaviorism" with which no one could disagree. What has happened, clearly, is that the terms "conditioned response" and "association" have joined "reinforcement" and "behaviorism" as terms with a merely ritual function, virtually deprived of substantive content, so far as I can see.

In *Word and Object* Quine introduced the notion of an innate quality space with a distance measure, to explain induction. In commenting on this, I noted (in the issue of *Synthese* just mentioned) that the examples Quine gives associate the qualities of the innate quality space with dimensions that have some simple physical correlate such as hue or brightness, with distance defined in terms of these physical correlates. I suggested that one could develop a substantive doctrine by making this association explicit. Quine, however, rejects this interpretation and rightly so, for it would make the doctrine false. He holds the "denizens of the quality spaces" to be "stimulations, any and all, with no prior imposition of dimensions." No further conditions are given. The concept is therefore quite vacuous. There is, for example, no objection to a quality space with a dimensional structure so abstract that any two sentences of English are "closer" than a sentence of English and a sentence of any other language, so that a person innately endowed with this quality space could learn all of English, by induction, from a presentation of a single sentence. He could, that is, generalize properly from such a presentation to full knowledge of all sentences of English with the situations in which they are appropriate. I take it that with this conclusion the notion "quality space" joins "reinforcement," "conditioned response," and "behaviorism."

This conclusion may be premature, however, for in *Language and Philosophy* (p. 97) Quine asserts that the quality space, thought not restricted to dimensions with simple physical correlates, nevertheless

[6] 19(1968):274–283.

still permits only what he calls "induction," and not the move to the "analytical hypotheses" which, he holds, must be developed somehow by the language-learner. I quote: "The as yet unknown innate structures, additional to mere quality space, that are needed in language-learning, are needed specifically to get the child over this great hump that lies beyond ostension, or induction." That is, they are needed to get the child to "analytical hypotheses." Quine also insists (in the *Synthese* article) that "generative grammar is what mainly distinguishes language from subhuman communication systems" — correctly, I am sure. Perhaps it is proper, then, to regard the principles and rules of the internalized generative grammar as among the "analytical hypotheses" that are arrived at by the "as yet unknown innate structures, additional to mere quality space, that are needed in language-learning . . . to get the child over this great hump that lies beyond ostension, or induction." Personally, I would find this interpretation of Quine's position quite congenial. The task of the linguist, in this interpretation, is to arrive at the internalized generative grammar using the evidence of language use. He will call the rules and principles of this grammar "analytical hypotheses," insofar as they are not determined through induction by the mechanism provided by the quality space of unknown dimensions. He will then seek to determine the nature of the quality space and the unknown innate structures, additional to the quality space, that are needed to account for the construction of this generative grammar by the language-learner. I believe that this is a fair rendition of Quine's most recent formulations. I think it is fair to describe this as an abandonment of behaviorism. I would only suggest that we now also abandon the terms "association," "conditioning," "reinforcement," and "behaviorism," now that they have been deprived of whatever content they have in the psychological literature, and now that all of the characteristic assumptions of behaviorism have been abandoned.

One last remark on Quine's behaviorist theory of language. In *Word and Object* Quine defines a language as a "complex of present dispositions to verbal behavior, in which speakers of the same language have perforce come to resemble one another." Surely he must also abandon this view. Presumably, a complex of dispositions is representable as a set of probabilities for utterances (responses) in certain definable circumstances or situations. Suppose that knowledge of language is represented by such a description. Then assuming 'circumstances' and 'situations' to be defined in terms of objective criteria, as Quine insists, it is surely the case that almost all entries in the situation-response matrix are null. That is, in any objectively definable situation, the probability of my producing any given sentence of English is zero, if probabilities are assessed on empirical grounds; in any event, it is not detectably different from the probability of my producing some sentence of, say, Japanese, if probabilities are assessed on empirical grounds. Thus knowledge of English is not differentiable from knowledge of Japanese. Clearly the whole approach is untenable and should be simply abandoned. If it is a generative grammar that mainly distinguishes language from subhuman communication systems, as Quine holds, then a language cannot be defined as a complex of dispositions to respond, since a generative grammar cannot be characterized in these terms.

To summarize: as a first approximation it is fair to assume, not as a matter of logical necessity, but as a plausible hypothesis, that the 'state of mind' of a person who knows a language is characterized by a generative grammar, a system of rules and principles that determines a sound-meaning connection for an infinite set of sentences. A deeper problem will be to investigate the 'state of mind' that is innate to the organism, that makes it possible for the grammar to be constructed in the specific way it is on the basis of experience. Thus far there appears to be near agreement between Quine, Harman, and myself. Agreement does not prove correctness, of course. To show that this approach is a valid one, one must demonstrate empirical successes in accounting for some interesting phenomena. But agreement does indicate, I think, the difficulty of conceiving a coherent alternative.

Let me consider a number of other apparent alternatives to this approach. In a brief essay in *Language and Philosophy*,[7] Nelson Goodman suggests that even if certain principles are discovered that characterize what might be called "the workings of the mind" in such a way as to meet the empirical conditions of description and explanation, there is no necessity to regard these principles as "in the mind" except in the sense that they can be inferred from what the mind does. That is, these principles may be nothing "more than descriptions, by an observer, of the resulting organization." They "need no more be in the minds in question than the theory of gravitation need be in bodies." (I should

[7] "The Emperor's New Ideas," pp. 138–142.

add, for clarification, that Goodman dislikes this whole terminology and accepts it only for purposes of discussion.) Presumably this argument holds both of the 'final state of mind' and the 'initial state of mind' of the language-learner, both of the mind of one who knows the language, and the mind of one equipped to learn it. In either case, following Goodman, we may say that the linguist's theory of the two states says nothing more about the organization of mind than the theory of gravitation says about the internal structure of bodies. That is, there is no reason to say anything other than that minds, in their initial and final states, are governed by these rules and principles. Just as the laws of gravitation are not represented internally in bodies governed by these laws, so the rules and principles of grammars are not represented in the minds governed by these rules and principles.

The argument is a bit unfair to Newton, who, after all, was not quite unconcerned with the question of how the behavior of a body under the laws of gravitation was affected by its form and internal structure. But putting that aside, suppose we were to apply Goodman's argument to a scientist who is presented with some device, say an automobile engine, which he cannot take apart and can investigate only by studying its behavior and 'input-output' relations. Suppose that the scientist cleverly hit upon a theory that worked quite well, namely, the theory that makes use of components with the properties of cylinders, spark plugs, and so on. He might hesitate to speculate about the precise physical realization of these concepts, and leave it as an abstract theory of the device, with various components, a certain kind of interaction between them, and so on. Observing Goodman's scruples, he should say only that the device behaves according to principles enunciated in his theory, that these principles describe its behavior only in the sense that the laws of gravitation describe its behavior. If the scientist were to propose that the device is actually constructed in accordance with the principles outlined, that the hypothesized components are embodied in the device in some fashion or other, he would be going beyond what Goodman regards as proper. Goodman's strictures are a bit vague, but if I understand him, he is saying that the scientist may say *that* the device obeys certain principles, but he may not ask *why* it does so; he may not speculate on the internal structure and organization that lead to its behaving in accordance with these principles. If this is his intention, the reference to the theory of gravitation is incorrect, since, as noted, Newton was quite concerned to explain why a complex object would behave like a point mass. But in any event, the scruples, so interpreted, simply amount to a lack of curiosity, an unwillingness to pursue questions beyond a certain arbitrary point.

Consider the remark that the principles may be nothing "more than descriptions, by an observer, of the resulting organization." To say that certain principles are descriptions of the resulting organization does not seem to me any different from saying that the system is organized in the way described by the principles in question. In this case Goodman's strictures appear to be merely terminological.

In any event I fail to see that he has offered any objection to the conclusion that a generative grammar is somehow internally represented in the mind of the person who knows the language, and that innate structures to which we are led in the study of language acquisition are internally represented in the mind of the language-learner; or, if you like, that the final state of mind of the language-learner is correctly described by a generative grammar, and the initial state by the postulated innate structures and principles. There is no difference, so far as I can see, between these terminologies.

Goodman does, however, have a different view with respect to language-learning. He argues that acquisition of language is a case of second-language learning, and that it is facilitated by the possession of a language of gestures and perceptual signs that can be used perhaps as analogues, or perhaps in explanation and instruction. This is pure hand-waving. In answer to the question how the specific structures of natural language are derived from systems of gestures and perceptual signs, he says nothing; or, to be more precise, he offers another analogy, namely, that just as tools can be used to make a clock, so prelinguistic symbolic systems can be used to explain — let us say — the principle of cyclic application of grammatical transformations. But this clearly tells us nothing about how the latter can be derived as an analogue to the former.

Instead of trying to deal with these problems, Goodman offers still another analogy. He points out that "the features that identify a picture as by a certain artist or of a certain school or period are in some sense deep (or obscure). Yet we learn with rather few examples to make some of the latter rather subtle distinctions. Must the mind therefore have been endowed at birth with a 'schematism' of artistic styles?" Apparently he regards it as self-evident that the answer to this rhetorical question is

313

negative. The analogy is entirely without force. To explain how a person learns to make subtle distinctions with few examples, we must attribute to the mind an innate structure rich enough to achieve this result, and not so rich as to be falsified by the actual range of such 'input-output' relations. If a postulated 'schematism of artistic styles' will meet these empirical conditions, there is no reason to be at all surprised, or to regard this as an objectionable hypothesis. Goodman seems amazed at the idea that one should seek to study human mentality in exactly the way one would approach any organism, or an inanimate device of unknown properties that modifies its state through time. The recent literature on this question contains many other rhetorical questions like the one Goodman asks, as if it were somehow absurd, prima facie, to construct a theory of the structure of mind that accounts for empirical facts, attributing to it as much structure as is necessary to do so. If we are forced to attribute to the mind separate 'faculties' in order to account for the empirical facts, then so be it. There is no reason for any a priori attitude as to the rightness or wrongness of such a result. If we find generalizations governing several such 'faculties', or more general structures that can account for many types of learning, well and good. Again, there is no reason for dogmatism in this matter.

When we try to characterize the state of mind of a person who knows a language, taking account of his ability to use and understand an indefinite range of sentences, each with its phonetic form and meaning-potential determined in a specific way, we are led to certain empirical hypotheses, specifically, to the construction of a generative grammar, a system of rules and principles that establishes a certain sound-meaning relationship. We may then say that this theory describes the organization of mind or that the mind is organized in accordance with this description. I see no difference. In either case we can then go on to ask how this organization developed through an interaction of experience and innate structure, and can seek to determine the specific innate endowment that makes this achievement possible.

So far I have said nothing of 'knowledge of language'. I think that it would be quite reasonable to suggest a characterization of 'knowledge of language' in the terms just given; to say, that is, that to know a language is to have internalized a generative grammar (equivalently, to have developed a state of mental organization as described by a generative grammar). It might be argued that this proposal does violence to the concept 'knowledge'. The latter concept seems to me sufficiently obscure so that I do not know whether or not this criticism is just. It seems to me that our concept of knowledge fades into obscurity at the point where we consider what Leibniz referred to as the principles that "enter into our thoughts, of which they form the soul and the connection," principles as necessary to thought "as the muscles and sinews are for walking, although we do not at all think of them." In the past I have tried to avoid, or perhaps evade the problem of explicating the notion 'knowledge of language' by using an invented technical term, namely, the term 'competence' in place of 'knowledge'. However, the term 'competence' suggests 'ability', 'skill', and so on, through a chain of association that leads direct to much new confusion. I do not think that the concepts of ordinary language suffice for the purpose at hand; they must either be sharpened, perhaps somewhat arbitrarily, or replaced by a new technical terminology. Either approach has a familiar disutility.

Suppose that we were to propose that to know a language is to have constructed, to be sure unconsciously, a specific generative grammar. A familiar argument against this proposal is that I can tell whether someone knows English, but I know nothing of the internal workings of his mind. The argument seems to me weak, because it begs a question rather like the one under discussion. If we are prepared to admit that the mind can incorporate unconscious theories, systems of principles and rules that we might describe as unconscious knowledge, then it is at least conceivable that I have an unconscious theory that attributes to other persons minds of a certain character, and that assigns to them mental states by virtue of certain actions that they perform. Suppose, furthermore, that this theory relates to my unconscious theory of English in such a way that I believe someone to know English when I attribute to him the mental state described by (or incorporating) the rules and principles of English grammar, arriving at this conclusion on the basis of his behavior. I see no incoherence in this formulation, which would support the conclusion that my concept 'knowledge of a language' is directly related to the concept 'internalization of the rules of grammar'.

Perhaps one might approach the analysis of "knowledge of a language" in a simpler and more direct way, supposing that to know a language is to know how to speak and understand, the latter being a dispositional concept of some sort. This leads us nowhere, so far as I can see. The problem immediately arises of characterizing the relevant dispositions, determining how they are related, how this complex is organized and developed, why certain dispositions are excluded from it and others not. Furthermore, we must face the fact that two people may have very different dispositions, may be inclined to say very different things in given circumstances and to interpret what is said quite differently, and yet we may attribute to them exactly the same knowledge of language. We are back where we were, so far as I can see.

There are difficulties, perhaps not insuperable but worth mention nevertheless, in attempting to account for knowledge of language in terms of 'knowing how'. To know English is, let us say, to know how to talk grammatically, how to understand what is said to us in English, and so on. To say that someone knows how to do such things is to say that when the relevant acts are performed, they are performed intelligently. To act intelligently, as Ryle put it long ago, is not merely to satisfy certain criteria but also to apply them. But what does it mean "to apply criteria"; how do we know which criteria to apply, how are these criteria related and organized. Again, we run into a barrier that can be overcome, I think, only by introducing some concept like generative grammar, and mental representation of generative grammar.

In fact, the discussion of 'knowing how' often takes a different, and I think misleading course, at exactly this point. It is said that to know how, to be able to act intelligently, is to have a skill, a disposition or complex of dispositions to act in a certain way. Consider the standard example, knowing how to play chess. This is sometimes described as a skill acquired by training, a complex of dispositions to make certain appropriate moves, a skill that improves with practice. A person is said to know how to play chess if he normally makes the permitted moves, avoids the wrong moves, and so on (Ryle). But there is a possible source of confusion here. Consider the difference between the two questions: "Does John know how to play chess?"; "How well does John (know how to) play chess?" It is the latter, not the former, that asks about a skill and how well it is exercised. The former asks about possession of knowledge, which can be partial or improved only in the way in which knowledge of facts can be partial or improved.

Knowing how to ride a bicycle may be a matter of having certain habits and reflexes, and knowing how to drive a car may be a matter of having certain skills. Similarly, being able to play chess well may be a matter of having certain skills, acting thoughtfully, and so on. But knowing how to play chess seems to me more like knowing how to get from New York to Chicago than like knowing how to ride a bicycle or drive a car. If a person does not know some specific rule of chess, say the rule of castling, he may still play chess very well; he may (in principle) be world champion, although it would still be correct to say that he does not know how to play chess in the way the rankest amateur who loses every game may very well know how to play chess. We may say that this chess champion has partial knowledge, but his defect of knowledge is very different from the defects in knowing how to play chess on the part of the poor player who knows all the rules (perhaps unconsciously, without ability to formulate them). Discussions of skills, dispositions, thoughtful action, and so on, seems to me appropriate in relation to ability to play well, but not in relation to knowing how to play; the latter concept has an irreducible intellectual component.

It is true that we say that a person knows how to play if he normally makes the right moves, etc. (as Ryle puts it); that is, his actions provide the evidence that leads us to attribute to him the knowledge how to play chess — perhaps erroneously. But to attribute to him this knowledge is not the same thing at all as attributing to him the ability to play well, perhaps on the same evidence. The concepts are entirely different, though the evidence for applying them may overlap.

Knowledge of a language, knowledge of classical mathematics, knowledge of chess, seem to me to be instances of 'knowing how', if at all, only in the sense in which knowing how to get from New York to Chicago is an instance of 'knowing how'. The latter implies no particular ability to get from New York to Chicago. That requires other talents apart from knowing how to get from New York to Chicago. And, as in the case of playing chess, the person most successful in getting from New York to Chicago, repeatedly and characteristically, may not know how to get from New York to Chicago as completely, as fully, as perfectly, as someone who repeatedly fails. Similarly, two people may know English in exactly the same way, they may have exactly the same knowledge of language, and yet differ in their dispositions, in how well they speak and understand the language that both know; as they might in principle match perfectly in these dispositions, and yet differ in their knowledge of language. Ability to speak and understand involves not

317

only knowledge of language, in the strict sense, but much else as well. The concepts should be kept quite separate.

There is a further problem that should be mentioned. Knowledge of language can be partial only insofar as there is an external standard. Let us assume, for simplicity, a perfectly homogeneous adult speech community. Customarily, one speaks of a child as having partial knowledge of a language, taking as the external standard the grammar of the adult language, of the speech community to which he belongs. In another sense the child has, by definition, a perfect knowledge of his own language. There is no contradiction here; rather, there are two notions that must be distinguished. The study of language should be concerned, in the first place, with the speaker's perfect knowledge of his own language. The notion of 'language' as a common property of a speech community is a construct, perfectly legitimate, but a higher order construct. In the real world there are no homogeneous speech communities, and no doubt every speaker in fact controls several grammars, in the strict sense in which a grammar is a formal system meeting certain fixed conditions. A person who is capable of learning a language in the real world of complex and overlapping dialects has a certain ability which would enable him to learn the language of a homogeneous speech community. It is this ability that we must seek to capture and understand, abstracting away from much real world complexity. Given some understanding of this ability, we may try to come to grips with the concrete problem of 'knowing how to speak and understand', of acquiring and using knowledge of language, in the real world of heterogeneous and overlapping styles and dialects.

One last problem. Suppose that one is willing to accept the characterization of knowledge of a language in terms of possession, internal representation of a generative grammar. Clearly the rules and principles of this grammar are not accessible to consciousness in general, though some undoubtedly are. I think that what we discover, empirically, is that those principles and rules that are accessible to consciousness are interspersed in some obscure and apparently chaotic way among others that are not, the whole complex of rules and principles constituting a system of a very tight and intricate design, and meeting stringent and restrictive general conditions of a sort that I will try to illustrate in subsequent lectures. Suppose that one is prepared to apply the notion 'knowledge' in this case — that is, to accept the locution "so-and-so knows the grammar of his language, its rules and principles," including those that lie beyond awareness. Suppose it is true, as I believe, that further investigation leads us to the conclusion that this knowledge is acquired on the basis of certain innate principles of what might be called 'universal grammar'. I mean by 'universal grammar' a certain fixed language-independent schematism that determines what counts as linguistic experience and what knowledge is acquired, what grammars are constructed, on the basis of this experience. Would we want to say, as well, that the child 'knows' the principles of universal grammar?

It seems to me that very little turns on the answer given to this question. It is also unclear to me whether the concept 'knowledge' is sufficiently clear to guide us in making a decision. It would be quite useful to have a concept X such that a person who knows English, and thus knows certain facts of English, also X's these facts; that furthermore, he X's the principles and rules of his internalized grammar, both those that can be brought to awareness and those that are forever hidden from consciousness; and that he X's the principles that underlie the acquisition of language in the first place. We have no clear concept such as X. My guess is that Leibniz would have been happy to extend the concept 'know' to have the meaning of X. It seems a natural enough step to me, though I do not want to press the point. If we were to adopt this way of speaking, extending familiar terminology in what seems to me a natural direction, we would then speak of knowledge, unconscious knowledge, and innate knowledge. A person who knows English would no doubt have conscious knowledge of certain facts, say that bachelors are unmarried or that "is" goes with singular and not plural subjects. He would have unconscious knowledge of the fact that the passive transformation is ordered in a certain way with respect to the transformation that inverts direct and indirect object. He would have innate knowledge of the fact that transformations apply in a cyclic ordering. We might refer to his innate knowledge as 'knowledge of universal grammar'. Alternatively, we may use the term 'know' in a narrower way, restricting it to conscious 'knowledge that', and to knowing how, why, who, and so on. 'Knowledge of' in the sense of 'knowledge of language' will then be explicated in terms of a new technical terminology: 'internal representation of a grammar', '. . . of an innate schematism,' and so on. In

319

this usage what is 'known' will be a somewhat ill-defined and, perhaps, a rather scattered and chaotic subpart of the coherent and important structures and systems that are described in terms of the new, more forbidding terminology. As long as we are clear about what we are doing, either approach seems quite all right.

Language, Rules, and Complex Behavior

Prominent among recent additions to psychology's store of theoretical terms are the terms "rule," "tacit knowledge of rules," and "rule guidance." Language, Noam Chomsky tells us,[1] is a rule system and what every speaker of a language has mastered and internalized is the system of rules that comprise his language. What speakers do when they speak and understand the language is done in accordance with the rules. In speaking and understanding a language, we interpret, execute, and follow rules.

Looking beyond language and verbal behavior to other forms of intelligent behavior, J. A. Fodor has maintained that "the paradigmatic psychological theory is a list of instructions for producing behavior."[2] What the rule theorist advances is what Fodor calls an "intellectualist" theory of psychological explanation. For the rule theorist, performing some bit of complex behavior involves employing rules. The explication of these rules is tantamount to a specification of how to perform the behavior.

In saying that performance involves rules or that the behavior is guided by rules, the rule theorist means to say more than that the rules describe regularities in the behavior. There are many alternative ways of describing complex behavior and one way is to have the description take the form of a rule or instruction that indicates how to behave from moment to moment. However, the claim that a system employs rules or that some bit of its behavior is guided by a rule presumably implies more than that the rule describes the behavior. There are regularities in the behavior of falling bodies and in the behavior of gases, but, of course,

AUTHOR'S NOTE: Portions of this paper were presented at the Third Annual Conference on Structural Learning, Philadelphia, March 1972, and at the Pacific Division Meetings of the American Philosophical Association, San Francisco, March 1973.
[1] Noam Chomsky, *Aspects of the Theory of Syntax* (Cambridge, Mass.: M.I.T. Press, 1965).
[2] J. A. Fodor, "The Appeal to Tacit Knowledge in Psychological Explanation," *Journal of Philosophy*, 65, no. 20:630.

almost no one wants to say that the behavior of bodies in free fall and of gases under increased pressure are rule-governed activities. Similarly there are regularities in the light emitted by a firefly during a brief flight, regularities in the feeding reactions of thrush nestlings, and in the zigzag dances of male sticklebacks. In man there are regularities in salivating and hand-out-of-the-fire responses, and there is a regularity in the upward kick of a crossed leg when a mallet is applied briskly to a spot under the kneecap. Yet we would not be prepared to say in all of these cases that the response or the behavior is guided by a rule.

Presumably when a rule theorist argues that speaking a language and doing arithmetic involve rules, he means that the rules play a role in the etiology of that behavior. The rules somehow cause or enable us to do what we do. Rules like beliefs can be the determinants of behavior.

However, when the rule theorist says that rules determine a subject's behavior, he doesn't mean that the subject is always or ever aware of these rules. The rule theorist denies that everyone who employs a rule for summing arithmetic series is able to answer the question, "How does one find the sum of an arithmetic series?" That is, a rule may guide a subject's behavior and the subject may not be conscious of the rule. In fact the rules that guide our behavior are often rules of which we are not aware.

What the rule theorist doesn't tell us, however, is how to justify talk about guidance by rules. That the rules describe the behavior is obviously not enough. What is needed is some sharper analysis of the notion of someone's behavior being guided by a rule. What is the difference between a rule's fitting someone's behavior and its guiding that behavior? As George Miller has said: "In spite of our increasing reliance on rules as explanations of thought and behavior, I do not know of any clear account of what rules are and how they function."[3]

In sections II and III of this essay I take up the question, "How does one justify a claim that what an organism does it does because it is guided by a rule?" Rule-guided behavior, I argue, displays a pattern of organization that is characteristic of language. The behavioral events that make up a rule-guided sequence of behavior are not selected in serial order. There are dependencies between distant events similar to the dependencies between distant words in a sentence. Most important:

---

[3] George A. Miller, "Four Philosophical Problems of Psycholinguistics," *Philosophy of Science*, 37, no. 2:191.

if there is no limit on the distance between dependent events, the etiology of the sequence must include a rule and we say that the behavior is rule guided. Here we have a sufficient condition for ascription of "rule guidance."

Language behavior is most often offered as the clearest example of rule-guided behavior. However, one result of our discussions of rule guidance is that the ability to speak and understand a language turns out to be more like our other abilities than first thought. Rules guide speech but they guide many of our other performances as well.

In section IV, I consider Chomsky's claims that man's ability to speak and understand a natural language represents a unique type of intellectual organization that manifests itself in what he refers to as the "creative aspect of language use." Language use is creative, Chomsky tells us, because language is unbounded in scope and stimulus-free. But the features that mark creative use are only coarsely drawn in these discussions. Once they are drawn more clearly, we see that, despite Chomsky's claims, language is not unique in having a creative use. Speaking a language again turns out to be a less distinctive activity than first thought.

I

Psychologists assume that behavior can be analyzed and that fixed, elementary units can be singled out. Descriptions of behavior are descriptions on this vocabulary of elementary units. However, two important questions remain, and of late these questions have received considerable attention: (a) what are the elementary units, and (b) on what principles of concatenation are complex behaviors built? Many of the interesting results that shed some light on (b) have come from linguistics.

Linguists have shown that there is a structure to most sentences that is independent of their linear structure. That is, two linguistic expressions may contain the same morphemes and the morphemes may occur in the same linear order and yet the expressions may differ in syntactically significant ways, for example, the expressions "*old men and women*" and " *old men and women*." Thus sentences like

(1)   John gave ten dollars to Harry and Phil
(2)   Intriguing women can be a drag

are ambiguous and the ambiguity does not arise because the meanings of some of the words are ambiguous, but because different syntactically significant structures can be associated with the same strings.

What has not been as often noticed is that these linguistic results can be easily generalized and quite naturally applied to nonlinguistic behavior as well. Most forms of complex nonverbal behavior have a non-linear, inner structure of their own and the significance of the behavior for the organism depends on this structure. Among the first American psychologists to make this point was Karl Lashley.

From such considerations, it is certain that any theory of grammatical form which ascribes it to direct associative linkage of the words of the sentence overlooks the essential structure of speech. The individual items of the temporal series do not in themselves have a temporal "valence" in their associative connections with other elements . . .

This is true not only of language, but of all skilled movements or successions of movement. In the gaits of a horse, trotting, pacing, and single footing involve essentially the same pattern. . . . The order in which the fingers of the musician fall on the keys or fingerboard is determined by the signature of the composition; this gives a set which is not inherent in the associations of the individual movements.[4]

Nonverbal behavior, then, may be ambiguous in much the same way that sentences are ambiguous. That is, two superficially identical bits of behavior may be derived from two different underlying cognitive configurations. Two bits of complex behavior may include the same behavioral events and the events may occur in the same temporal order and yet the bits of behavior may differ in cognitively significant ways. Consider, for example, the following sequences of behavioral events: B walks out the front door of his house, walks to the end of the block, notices a newsstand, buys an evening paper, stops and speaks to a passerby, turns, walks back down the block, and enters his house through the front door. This description may equally well fit the following three cases: (a) B makes it a practice always to return home by the same door from which he leaves; (b) B was out for a stroll and bought a newspaper as a second thought; and (c) B went out in order to buy a newspaper. If (a) or (c), then there is an order to the previously specified sequences of behavioral events that is independent of the linear order. There are significant relations between events that are not tem-

---

[4] K. S. Lashley, "The Problem of Serial Order in Behavior," in *Cerebral Mechanisms in Behavior*, ed. L. A. Jeffress (New York: Wiley, 1951), p. 116.

porally adjacent. In order to give a more complete and unambiguous description of B's activities, we need to take account of underlying structure. A description sufficient to distinguish (a) from (c) will need to include a description of hierarchical relations between the events. In short, B's walk to buy a newspaper is hierarchically ordered in the way in which the sentences of his language are hierarchically ordered, and while we need to appeal to that ordering in order to disambiguate many of those sentences, so we need to appeal to underlying structure and hierarchies in order to disambiguate B's nonverbal behavior, his walk and purchase of a paper.

We can draw the linguistic analogy even farther, for we can represent behavioral organization in the same manner in which we represent the constituent structure of linguistic expressions, for example, tree diagrams or labeled bracketings. In an unambiguous description of B's behavior, we need to indicate which events depend on the character of other events in the sequences, and we can do this for nonadjacent events by connecting the dependent events to a node that lies outside the linear sequence itself. The event of B's walking out the front door of his house, e_1, and the event of B's walking back in by way of the front door of his house, e_8, are dominated by a common node

and are thus coordinated even though they are a considerable distance from one another in the temporal chain of behavioral events. What a tree diagram for a temporally ordered sequence of behaviors will indicate is that there are psychologically significant relations between events that are independent of their temporal order. What the node represents is an integrative function that is independent of the associations between temporally adjacent events. As the example illustrates, a subject's selection of one event will often determine or restrict the choice of proceeding events that are far removed in space and time. Again the parallel with the structure of natural language is obvious. In English choice of verb determines or restricts choice of complement and, similarly, it often determines or restricts the choice of subject. We say (3) but not (4) and we say (5) but not (6).

(3)   John broke the bottle

(4)   *John broke the water
(5)   John who loves dogs also love cats
(6)   *John who love dogs also love cats

In *Syntactic Structures* Chomsky noted that dependencies between morphemes, between elements in a terminal string, often stretch across a number of other words and that there is no upper limit on that number. There are, for example, English sentences of the form

(7)   If S_1, then S_2

and these sentences include a dependency between "if" and "then," and yet S_1 and S_2 may in turn have other sentences within them and those in turn sentences within them and so on.

A similar case can be made out for most forms of complex nonverbal behavior. Complexity, we noted, frequently takes the form of a hierarchy. The behavior involved in leaving one's house in order to purchase a newspaper is hierarchically ordered. Moreover, there is no upper limit on the number of behavioral events across which these motor coordinations or behavioral dependencies can stretch.

Let me illustrate this last by way of another example. You learn to knit and then undertake to knit your husband or wife a sweater. Your plan is first to knit the left sleeve, next to knit the front and back panels, and then to knit the right sleeve. You proceed to execute your plan and knit the left sleeve to some desired length n and proceed to complete the center panels. Finally, you turn to the right sleeve. How you knit this last sleeve will depend on how you knitted the first. You want the patterns, the stitches, the colors, and length to be the same for each. Although the sequence of events that constituted the knitting of the first sleeve is far removed from the events that need to be included in the knitting of the last, they depend upon one another. You want to coordinate what you do now with what you did earlier. Of course, you could return to the left sleeve, rehearse your earlier behavior, and follow each "operation" on the first sleeve with an appropriate bit of knitting behavior on the second. However, this is a needlessly onerous and time-consuming strategy. It is likely that you can coordinate the knitting of the two sleeves with very little explicit cross-reference. If indeed you are following a rule, it is the rule that provides the basis for the coordination and not the individual units of behavior. In any case, knitting the sweater includes coordination of events that are discontinuous and the events that are distant need to be coordinated. Further, there is no upper limit on that distance. You can interrupt knitting the sleeve in order to complete the front and back panels, but can also take time out to write a book about knitting, brush your teeth, prepare dinner, or sail to Calais. There is no longest interruption.

There are then nonverbal behaviors of the form

(8) $\quad X_1 + X_2 + X_3$

and these behaviors include dependencies between events in the sequence X_1 and events in the sequence X_3. More importantly, the dependencies cut across the sequences of events X_2 and yet this last sequence of events may itself include behavior of the form (8) and so on. Thus knitting behavior can include an unlimited number of what linguists call "nested dependencies" and a structural tree for the behavior involved in

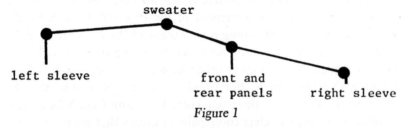

Figure 1

knitting a sweater might look like the diagram in Figure 1, and this, of course, resembles the configuration assigned to sentences of the form of (7).

II

It is possible to make the parallels between the structure of language and the structure of many forms of complex behavior more precise. Specifically, we can define a relation on hierarchically ordered behavioral events that corresponds to a relation that Chomsky[5] has defined on elements in terminal strings.

Suppose we have a natural and antecedently agreed-upon taxonomy of behavioral events. For example, we might count drawing as a type of behavior and each of the following as (1) a type of behavior and (2) a proper part of the type of behavior that precedes it: drawing a face, John's drawing a face, John's drawing a face with prominent features.

[5] Noam Chomsky, "Three Models for the Description of Language," I.R.E. Transactions on Information Theory, IT-2, no. 3, pp. 113–124.

Or consider: entering and leaving, John's entering and leaving, John's entering and leaving by the same door.

Let K be a class of behavioral sequences each of which (a) a subject S is able to perform and (b) counts as an instance of the same type of behavioral event. Let $Y_1, Y_2, \ldots, Y_n = X$ and X be a sequence in K. We say that X has an (i,j)-dependency with respect to K if and only if:

(i) $1 \leq i$ and $i + m < j \leq n$ where $m \geq 1$,

(ii)   there are behavioral events Z_i, Z_j which S is able to perform and $Z_i \neq Y_i$ and $Z_j \neq Y_j$,

(iii) Z_i, Z_j have the property that X' is not while X'' is in K where X' is formed from X by replacing the ith behavioral event in X (namely, Y_i) by Z_i and X'' is formed from X' by replacing the jth behavioral event in X' (namely, Y_j) by Z_j.

In other words, a behavioral sequence X has an (i,j)-dependency with respect to K if and only if replacement of the ith event Y_i of X by Z_i ($Z_i \neq Y_i$) requires a corresponding replacement of the jth event Y_j of X by Z_j ($Z_j \neq Y_j$) for the resulting sequence to belong to K. Now let us assume that sequence X and every other sequence included in K is (i,j)-dependent with respect to K. For simplicity we might assume that K consists of X and all and only those sequences that result from X by means of the following steps: we select those pairs of events that are not identical with Y_i, Y_j but whose substitution for this last pair will preserve membership in K; we replace Y_i, Y_j with each pair of these events. I assume that there are only finitely many such pairs.

I wish to consider a subject, S, who has the ability to perform every sequence in K and who has the ability to perform them in circumstances in which it is possible for S to perform sequences like X'. That is, I wish to rule out cases in which S is able to perform all of the sequences in K but only because performance of any other sequence, e.g., X', is excluded by external circumstances. An example here would be helpful. Consider those behavioral sequences that we would call "entering and leaving by the same door." K consists of all and only the sequences that satisfy this description. Each of these sequences will be (i,j)-dependent with respect to K. We will let X here be a sequence in which one enters by the front door and leaves by the front door. Event Y_i is entering by the front door and Y_j is leaving by the front door. Events Z_i and Z_j can be entering by the back door and leaving by the back door. Sequence

X' will not be in K while X'' will. The subject I wish to consider is one who is always able to enter and leave by the same door but one who is able to do this in quarters that contain more than one door, in quarters that allow him to enter and leave by different doors.

In these circumstances, where the subject has a choice in his enterings and leavings, the ability to perform all of the sequences in K requires that the subject's choice of an ith event determine his choice of the jth event. When S goes to leave he cannot leave by just any door and of course he cannot decide willy-nilly to leave by the window. Our subject has to leave as he entered and this requires an agreement between i^{th} and j^{th} events.

Finally let us assume that there is no limit on m in the expression "$i + m < j \leqq n$" and that there is no longest sequence in K. We will assume that K is an infinite class of (i,j)-dependent behavioral sequences and S can perform each of the sequences in K. To return to our example of entering and leaving by the same door: to say now that S can perform all of the sequences in this class is to say that he can enter by the front door, walk up and down a flight of stairs, and still manage to leave by the same door. To say here that there is no limit on m is to say that in principle there is no limit on the number of times that S can repeat the walk-up-and-down-the-stairs cycle. Indefinitely many trips up and down the stairs may intervene between the ith and j^{th} events, between entering by the front door and leaving by the front door. Nevertheless S in being able to perform all of the sequences in K is always able to get his enterings and leavings to agree. The recursion of trips up and down the stairs results in infinitely many sequences that begin with an entrance and end with a departure. S is able to perform them all. This is no simple task. It requires the employment of what I want to call "a rule."

III

We return now to the question of rule guidance. How do we justify a claim that a rule is guiding an organism's behavior? What is the cash value of such a claim? The notions introduced in the last section provide the basis for an answer; they provide us with a principle by which inferences to rule guidance can be licensed. The principle goes thus: if an organism is capable of producing a certain type of behavior and this behavior consists of (i,j)-dependent sequences and there is no limit on the distance between i^{th} and j^{th} events, then in producing these sequences the organism is being guided by a rule.[6] Thus if we know that an organism's ability is an ability to produce a class of behavioral sequences like the sequences in K, then by this principle we can conclude that in the production of each sequence the organism is being guided by a rule.

There are, I believe, a number of good reasons for accepting this principle and each reason depends on features built into class K, a class of (i,j)-dependent sequences with no limit on m, no limit on the distance between i^{th} and j^{th} events. The behavioral sequences in this class have the property that they cannot all be produced from left to right in a single pass. The processes that are involved in their production have to be hierarchical processes. With hierarchical processes the linear order of behavioral events may not reflect the order in which the events are actually produced. Control over production may shift forward and backward across the sequence. An initial event may determine the character of an event that occurs only much later in the sequence. In the case of our (i,j)-dependent sequences not every k^{th} event $(1 \leq k \leq n)$ in the sequence can be selected on the basis of the $k\text{-}1^{th}$ event alone. Selection of the jth event is determined by the ith event and $i \neq j\text{-}1$. What we require for the production of all of the sequences in K is a special pattern of control: at various times control over the subject's behavior has to shift from the preceding event to a state or structure outside of the sequence that stores the location of the ith event and that selects the appropriate event for j.

Class K shares a feature with languages that include sentences of the form of (7) where there is no limit on the length of S_1. The sentences of these languages and the sequences in K cannot be produced by a Markov process, by a process that moves only from left to right and in which the occurrence of each element that the process produces is determined by the immediately preceding element or series of elements. In sentences like (7) choice of a word is determined by a word that occurs much earlier in the sentence. While in sequences like those in K choice of a behavioral event is determined by an event that occurs much earlier in the sequence. A process that produces these sentences and a process that produces the sequences in K must employ elements or structures that never appear in any of the sequences or sentences. In the case of

---

[6] We again assume that our subject can perform sequences like X', that the circumstances are such, for example, that he can enter and leave by different doors.

language these elements are grammatical constructions or categories that label the nodes on the grammarian's tree. In the nonlinguistic case they have a quite similar role. Their role in both the production of sentences and in the production of the sequences in K is essentially integrative. They impose the order on the sequences of skilled movements of which Lashley spoke. They provide a "generalized schemata of action which determine the sequence of specific acts, acts which in themselves or in their associations seem to have no temporal valence."[7] Given their independence of motor events and integrative function, an apt description of these structures is "rules." That is, (1) having an integrative function and (2) being independent of the elements they integrate are the two properties that we intuitively assign to rules. A rule, I suggest, is any structure that is somehow realized in an organism or machine, that controls the order in which a sequence of operations or events is to be performed, and yet that is independent of those operations or events. A rule is guiding some bit of an organism's behavior when it is the rule that determined the sequence of operations the organism is carrying out.

Consider again a subject B who goes out to buy a newspaper and makes it a practice to depart and return by the same door. $X = e_1$, $e_2, \ldots, e_8$ where e_1 is B's walking out the front door, e_2 is his walking to the end of the block, e_3 is his noticing a newspaper stand, e_4 is his buying an evening newspaper, e_5 is his stopping and speaking to a passerby, e_6 is his turning, e_7 is his walking back down the block, and e_8 is his entering by the front door. X is an instance of leaving-and-entering-by-the-same-door behavior and B performs X. Moreover, *ex hypothesi*, there is an (i,j)-dependency between events e_i and e_8. These events must "agree" with one another, the door he leaves and enters by must be the same door. If B were to choose to depart by a different door, by the back door, his choice somewhat later on, his choice of the 8th event would have to be of a different door as well. Moreover, B can always lengthen the sequence by iterating walks up and down the block. He may always walk up and down another block before he reenters his home. So long as these sequences are hierarchically ordered this poses no special problem. Even in the short sequences agreement between the first and last events, between e_1 and e_n, was in no way dependent on the e_{n-1}^{th} event.

Figure 2 reveals the cognitive relations and structure that underlie X.

---

[7] K. S. Lashley, "The Problem of Serial Order in Behavior," p. 122.

Michael D. Root

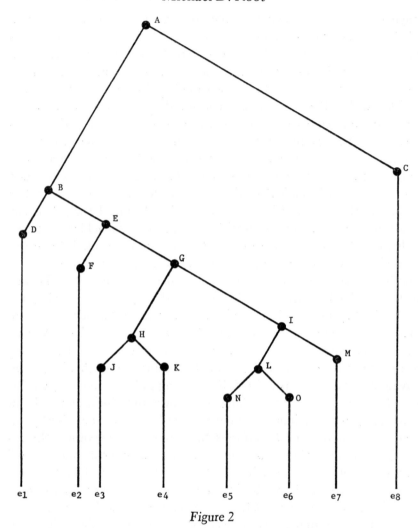

Figure 2

The entire sequence is dominated by the node marked "A." Each node, i.e., A to O, is associated with a class of behavioral events. For example, node A is associated with all and only the events that fit the description "leaving and entering by the same door." The lines that connect the nodes stand for relations between the sets. The lines that connect nodes B and C to A indicate that the set associated with A is the Cartesian product of B and C. All of the sequences of events that belong to A, all leaving-and-entering-by-the-same-door events, consist of an ordered pair

332

of sequences; the first member of the pair is a sequence drawn from B; the second member is a sequence drawn from C. The lines that connect the nodes to the "terminal" symbols, to e_1 to e_8, stand for the membership relation. Event e_1 is a member of the set of events associated with D, e.g., the set of all departures through a door. The relation between events e_1 and e_8, between our two (i,j)-dependent events, is represented by the lines that connect e_1 to node D, e_8 to C, D to B, B to A, and C to A.

Looked at from the point of view of production and control, each node plays a special role in the production of the final sequence. Node A, for example, constrains the sequence by requiring that it consist of sequences drawn from B and C. In addition sequences from B and C are so selected that the final sequence of events, e.g., events $e_1 \ldots ,e_8$, belongs to the set associated with A. Looked at this way each node has the function of a rule.

My claim has been this: for an organism to have some behavioral capacities, for an organism to have the capacity to perform each sequence in a class K of behavioral sequences, is for him to be guided by a rule. By "rule" I mean something that (1) is independent of the individual events that make up the sequences in K and (2) still serves to integrate the events and to preserve various sorts of agreement between them. But I need to add a note of caution here. The term that I have been discussing is the term "rule guidance." This term is one of a family of terms that have been introduced into talk about language and into talk about complex behavior. What I have said about rule guidance suggests that the process of being guided by a rule is neither a distinctively human nor a particularly mysterious process. In fact a general purpose computer is on my analysis a paradigm case of a system whose behavior is guided by a rule. But I have said nothing about some of the more exotic members of this favored circle of terms. I have said nothing about knowledge of rules or the expression "tacit knowledge of rules," the expression that Noam Chomsky so often uses when talking about language. On my analysis the expression "guided by a rule" has broad application. It applies to the behavior of systems to which we are not inclined to attribute knowledge or beliefs. On my analysis of "guided by a rule" the following is not a valid inference: S's behavior is guided by a rule; therefore S has knowledge (albeit tacit) of a rule. S's behavior may be guided by a rule and S may have no knowledge at all.

333

Michael D. Root

On the other hand, a subject may know a rule and his behavior may not be guided by it. In fact here lies one of the paradoxes of Chomsky's distinction between language performance and language competence. Chomsky attributes knowledge of a grammar to a speaker but denies that the speaker puts the grammar to use in the perception and production of speech.[8] The point seems to be that while the speaker knows these rules (in some sense of "knows") he is not guided by them when he actually speaks and understands the sentences of his language. This of course makes the speaker's possession of the rules much like a wheel that doesn't turn anything, and it makes Chomsky's talk about knowledge of these rules quite mysterious. An investigation of this mystery, however, will have to await another day. (See my paper "Language, Rules and Psychological Explanation," *Biosciences Communications* [in press].)

IV

Chomsky believes that every speaker of a natural language has mastered and internalized a system of rules that express his ability to speak and understand his language. He also believes that any fully adequate theory of a language must assign to each of infinitely many sentences a structural description that indicates how the sentence is understood by the speakers of the language. In sections II and III, I gave reasons for believing that you and I have other abilities that are somewhat similar. We have the ability to perform sequences of complex nonverbal behavior, and we have seen reasons for saying that this behavior is rule guided. Thus in mastering these sequences we master and internalize a system of rules. These rules subsequently guide our behavior and describe or express our ability to perform each of infinitely many sequences of complex nonverbal behavior. Moreover, a fully adequate theory of this behavior must assign to each of the sequences a structural description that indicates the significance of the sequence for the subject.

Among the claims that Chomsky makes about our ability to speak and understand a natural language is that it is species-specific and consists of a unique type of intellectual organization that manifests itself in the "creative" aspect of ordinary language use: "man has a species-specific capacity, a unique type of intellectual organization which can-

---

[8] Noam Chomsky, *Aspects of the Theory of Syntax*, p. 9.

not be attributed to peripheral organs or related to general intelligence and which manifests itself in what we may refer to as the 'creative aspect' of ordinary language use — its property being both unbounded in scope and stimulus-free."[9] According to Chomsky the properties of being unbounded in scope and being stimulus-free are independent. A system may have either one of these properties and not the other. A tape recorder has the ability to record speech. In principle there is no limit on the number of sentences it can record. Given another reel of tape and neglecting the mortality of the machine, the tape recorder can always record one more sentence. Since there is no last sentence that the machine can record, we can say that the recorder's ability to record speech is unbounded in scope. However, the recorder's ability to record speech is not stimulus-free. There is a 1–1 mapping from what we say to what it records.

On the other hand, a machine that has only two responses that are produced randomly is a machine whose responses are stimulus-free but whose ability to respond is obviously not unbounded. In short, devices that are capable of producing infinitely many responses manifest unboundedness. Devices whose responses are not determined by any stimulus manifest stimulus-freedom.

One way of reading Chomsky on the issue of "creativity" might be as follows: while many abilities may manifest either one or the other of these properties, namely, unboundedness and freedom from stimuli, only the ability to speak and understand a natural language manifests both: "Animal behavior is typically regarded by the Cartesians as unbounded, but not stimulus-free, and hence not 'creative' in the sense of human speech."[10] Despite the textual evidence, this is not the way to read Chomsky. The ability to use a natural language is obviously not the only ability that manifests both unboundedness and stimulus-freedom. There are machines with simple nonlinguistic abilities that manifest both of these properties. A machine that randomly prints out the integers manifests both. The machine can print any integer, but which integer it prints at any moment is selected randomly. That is, each integer has an equal probability of being selected and printed at each moment in time as any other. If the only properties that mark "creative"

[9] Noam Chomsky, *Cartesian Linguistics* (New York: Harper and Row, 1966), pp. 4–5.
[10] *Ibid.*, p. 77.

335

use are stimulus-freedom and unboundedness, a machine that randomly prints the integers is creative.

However, these are not the only properties that mark the "creative" aspect of ordinary language use. As Chomsky points out, language not only provides a means for expressing indefinitely many thoughts, i.e., speaking and understanding an unbounded number of sentences, but also a means for reacting appropriately in an indefinite range of new situations. Our linguistic responses are not only unbounded and stimulus-free, but are also appropriate to the contexts in which we make them. While the contexts do not determine the responses, the responses are appropriate to the contexts: "The essential difference between man and animal is exhibited most clearly by human language, in particular, by man's ability to form new statements which express new thoughts and which are appropriate to new situations."[11] The sentences (responses) a fluent English speaker produces are not determined by the context in which they are uttered. The speaker could have always produced a different sentence in the same context. Nevertheless the sentences are all appropriate to the context.

Consider, again, the machine that randomly generates the integers. The machine can print infinitely many integers but, since each integer has an equal probability of being selected and printed at each moment, the machine's responses cannot properly be said to be appropriate to the contexts in which they are made. While the machine can produce an unlimited number of responses and the English speaker can produce an infinite number of sentences, the speaker's sentences are appropriate to the contexts in which they are produced and the machine's are not. Our ability to speak English involves more than unboundedness and stimulus-freedom, and this something more, namely, appropriateness to the context of use, is apparently not a property shared by our machine.

But perhaps we are selling our machine a bit short. There are contexts in which printing out any integer would be quite appropriate. I could ask the following questions: "What is an example of an integer?" or "What is an example of a whole number?" or "Can you give me an example of a number?" In these contexts printing out any integer would be to produce an appropriate response. That is to say, there are some contexts in which the machine is able to produce infinitely many re-

[11] *Ibid.*, p. 3.

sponses each of which is appropriate to, but undetermined by, the context. On the other hand, there are some contexts in which an English speaker is not able to produce more than a small finite number of appropriate responses. If I ask you to tell me in six words or less why you read and study philosophy, to reply with a sentence of ten words would, presumably, not be to produce an appropriate response. But if the distinction between man's language abilities and the machine's abilities is not that the one's responses are always appropriate and the other's never appropriate, how is the distinction to be made?

If this objection is at all appealing, it is only because of some unclarity over the notion of one's responses being appropriate to the context of use. We can provide the machine that prints the numbers with some contexts in which each of its responses may appear appropriate. But we provide the machine with the contexts. We bring about the appropriateness. We select the context to fit the response. If we change the context, there will be no appropriate change in the machine's response. If we ask "What is an example of an integer between 4 and 6?" we can expect the same response as if we were to ask "Can you give me an example of an integer?" If the machine's responses are appropriate to the context, it is presumably because of something we do. If the machine's responses fit, it is due to our abilities and our sapience rather than the machine's. The case is quite different with language. A speaker of English is able to produce sentences that are appropriate to novel contexts. Moreover, if the response is appropriate to the context, it is because of something the speaker is able to do. The speaker brings it about that his response fits the context. Part of what the speaker of English is able to do when he is able to speak and understand English is to choose sentences that are appropriate to the context of use. On the other hand, this is not part of what a machine is able to do when it is able to randomly print the integers.

While I think this reply is apt, more needs to be said about a sentence's being appropriate to the context of use. As G. P. Grice has pointed out,[12] the use of a linguistic expression may be inappropriate for any number of reasons. An expression may be inappropriate because it fails to correspond to the world in some favored way, because it is pointless, or because it violates some general principles governing communica-

---

[12] G. P. Grice, "Logic and Conversation." William James Lectures, Harvard University, 1967–1968. Unpublished.

tion or rational behavior. Chomsky's point that an essential property of language is that it provides the means for producing novel but appropriate sentences and for producing sentences that are appropriate to novel contexts suffers, because he says so little about the conditions under which sentences can be said to be appropriate and inappropriate to the contexts of their use.

Grice's work is helpful here. Most linguistic behavior is purposive and Grice has formulated a rough general principle that speakers of a language expect one another to observe and that sets some conditions on what counts as appropriate: make your conversational contribution such as is required, at the stage at which it occurs, by the accepted purpose or direction of the talk-exchange in which you are engaged. One purpose that underlies the use of some linguistic expressions is to produce an effective exchange of information. Given this purpose, users of these expressions share some expectations and presumptions concerning whether their use is appropriate in various contexts. If a speaker uses a sentence in a context that does not satisfy the purpose of the talk-exchange, that sentence is inappropriate.

It is clear from this that the responses of the number generator (the machine that randomly prints out the integers) are neither inappropriate nor appropriate to the contexts in which they are produced. The behavior of the machine is not purposive. Accordingly, it does not have a set of expectations or presumptions concerning how it ought to respond in order to achieve its ends and its purposes, and, accordingly, it cannot meet or fail to meet some shared expectations. The ability of the machine, then, is quite different from the abilities of language speakers and the machine does not manifest all the properties that Chomsky identifies with the creative aspect of language use. Where I take issue with Chomsky is over whether other human intellectual abilities have the properties that he associates with the creative aspects of ordinary language use. I believe that man has many nonlinguistic abilities that are "creative" in the sense that linguistic processes and abilities are.

The rules that express a speaker's ability to speak and understand his language can iterate and thereby generate an indefinitely large number of linguistic structures. In sections II and III, I argued that there are rules that express man's ability to perform nonlinguistic behaviors. We saw that these rules can similarly iterate to generate an indefinitely large number of structures. So man has the ability to perform an unbounded number of nonlinguistic behaviors. But we also saw that these behaviors are determined not by stimuli or by previous responses but by a generalized schemata of action, i.e., rules. These rules are independent of the acts that they determine but serve to determine the order in which these acts are carried out. It remains to show that these abilities can include the ability to perform acts that are appropriate to the contexts in which they are performed and which are appropriate to new and novel situations.

Grice believes that language use, talking, is just a special case of purposive or rational behavior. Further he believes that some of the purposes that underlie the use of language underlie the use of nonlinguistic behaviors as well. One end for which language is used is for the exchange of information. For example, we can demonstrate our displeasure with another's conduct by saying, "What you have done displeases me." But there are also exchanges that don't involve talk that accomplish this same end. Participants in these exchanges share common purposes. Like those who participate in talk-exchanges, they also share some expectations and presumptions concerning whether some bit of behavior is appropriate to the context in which it is performed. In this way the terms "appropriate" and "inappropriate" can be applied to sequences of nonverbal behavior. If you are assisting me to fix my automobile and I need a wrench, I don't expect you to hand me a feather duster. If you share with me the end that I get the wrench and, nevertheless, hand me the duster, then, on its face, your behavior is inappropriate.

Earlier I argued that there are classes of nonlinguistic behaviors which each of us is able to perform and which require rules for their production. We are capable of performing any one of indefinitely many sequences of behavior each of which require some agreement between nonadjacent events, and there is no limit on how far removed these events may be from one another. All of these behaviors are purposive in at least one sense: changes in the jth event in a sequence are for the purpose of maintaining agreement with th i^{th} event. The subject makes it a practice or girds himself to enter the same door from which he leaves. Under this description, his entering by the back door is purposive. The subject enters by the back door in order to bring it about that he enters and leaves by the same door.

Nonlinguistic behavior, then, like linguistic behavior can be purposive. Both can be purposive because we can perform either kind of behavior in order to bring about some desired result. We knit a sweater in order to please a friend. We solve a mathematics problem in order to determine the cost of a new driveway. In addition there are some rough general conditions that persons who engage in purposive behavior expect each other to observe: make your performance such as is required, at the stage at which it occurs, by the accepted purpose or direction of the activity in which you are engaged. And a performance or some bit of behavior is appropriate to a context just in case it is what is required in the context by whatever is the accepted purpose or direction of the activity.

Someone who makes it a practice to enter and leave by the same door, who coordinates his entering and leaving regardless of how removed these events may be, has an ability that can only be expressed by a rule. However, given this practice, a subject's entering by the back door can be said to be appropriate to the context in which it occurs just in case he first leaves by that door. If the subject makes it a practice to enter and leave his home by the same door, it would be inappropriate for him to leave by the front door and enter by the back door. If he resolves or makes it a practice to enter and leave his home by the same door but not before walking around the block four times, it would be inappropriate for him to enter after having walked around the block twice. A subject who is able to coordinate his arrivals and departures around an unlimited number of trips around the block and who is able to carry out this coordination in order to satisfy some resolve, has the means for fitting his behavior to the context of its use. However, not only is the subject able to make his performances appropriate to their contexts, he is able to form new behavioral sequences that are appropriate to new and novel situations. He has the means for performing an unlimited number of behaviors and for reacting appropriately in an unlimited number of new or novel situations. In our present example, for the subject to act appropriately is for him to enter and leave by the same door. But a subject who can coordinate his enterings and leavings around an unlimited number of other behavioral elements can react appropriately in new or novel contexts. Specifically, he can react appropriately in a situation in which, in order to enter and leave by the same door, he must walk n times around the block for any integer value of n. In other words, walks around the block interrupt his leaving and entering. Since there are infinitely many such situations, i.e., one situation for each value of n and infinitely many values for n, the subject can act appropriately in an unlimited number of new or novel situations.

In short, the ability to coordinate one's leavings and enterings has the following properties: it is unbounded in scope, i.e., there is no longest sequence of entering-and-leaving-by-the-same-door behavior that a subject who has this ability is able to perform; the performances that result when the subject exercises this ability are stimulus-free, i.e., the sequences of behaviors that he performs are mediated by rules; and finally the ability to coordinate one's leavings and enterings includes the ability to make each of the sequences appropriate to the contexts in which they are performed even when the contexts are new or novel, i.e., a person who is able to coordinate his leavings and enterings can enter and leave by the same door when to do so requires the performance of a sequence of behavior unlike those which he has previously performed. If I am right, then there are nonlinguistic abilities and processes that are "creative" in the same sense that, according to Noam Chomsky, linguistic processes are. The type of intellectual organization that manifests itself in language use and in the ability to speak and understand a natural language is not unique. This same type of intellectual organization manifests itself in nonlinguistic performances and in the ability to perform many types of nonlinguistic behavior of which the ability to coordinate one's leavings and enterings is but a simple example.

V

I have been arguing that some of the notions and results that have arisen from the study of natural languages can be generalized and the generalizations applied to nonlinguistic behavior. In sections II and III, I pointed out that a good deal more behavior than verbal behavior is hierarchically ordered and that this order is significant for both the organism and the specification of his behavior. In section IV, I argued that if there is a creative aspect of ordinary language use, there is a creative aspect in the performance of some nonlinguistic behavior as well. However, the analogies between verbal and nonverbal abilities can be pushed too far, and I wish to turn my final comments to some of the differences.

The system of rules that express a speaker's abilities to speak and understand his language can be analyzed (at least on Chomsky's view) into three major components: syntactic, phonological, and semantic rules. The syntactic component, in turn, includes a base subcomponent and a transformational subcomponent. The former consists of rules that generate a set of basic strings each with an associated structural description. The latter consists of rules that map strings and their associated structural descriptions onto strings with associated structural descriptions. There is, to my knowledge, nothing corresponding to each of these components in the nonlinguistic case. Though there is some evidence that the notion of grammatical transformation can be generalized and applied to phenomena in visual perception, there is presently no good reason to believe that for each of the phenomena the linguist describes or for each element in his description there corresponds a similar phenomena or element in some nonlinguistic dimension.

However, I think I can make a more interesting point. Where there are (i,j)-dependencies between elements in linguistic strings, in strings of linguistic formatives, there are also always selectional restrictions or restrictions of co-occurrence that constrain the choice of elements that can be nested within i^{th} and j^{th} elements. A speaker of English can speak and understand the sentence

(9)   Blonds are abundant in Sweden

and infinitely many other sentences in which the subject of the sentence is "blonds" and "are abundant in Sweden" its predicate. In each sentence there will be an (i,j)-dependency between "blonds" and "are abundant." There is no longest such sentence because we can iterate relative clauses between the ith and j^{th} elements. For example,

(10)   Blonds who love children are abundant in Sweden,
(11)   Blonds who love children who love dogs are abundant in Sweden,

but not

(12)   *Blonds which boil at 212 degrees Fahrenheit are abundant in Sweden.

What the examples show us is that there is no limit on the number of relative clauses that can be nested between subject and predicate, but the relative clauses cannot be freely selected. Within the language there are restrictions on what clauses can modify what nouns. Consequently, there are restrictions on what we can iterate between i^{th} and j^{th} elements. The clause "who love children" is permitted, but the clause "which boils at 212 degrees Fahrenheit" is not.

On the other hand, in at least some of the nonlinguistic cases I considered in sections II and III, there were no such restrictions. That is, there were no restrictions on the behavioral elements that we could nest between i^{th} and j^{th} behavioral events. Where there are (i,j)-dependencies between elements in nonlinguistic strings, e.g., entering and leaving by the same door, there need be nothing that corresponds to selectional or co-occurrence restrictions on the embedded elements. One can walk up and down the block, write a book about knitting, brush his teeth, prepare dinner, or sail to Calais. In this nonlinguistic case one's choice is unrestricted by the rules that express the ability to perform the sequences in K, e.g., to enter and leave by the same door. That is, the intervening behavior may have nothing to do with the ith and j^{th} events.

In short, linguists have erred in overestimating the distinctiveness of language. Walking and talking are more alike than the Cartesians think. On the other hand, it is easy to err on the side of parity. It would be a mistake to allow the similarities between sweater-making and English-speaking to mask the interesting differences. Restrictions on occurrence may be one difference. I leave it to the linguist to remind us of the rest.

A Taxonomy of Illocutionary Acts

I. Introduction

The primary purpose of this paper is to develop a reasoned classification of illocutionary acts into certain basic categories or types. It is to answer the question: How many kinds of illocutionary acts are there? Since any such attempt to develop a taxonomy must take into account Austin's classification of illocutionary acts into his five basic categories of verdictive, expositive, exercitive, behabitive, and commissive, a second purpose of this paper is to assess Austin's classification to show in what respects it is adequate and in what respects inadequate. Furthermore, since basic semantic differences are likely to have syntactical consequences, a third purpose of this paper is to show how these different basic illocutionary types are realized in the syntax of a natural language such as English.

In what follows, I shall presuppose a familiarity with the general pattern of analysis of illocutionary acts offered in such works as Austin, *How to Do Things with Words*, Searle, *Speech Acts*, and Searle, "Austin on Locutionary and Illocutionary Acts."[1] In particular, I shall presuppose a distinction between the illocutionary force of an utterance and its propositional content as symbolized as $F(p)$. The aim of this paper then is to classify the different types of F.

II. Different Types of Differences between Different Types of Illocutionary Acts

Any taxonomical effort of this sort presupposes criteria for distinguishing one (kind of) illocutionary act from another. What are the criteria by which we can tell that of three actual utterances one is a report, one

---

[1] J. L. Austin, *How to Do Things with Words* (Oxford: Clarendon Press, 1962); J. R. Searle, *Speech Acts: An Essay in the Philosophy of Language* (London: Cambridge University Press, 1969); and J. R. Searle, "Austin on Locutionary and Illocutionary Acts," *Philosophical Review*, 1968.

a prediction, and one a promise? In order to develop higher order genera, we must first know how the species *promise, prediction, report,* etc., differ one from another. When one attempts to answer that question one discovers that there are several quite different principles of distinction; that is, there are different kinds of differences that enable us to say that the force of this utterance is different from the force of that utterance. For this reason the metaphor of force in the expression "illocutionary force" is misleading since it suggests that different illocutionary forces occupy different positions on a single continuum of force. What is actually the case is that there are several distinct crisscrossing continua. A related source of confusion is that we are inclined to confuse illocutionary verbs with types of illocutionary acts. We are inclined, for example, to think that where we have two nonsynonymous illocutionary verbs they must necessarily mark two different kinds of illocutionary acts. In what follows, I shall try to keep a clear distinction between illocutionary verbs and illocutionary acts. Illocutions are a part of language as opposed to particular languages. Illocutionary verbs are always part of a particular language: French, German, English, or whatnot. Differences in illocutionary verbs are a good guide but by no means a sure guide to differences in illocutionary acts.

It seems to me there are (at least) twelve significant dimensions of variation in which illocutionary acts differ one from another and I shall — all too briskly — list them:

1. *Differences in the point (or purpose) of the (type of) act.* The point or purpose of an order can be specified by saying that it is an attempt to get the hearer to do something. The point or purpose of a description is that it is a representation (true or false, accurate or inaccurate) of how something is. The point or purpose of a promise is that it is an undertaking of an obligation by the speaker to do something. These differences correspond to the essential conditions in my analysis of illocutionary acts in *Speech Acts.*[2] Ultimately, I believe, essential conditions form the best basis for a taxonomy, as I shall attempt to show. It is important to notice that the terminology of "point" or "purpose" is not meant to imply, nor is it based on the view, that every illocutionary act has a definitionally associated perlocutionary intent. For many, perhaps most, of the most important illocutionary acts, there is no essential perlocutionary intent associated by definition with the corre-

[2] Searle, *Speech Acts,* chap. 3.

sponding verb, e.g., statements and promises are not by definition attempts to produce perlocutionary effects in hearers.

The point or purpose of a type of illocution I shall call its *illocutionary point*. Illocutionary point is part of but not the same as illocutionary force. Thus, for example, the illocutionary point of a request is the same as that of a command: both are attempts to get hearers to do something. But the illocutionary forces are clearly different. In general, one can say that the notion of illocutionary force is the resultant of several elements of which illocutionary point is only one, though, I believe, the most important one.

2. Differences in the direction of fit between words and the world. Some illocutions have as part of their illocutionary point to get the words (more strictly, their propositional content) to match the world, others to get the world to match the words. Assertions are in the former category, promises and requests are in the latter. The best illustration of this distinction I know of is provided by Miss Anscombe.[3] Suppose a man goes to the supermarket with a shopping list given him by his wife on which are written the words "beans, butter, bacon, and bread." Suppose as he goes around with his shopping cart selecting these items, he is followed by a detective who writes down everything he takes. As they emerge from the store both shopper and detective will have identical lists. But the function of the two lists will be quite different. In the case of the shopper's list, the purpose of the list is, so to speak, to get the world to match the words; the man is supposed to make his actions fit the list. In the case of the detective, the purpose of the list is to make the words match the world; the man is supposed to make the list fit the actions of the shopper. This can be further demonstrated by observing the role of a "mistake" in the two cases. If the detective gets home and suddenly realizes that the man bought pork chops instead of bacon, he can simply erase the word "bacon" and write "pork chops." But if the shopper gets home and his wife points out he has bought pork chops when he should have bought bacon, he cannot correct the mistake by erasing "bacon" from the list and writing "pork chops."

In these examples the list provides the propositional content of the illocution, and the illocutionary force determines how that content is supposed to relate to the world. I propose to call this difference a dif-

[3] G. E. M. Anscombe, *Intentions* (Oxford: Blackwell, 1957).

ference in *direction of fit*. The detective's list has the *word-to-world* direction of fit (as do statements, descriptions, assertions, and explanations); the shopper's list has the *world-to-word* direction of fit (as do requests, commands, vows, promises). I represent the word-to-world direction of fit with a downward arrow thus $\downarrow$ and the world-to-word direction of fit with an upward arrow thus $\uparrow$. Direction of fit is always a consequence of illocutionary point. It would be very elegant if we could build our taxonomy entirely around this distinction in direction of fit, but though it will figure largely in our taxonomy, I am unable to make it the entire basis of the distinctions.

3. *Differences in expressed psychological states.* A man who states, explains, asserts, or claims that p *expresses the belief that p*; a man who promises, vows, threatens, or pledges to do a *expresses an intention to do a*; a man who orders, commands, requests H to do A *expresses a desire (want, wish) that H do A*; a man who apologizes for doing A *expresses regret at having done A*; etc. In general, in the performance of any illocutionary act with a propositional content, the speaker expresses some attitude, state, etc., to that propositional content. Notice that this holds even if he is insincere, even if he does not have the belief, desire, intention, regret, or pleasure which he expresses, he nonetheless expresses a belief, desire, intention, regret, or pleasure in the performance of the speech act. This fact is marked linguistically by the fact that it is linguistically unacceptable (though not self-contradictory) to conjoin the explicit performative verb with the denial of the expressed psychological state. Thus one cannot say "I state that p but do not believe that p," "I promise that p but I do not intend that p," etc. Notice that this only holds in the first person performative use. One can say, "He stated that p but didn't really believe that p," "I promised that p but did not really intend to do it," etc. The psychological state expressed in the performance of the illocutionary act is the *sincerity condition* of the act, as analyzed in *Speech Acts*, chapter 3.

If one tries to do a classification of illocutionary acts based entirely on different expressed psychological states (differences in the sincerity condition), one can get quite a long way. Thus *belief* collects not only statements, assertions, remarks, and explanations, but also postulations, declarations, deductions, and arguments. *Intention* will collect promises, vows, threats, and pledges. *Desire* or *want* will collect requests, orders, commands, askings, prayers, pleadings, beggings, and entreaties. *Pleasure* doesn't collect quite so many — congratulations, felicitations, welcomes, and a few others.

In what follows, I shall symbolize the expressed psychological state with the capitalized initial letters of the corresponding verb, the "*B*" for "believe," "*W*" for "want," "*I*" for "intend," etc.

These three dimensions — illocutionary point, direction of fit, and sincerity condition — seem to me the most important, and I will build most of my taxonomy around them, but there are several others that need remarking.

4. *Differences in the force or strength with which the illocutionary point is presented.* Both "I suggest we go to the movies" and "I insist that we go to the movies" have the same illocutionary point, but it is presented with different strengths, analogously with "I solemnly swear that Bill stole the money" and "I guess Bill stole the money." Along the same dimension of illocutionary point or purpose there may be varying degrees of strength or commitment.

5. *Differences in the status or position of the speaker and hearer as these bear on the illocutionary force of the utterance.* If the general asks the private to clean up the room, that is in all likelihood a command or an order. If the private asks the general to clean up the room, that is likely to be a suggestion or proposal or request but not an order or command. This feature corresponds to one of the preparatory conditions in my analysis in *Speech Acts,* chapter 3.

6. *Differences in the way the utterance relates to the interests of the speaker and the hearer.* Consider, for example, the differences between boasts and laments, between congratulations and condolences. In these two pairs one hears the difference as being between what is or is not in the interests of the speaker and hearer respectively. This feature is another type of preparatory condition according to the analysis in *Speech Acts.*

7. *Differences in relations to the rest of the discourse.* Some performative expressions serve to relate the utterance to the rest of the discourse (and also to the surrounding context). Consider, e.g., "I reply," "I deduce," "I conclude," and "I object." These expressions serve to relate utterances to other utterances and to the surrounding context. The features they mark seem mostly to involve utterances within the class of statements. In addition to simply stating a proposition, one may state it by way of objecting to what someone else has said, by way of replying

to an earlier point, by way of deducing it from certain evidentiary premises, etc. "However," "moreover," and "therefore" also perform these discourse-relating functions.

8. *Differences in propositional content that are determined by illocutionary-force indicating devices.* The differences, for example, between a report and a prediction involve the fact that a prediction must be about the future whereas a report can be about the past or present. These differences correspond to differences in propositional content conditions as explained in *Speech Acts.*

9. *Differences between those acts that must always be speech acts, and those that can be, but need not be, performed as speech acts.* One may classify things, for example, by saying "I classify this as an A and this as a B." But one need not say anything at all in order to be classifying; one may simply throw all the A's in the A box and all the B's in the B box. Similarly with estimate, diagnose, and conclude. I may make estimates, give diagnoses, and draw conclusions in saying "I estimate," "I diagnose," and "I conclude," but in order to estimate, diagnose, or conclude it is not necessary to say anything at all. I may simply stand before a building and estimate its height, silently diagnose you as a marginal schizophrenic, or conclude that the man sitting next to me is quite drunk. In these cases, no speech act, not even an internal speech act, is necessary.

10. *Differences between those acts that require extra-linguistic institutions for their performance and those that do not.* There are a large number of illocutionary acts that require an extra-linguistic institution, and generally a special position by the speaker and the hearer within that institution in order for the act to be performed. Thus in order to bless, excommunicate, christen, pronounce guilty, call the base runner out, bid three no-trump, or declare war, it is not sufficient for any old speaker to say to any old hearer "I bless," "I excommunicate," etc. One must have a position within an extra-linguistic institution. Austin sometimes talks as if he thought all illocutionary acts were like this, but plainly they are not. In order to make a statement that it is raining or promise to come and see you, I need only obey the rules of language. No extra-linguistic institutions are required. This feature of certain speech acts, that they require extra-linguistic institutions, needs to be distinguished from feature 5, the requirement of certain illocutionary acts that the speaker and possibly the hearer as well have a certain status. Extra-

349

linguistic institutions often confer status in a way relevant to illocutionary force, but not all differences of status derive from institutions. Thus an armed robber in virtue of his possession of a gun may *order* as opposed to, e.g., request, entreat, or implore victims to raise their hands. But his status here does not derive from a position within an institution but from his possession of a weapon.

11. Differences between those acts where the corresponding illocutionary verb has a performative use and those where it does not. Most illocutionary verbs have performative uses — e.g., 'state', 'promise', 'order', 'conclude'. But one cannot perform acts of, e.g., boasting or threatening, by saying "I hereby boast," or "I hereby threaten."⁴ Not all illocutionary verbs are performative verbs.

12. Differences in the style of performance of the illocutionary act. Some illocutionary verbs serve to mark what we might call the special *style* in which an illocutionary act is performed. Thus the difference between, for example, announcing and confiding need not involve any difference in illocutionary point or propositional content but only in the *style* of performance of the illocutionary act.

III. Weaknesses in Austin's Taxonomy

Austin advances his five categories very tentatively, more as a basis for discussion than as a set of established results. "I am not," he says, "putting any of this forward as in the very least definitive."⁵ I think they form an excellent basis for discussion but I also think that the taxonomy needs to be seriously revised because it contains several weaknesses. Here are Austin's five categories:

Verdictives. These "consist in the delivering of a finding, official or unofficial, upon evidence or reasons as to value or fact so far as these are distinguishable." Examples of verbs in this class are: *acquit, hold,*

---

⁴ There are other verbs in English which sound odd in the first person present. Consider "lurk" and "skulk." It is odd in answer to the question "What are you doing?" to say "I am lurking in the bushes" or "I am skulking today." The reason may be that both verbs involve a negative assessment and it is odd to give a negative assessment of what one is doing while voluntarily doing it. Perhaps a similar explanation will work for "boast" and "threaten," since they both also seem to contain an element of negative assessment. Notice that they are acceptable if embedded in some apologetic form, e.g., "I hope you won't mind if I boast about my new motorcycle."

⁵ Austin, *How to Do Things with Words*, p. 151.

calculate, describe, analyze, estimate, date, rank, assess, characterize, and describe.

Exercitives. One of these "is the giving of a decision in favor of or against a certain course of action or advocacy of it," "a decision that something is to be so, as distinct from a judgment that it is so." Some examples are: order, command, direct, plead, beg, recommend, entreat, and advise. Request is also an obvious example, but Austin does not list it. As well as the above, Austin also lists: appoint, dismiss, nominate, veto, declare closed, declare open, as well as announce, warn, proclaim, and give.

Commissives. "The whole point of a commissive," Austin tells us, "is to commit the speaker to a certain course of action." Some of the obvious examples are: promise, vow, pledge, covenant, contract, guarantee, embrace, and swear.

Expositives "are used in acts of exposition involving the expounding of views, the conducting of arguments and the clarifying of usages and references." Austin gives many examples of these; among them are: affirm, deny, emphasize, illustrate, answer, report, accept, object to, concede, describe, class, identify, and call.

Behabitives. This class, with which Austin was very dissatisfied ("a shocker," he called it) "includes the notion of reaction to other people's behavior and fortunes and of attitudes and expressions of attitudes to someone else's past conduct or imminent conduct."

Among the examples Austin lists are: apologize, thank, deplore, commiserate, congratulate, felicitate, welcome, applaud, criticize, bless, curse, toast, and drink. But also, curiously: dare, defy, protest, and challenge.

The first thing to notice about these lists is that they are not classifications of illocutionary acts but of English illocutionary verbs. Austin seems to assume that a classification of different verbs is eo ipso a classification of kinds of illocutionary acts, that any two nonsynonymous verbs must mark different illocutionary acts. But there is no reason to suppose that this is the case. As we shall see, some verbs, for example, mark the manner in which an illocutionary act is performed, e.g., "announce." One may announce orders, promises, and reports, but announcing is not on all fours with ordering, promising, and reporting. Announcing, to anticipate a bit, is not the name of a type of illocutionary act, but of the way in which some illocutionary act is performed. An announcement is never just an announcement because "announcing" is not the name

351

of an illocutionary point. An announcement must also be a statement, order, etc.

Even granting that the lists are of illocutionary verbs and not necessarily of different illocutionary acts, it seems to me one can level the following additional criticisms against it.

1. First, a minor cavil, but one worth noting. Not all of the verbs listed are even illocutionary verbs. For example, "sympathize," "regard as," "mean to," "intend," and "shall." Take "intend": it is clearly not performative. Saying "I intend" is not intending; nor in the third person does it name an illocutionary act: "He intended . . ." does not report a speech act. Of course there is an illocutionary act of *expressing an intention*, but the illocutionary verb phrase is "express an intention," not "intend." Intending is never a speech act; expressing an intention usually, but not always, is.

2. The most important weakness of the taxonomy is simply this. There is no clear or consistent principle or set of principles on the basis of which the taxonomy is constructed. Only in the case of commissives has Austin clearly and unambiguously used illocutionary point as the basis of the definition of a category. Expositives, insofar as the characterization is clear, seem to be defined in terms of discourse relations (my feature 7). Exercitives seem to be at least partly defined in terms of the exercise of authority. Both considerations of status (my feature 5) as well as institutional considerations (my feature 10) are lurking in it. Behabitives do not seem to me at all well defined (as Austin, I am sure, would have agreed) but it seems to involve notions of what is good or bad for the speaker and hearer (my feature 6) as well as expressions of attitudes (my feature 3).

3. Because there is no clear principle of classification and because there is a persistent confusion between illocutionary acts and illocutionary verbs, there is a great deal of overlap from one category to another and a great deal of heterogeneity within some of the categories. The problem is not that there are borderline cases — any taxonomy that deals with the real world is likely to come up with borderline cases — nor is it merely that a few unusual cases will have the defining characteristics of more than one category. Rather a very large number of verbs find themselves smack in the middle of two competing categories because the principles of classification are unsystematic. Consider, for example, the verb "describe," a very important verb in anybody's theory of speech

acts. Austin lists it as both a verdictive and an expositive. Given his definitions, it is easy to see why: describing can be both the delivering of a finding and an act of exposition. But then any "act of exposition involving the expounding of views" could also in his rather special sense be "the delivering of a finding, official or unofficial, upon evidence or reasons." And indeed, a look at his list of expositives (pp. 161–162) is sufficient to show that most of his verbs fit his definition of verdictives as well as does "describe." Consider "affirm," "deny," "state," "class," "identify," "conclude," and "deduce." All of these are listed as expositives, but they could just as easily have been listed as verdictives. The few cases which are clearly not verdictives are cases where the meaning of the verb has purely to do with discourse relations, e.g., "begin by," "turn to," or where there is no question of evidence or reasons, e.g., "postulate," "neglect," "call," and "define." But then that is really not sufficient to warrant a separate category, especially since many of these — "begin by," "turn to," "neglect" — are not names of illocutionary acts at all.

4. Not only is there too much overlap from one category to the next, but within some of the categories there are quite distinct kinds of verbs. Thus Austin lists "dare," "defy," and "challenge" alongside "thank," "apologize," "deplore," and "welcome" as behabitives. But "dare," "defy," and "challenge" have to do with the hearer's subsequent actions; they belong with "order," "command," and "forbid" both on syntactical and semantic grounds, as I shall argue later. But when we look for the family that includes "order," "command," and "urge," we find these are listed as exercitives alongside "veto," "hire," and "demote." But these, again as I shall argue later, are in two quite distinct categories.

5. Related to these objections is the further difficulty that not all of the verbs listed within the classes really satisfy the definitions given, even if we take the definitions in the rather loose and suggestive manner that Austin clearly intends. Thus nominating, appointing, and excommunicating are not the "giving of a decision in favor of or against a certain course of action," much less are they "advocating" it. Rather they are, as Austin himself might have said, performances of these actions, not advocacies of anything. That is, in the sense in which we might agree that ordering, commanding, and urging someone to do something are all cases of advocating that he do it, we cannot also agree that nominating or appointing is also advocating. When I appoint you

chairman, I don't advocate that you be or become chairman; I *make* you chairman.

In sum, there are (at least) the following six related difficulties with Austin's taxonomy. In ascending order of importance, there is a persistent confusion between verbs and acts; not all the verbs are illocutionary verbs; there is too much overlap of the categories; there is too much heterogeneity within the categories; many of the verbs listed in the categories don't satisfy the definition given for the category; and, most important, there is no consistent principle of classification.

I don't believe I have fully substantiated all six of these charges and I will not attempt to do so within the confines of this paper, which has other aims. I believe, however, that my doubts about Austin's taxonomy will have greater clarity and force after I have presented an alternative. What I propose to do is take illocutionary point and its corollaries, direction of fit and expressed sincerity conditions, as the basis for constructing a classification. In such a classification, other features — the role of authority, discourse relations, etc. — will fall into their appropriate places.

IV. Alternative Taxonomy

In this section I shall present a list of what I regard as the basic categories of illocutionary acts. In so doing, I shall discuss briefly how my classification relates to Austin's.

Representatives. The point or purpose of the members of the representative class is to commit the speaker (in varying degrees) to something's being the case, to the truth of the expressed proposition. All of the members of the representative class are assessable on the dimension of assessment which includes *true* and *false*. Using Frege's assertion sign to mark the illocutionary point common to all the members of this class and the symbols introduced above, we may symbolize this class as follows:

$$\vdash \downarrow B(p)$$

The direction of fit is words-to-the-world; and the psychological state expressed is belief (*that p*). It is important to emphasize that words such as "belief" and "commitment" are here intended to mark dimensions; they are so to speak determinables rather than determinates. Thus there is a difference between *suggesting that p* or *putting it forward as a hypothesis that p* on the one hand, and *insisting that p* or *solemnly swear-*

ing that p on the other. The degree of belief and commitment may approach or even reach zero, but it is clear or will become clear that *hypothesizing that p* and *flatly stating that p* are in the same line of business in a way that neither is like requesting. Once we recognize the existence of representatives as a quite separate class, based on the notion of illocutionary point, then the existence of a large number of performative verbs denoting illocutions that seem to be assessable in the true-false dimension and yet are not just "statements" will be easily explicable in terms of the fact that they mark features of illocutionary force which are in addition to illocutionary point. Thus, for example, consider "boast" and "complain." They both denote representatives with the added feature that they have something to do with the interest of the speaker (feature 6 above). "Conclude" and "deduce" are also representatives with the added feature that they mark certain relations between the representative illocutionary act and the rest of the discourse or the context of utterance (feature 7 above). This class will contain most of Austin's expositives as well as many of his verdictives for the, by now I hope obvious, reason that they all have the same illocutionary point and differ only in other features of illocutionary force. The simplest test of a representative is this: can you literally characterize it (inter alia) as true or false. Though I hasten to add that this will give neither necessary nor sufficient conditions, as we shall see when we get to my fifth class.

These points about representatives will, I hope, be clearer when I discuss my second class which, with some reluctance, I will call

Directives. The illocutionary point of these consists in the fact that they are attempts (of varying degrees, and hence more precisely, they are determinates of the determinable which includes attempting) by the speaker to get the hearer to do something. They may be very modest "attempts," as when I invite you to do it or suggest that you do it, or they may be very fierce attempts as when I insist that you do it. Using the exclamation mark as the illocutionary-point indicating device for the members of this class generally, we have the following symbolism:

$$! \uparrow W \,(H \text{ does } A)$$

The direction of fit is world-to-words and the sincerity condition is want (or wish or desire). The propositional content is always that the hearer *H* does some future action *A*. Verbs denoting members of this class are

355

order, command, request, ask, question,[6] beg, plead, pray, entreat, and also *invite, permit,* and *advise.* I think also that it is clear that *dare, defy,* and *challenge* which Austin lists as behabitives are in this class. Many of Austin's exercitives are also in this class.

Commissives. Austin's definition of commissives seems to me unexceptionable, and I will simply appropriate it as it stands with the cavil that several of the verbs he lists as commissive verbs do not belong in this class at all, such as "shall," "intend," "favor," and others. Commissives then are those illocutionary acts whose point is to commit the speaker (again in varying degrees) to some future course of action. Using C for the members of this class, generally we have the following symbolism:

$$C \uparrow I \ (S \ \text{does} \ A)$$

The direction of fit is world-to-words and the sincerity condition is intention. The propositional content is always that the speaker S does some future action A. Since the direction of fit is the same for commissives and directives, it would give us a more elegant taxonomy if we could show that they are really members of the same category. I am unable to do this, because whereas the point of a promise is to commit the speaker to doing something (and not necessarily to try to get himself to do it), the point of a request is to try to get the hearer to do something (and not necessarily to commit or obligate him to do it). In order to assimilate the two categories, one would have to show that promises are really a species of requests to oneself (this has been suggested to me by Julian Boyd) or alternatively one would have to show that requests placed the hearer under an obligation (this has been suggested to me by William Alston and John Kearns). I have been unable to make either of these analyses work and am left with the inelegant solution of two separate categories with the same direction of fit.

A fourth category I shall call

Expressives. The illocutionary point of this class is to express the psychological state specified in the sincerity condition about a state of affairs specified in the propositional content. The paradigms of expressive verbs are "thank," "congratulate," "apologize," "condole," "deplore," and "welcome." Notice that in expressives there is no direction of fit.

---

[6] Questions are directives, since they are attempts to get the hearer to perform a speech act.

In performing an expressive, the speaker is neither trying to get the world to match the words nor the words to match the world; rather the truth of the expressed proposition is presupposed. Thus, for example, when I apologize for having stepped on your toe, it is not my purpose either to claim that your toe was stepped on nor to get it stepped on. This fact is neatly reflected in the syntax (of English) by the fact that the paradigm expressive verbs in their performance occurrence will not take *that* clauses but require a gerund nominalization transformation (or some other nominal). One cannot say:

> * I apologize that I stepped on your toe;

rather the correct English is

> I apologize for stepping on your toe.

Similarly, one cannot have

> * I congratulate you that you won the race

nor

> * I thank you that you paid me the money.

One must have

> I congratulate you on winning the race (congratulations on winning the race).

> I thank you for paying me the money (thanks for paying me the money).

These syntactical facts, I suggest, are consequences of the fact that there is in general no direction of fit in expressives. The truth of the proposition expressed in an expressive is presupposed. The symbolization of this class therefore must proceed as follows:

$$E \phi (P) (S/H + \text{property})$$

where E indicates the illocutionary point common to all expressives, ϕ is the null symbol indicating no direction of fit, (P) is a variable ranging over the different possible psychological states expressed in the performance of the illocutionary acts in this class, and the propositional content ascribes some property (not necessarily an action) to either S or H. I can congratulate you not only on your winning the race, but also on your good looks or on your son's winning the race. The property specified in the propositional content of an expressive must, however, be related to *S* or *H*. I cannot without some very special assumptions congratulate you on Newton's first law of motion.

It would be economical if we could include all illocutionary acts in these four classes and would lend some further support to the general pattern of analysis adopted in *Speech Acts*, but it seems to me it is still not complete. There is still left an important class of cases, where the state of affairs represented in the proposition expressed is realized or brought into existence by the illocutionary-force indicating device, cases where one brings a state of affairs into existence by declaring it to exist, cases where, so to speak, "saying makes it so." Examples of these cases are "I resign," "You're fired," "I excommunicate you," "I christen this ship the battleship *Missouri*," "I appoint you chairman," and "War is hereby declared." These cases were presented as paradigms in the very earliest discussions of performatives, but it seems to me they are still not adequately described in the literature and their relation to other kinds of illocutionary acts is usually misunderstood. Let us call this class

Declarations. It is the defining characteristic of this class that the successful performance of one of its members brings about the correspondence between the propositional content and reality; successful performance guarantees that the propositional content corresponds to the world: if I successfully perform the act of appointing you chairman, then you are chairman; if I successfully perform the act of nominating you as candidate, then you are a candidate; if I successfully perform the act of declaring a state of war, then war is on; if I successfully perform the act of marrying you, then you are married.

The surface syntactical structure of many sentences used to perform declarations conceals this point from us because in them there is no surface syntactical distinction between propositional content and illocutionary force. Thus "You're fired" and "I resign" do not seem to permit a distinction between illocutionary force and propositional content, but I think in fact that in their use to perform declarations their semantic structure is:

> I declare: your employment is (hereby) terminated.
>
> I declare: my position is (hereby) terminated.

Declarations bring about some alternation in the status or condition of the referred to object or objects solely in virtue of the fact that the declaration has been successfully performed. This feature of declarations

358

distinguishes them from the other categories. In the history of the discussion of these topics since Austin's first introduction of his distinction between performatives and constatives, this feature of declarations has not been properly understood. The original distinction between constatives and performatives was supposed to be a distinction between utterances which are sayings (constatives: statements, assertions, etc.) and utterances which are doings (performatives: promises, bets, warnings, etc.). What I am calling declarations were included in the class of performatives. The main theme of Austin's mature work, *How to Do Things with Words*, is that this distinction collapses. Just as saying certain things constitutes getting married (a "performative") and saying certain things constitutes making a promise (another "performative"), so saying certain things constitutes making a statement (supposedly a "constative"). As Austin saw, but as many philosophers still fail to see, the parallel is exact. Making a statement is as much performing an illocutionary act as making a promise, a bet, a warning, or what have you. Any utterance will consist in performing one or more illocutionary acts.

The illocutionary-force indicating device in the sentence operates on the propositional content to indicate among other things the direction of fit between the propositional content and reality. In the case of representatives the direction of fit is words-to-world; in the case of directives and commissives it is world-to-words; in the case of expressives there is no direction of fit carried by the illocutionary force because the existence of fit is presupposed. The utterance can't get off the ground unless there already is a fit. But now with the declarations we discover a very peculiar relation. The performance of a declaration brings about a fit by its very successful performance. How is such a thing possible?

Notice that all of the examples we have considered so far involve an extra-linguistic institution, a system of constitutive rules in addition to the constitutive rules of language, in order that the declaration may be successfully performed. The mastery of those rules which constitute linguistic competence by the speaker and hearer is not in general sufficient for the performance of a declaration. In addition there must exist an extra-linguistic institution and the speaker and hearer must occupy special places within this institution. It is only given such institutions as the church, the law, private property, the state, and a special position of the speaker and hearer within these institutions that one can excommunicate, appoint, give and bequeath one's possessions, or declare war.

359

The only exceptions to the principle that every declaration requires an extra-linguistic institution are those declarations that concern language itself,[7] as for example when one says, "I define, abbreviate, name, call, or dub." Austin sometimes talks as if all performatives (and in the general theory, all illocutionary acts) required an extra-linguistic institution, but this is plainly not the case. Declarations are a very special category of speech acts. We shall symbolize their structure as follows:

$$D \updownarrow \phi (p)$$

where D indicates the declarational illocutionary point; the direction of fit is both words-to-world and world-to-words because of the peculiar character of declarations; there is no sincerity condition, hence we have the null symbol in the sincerity condition slot; and we employ the usual propositional variable p.

The reason there has to be a relation-of-fit arrow here at all is that declarations do attempt to get language to match the world. But they do not attempt to do it either by describing an existing state of affairs (as do representatives) or by trying to get someone to bring about a future state of affairs (as do directives and commissives).

Some members of the class of declarations overlap with members of the class of representatives. This is because in certain institutional situations we not only ascertain the facts but we need an authority to lay down a decision as to what the facts are after the fact-finding procedure has been gone through. The argument must eventually come to an end and issue in a decision, and it is for this reason that we have judges and umpires. Both the judge and the umpire make factual claims: "you are out," "you are guilty." Such claims are clearly assessable in the dimension of word-world fit. Was he really tagged off base? Did he really commit the crime? They are assessable in the word-to-world dimension. But at the same time both have the force of declarations. If the umpire calls you out (and is upheld on appeal), then for baseball purposes you are out regardless of the facts in the case, and if the judge declares you guilty (on appeal), then for legal purposes you are guilty. There is nothing mysterious about these cases. Institutions characteristically require illocutionary acts to be issued by authorities of various kinds which have the force of declarations. Some institutions require

---

[7] Another rather special class of exceptions concerns the supernatural. When God says "Let there be light," that is a declaration.

representative claims to be issued with the force of declarations in order that the argument over the truth of the claim can come to an end somewhere and the next institutional steps which wait on the settling of the factual issue can proceed: the prisoner is released or sent to jail, the side is retired, a touchdown is scored. The members of this class we may dub "representative declarations." Unlike the other declarations, they share with representatives a sincerity condition. The judge, jury, and umpire can, logically speaking, lie, but the man who declares war or nominates you cannot lie in the performance of his illocutionary act. The symbolism for the class of representative declarations, then, is this:

$$D_r \downarrow \updownarrow B(p)$$

where D_r indicates the illocutionary point of issuing a representative with the force of a declaration, the first arrow indicates the representative direction of fit, the second indicates the declarational direction of fit, the sincerity condition is belief, and the p represents the propositional content.

V. Some Syntactical Aspects of the Classification

So far, I have been classifying illocutionary acts and have used facts about verbs for evidence and illustration. In this section I want to discuss explicitly some points about English syntax. If the distinctions marked in section IV are of any real significance, they are likely to have various syntactical consequences, and I now propose to examine the deep structure of explicit performative sentences in each of the five categories; that is, I want to examine the syntactical structure of sentences containing the performative occurrence of appropriate illocutionary verbs appropriate to each of the five categories. Since all of the sentences we will be considering will contain a performative verb in the main clause, and a subordinate clause, I will abbreviate the usual tree structures in the following fashion: The sentence, for example, "I predict John will hit Bill," has the deep structure shown in the accompanying diagram. I will simply abbreviate this as: "I predict + John will hit Bill." Parentheses will be used to mark optional elements or elements that are obligatory only for restricted classes of the verbs in question. Where there is a choice of one of two elements, I will put a stroke between the elements, for example, "I/you."

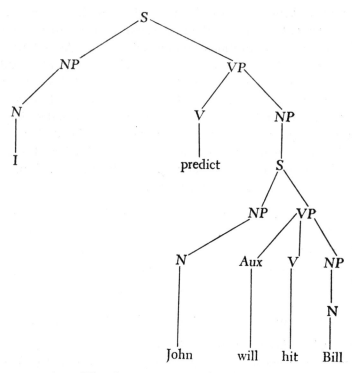

Representatives. The deep structure of such paradigm representative sentences as "I state that it is raining" and "I predict he will come" is simply,

I verb (that) + S.

This class, as a class, provides no further constraints; though particular verbs may provide further constraints on the lower node S. For example, "predict" requires that an Aux in the lower S must be future or at any rate cannot be past. Such representative verbs as "describe," "call," "classify," and "identify" take a different syntactical structure, similar to many verbs of declaration, and I shall discuss them later.

Directives. Such sentences as "I order you to leave" and "I command you to stand at attention" have the following deep structure:

I verb you + you Fut Vol Verb (NP) (Adv).

"I order you to leave" is thus the surface structure realization of "I order you + you will leave" with equi NP deletion of the repeated "you." Notice that an additional syntactical argument for my including "dare," "defy," and "challenge," in my list of directive verbs and objecting to Austin's including them with "apologize," "thank," "congratulate," etc., is that they have the same syntactical form as do the paradigm directive verbs "order," "command," and "request." Similarly, "invite" and "advise" (in one of its senses) have the directive syntax. "Permit" also has the syntax of directives, though giving permission is not strictly speaking trying to get someone to do something; rather it consists in removing antecedently existing restrictions on his doing it.

Commissives. Such sentences as "I promise to pay you the money" and "I pledge allegiance to the flag" and "I vow to get revenge" have the deep structure

I verb (you) + I Fut Vol Verb (NP) (Adv).

Thus "I promise to pay you the money" is the surface structure realization of "I promise you + I will pay you the money," with equi NP deletion of the repeated "I." We hear the difference in syntax between "I promise you to come on Wednesday" and "I order you to come on Wednesday" as being that "I" is the deep structure subject of "come" in the first and "you" is the deep structure subject of "come" in the second, as required by the verbs "promise" and "order" respectively. Notice that not all of the paradigm commissives have "you" as an indirect object of the performative verb. In the sentence "I pledge allegiance to the flag" the deep structure is not "I pledge to you flag + I will be allegiant." It is

I pledge + I will be allegiant to the flag.

Whereas there are purely syntactical arguments that such paradigm directive verbs as "order" and "command," as well as the imperative mood, require "you" as the deep structure subject of the lower node S, I do not know of any purely syntactical argument to show that commissives require "I" as the deep structure subject on their lower node S. Semantically, indeed, we must interpret such sentences as "I promise that Henry will be here on Wednesday" as meaning

I promise that *I will see to it* that Henry will be here next Wednesday,

insofar as we interpret the utterance as a genuine promise, but I know of no purely syntactical arguments to show that the deep structure of the former sentence contains the italicized elements in the latter.

363

Expressives. As I mentioned earlier, expressives characteristically require a gerund transformation of the verb in the lower node S. We say:

> I apologize for stepping on your toe,
> I congratulate you on winning the race,
> I thank you for giving me the money,

The deep structure of such sentences is:

> I verb you + I/you VP⇒Gerund Nom.

And, to repeat, the explanation of the obligatory gerund is that there is no direction of fit. The forms that standardly admit of questions concerning direction of fit, *that* clauses and infinitives, are impermissible. Hence, the impossibility of

> * I congratulate you that you won the race,
> * I apologize to step on your toe.

However, not all of the permissible nominalization transformations are gerunds; the point is only that they must not produce *that* clauses or infinitive phrases; thus, we can have either

> I apologize for behaving badly,

or

> I apologize for my bad behavior,

but not

> * I apologize that I behaved badly,
> * I apologize to behave badly.

Before considering declarations, I want now to resume discussion of those representative verbs which have a different syntax from the paradigms above. I have said that the paradigm representatives have the syntactical form

> I verb (that) + S.

But if we consider such representative verbs as "diagnose," "call," and "describe," as well as "class," "classify," and "identify," we find that they do not fit this pattern at all. Consider "call," "describe," and "diagnose," in such sentences as

> I call him a liar,

> I diagnose his case as appendicitis,
> I describe John as a Fascist.

and in general the form of this is

> I verb $NP_1 + NP_1$ be pred.

One cannot say

> * I call that he is a liar,
> * I diagnose that his case is appendicitis

(perversely, some of my students find this form acceptable),

> * I describe that John is a Fascist.

There seems, therefore, to be a very severe set of restrictions on an important class of representative verbs which is not shared by the other paradigms. Would this justify us in concluding that these verbs were wrongly classed as representatives along with "state," "assert," "claim," and "predict" and that we need a separate class for them? I have heard it argued that the existence of these verbs substantiates Austin's claim that we require a separate class of verdictives distinct from expositives, but that would surely be a very curious conclusion to draw since Austin lists most of the verbs we mentioned above as expositives. He includes "describe," "class," "identify," and "call" as expositives and "diagnose" and "describe" as verdictives. A common syntax of many verdictives and expositives would hardly warrant the need for verdictives as a separate class. But leaving aside Austin's taxonomy, the question still arises, do we require a separate semantic category to account for these syntactical facts? I think not. I think there is a much simpler explanation of the distribution of these verbs. Often, in representative discourse, we focus our attention on some topic of discussion. The question is not just what is the propositional content we are asserting, but what do we say about the object(s) referred to in the propositional content: not just what do we state, claim, characterize, or assert, but how do we describe, call, diagnose, or identify it, some previously referred-to topic of discussion. When, for example, there is a question of diagnosing or describing, it is always a question of diagnosing a person or his case, of describing a landscape or a party or a person, etc. These representative illocutionary verbs give us a device for isolating topics from what is said about topics. But this very genuine syntactical difference does not mark a semantic difference big enough to justify the formation of a separate category. Notice in support of my argument here that the actual sentences in which the describing, diagnosing, etc., is done are seldom of the explicit perfomative type, but rather are usually in the standard indicative forms which are so characteristic of the representative class.

Utterances of:

> He is a liar,
> He has appendicitis,
> He is a Fascist,

are all characteristically *statements*, in the making of which we call, diagnose, and describe, as well as accuse, identify, and characterize. I conclude then that there are typically two syntactical forms for representative illocutionary verbs: one of which focuses on propositional content, the other on the object(s) referred to in the propositional content, but both of which are semantically representatives.

Declarations. I mention the syntactical form

> I verb NP_1 + NP_1 be pred

both to forestall an argument for erecting a separate semantic category for them and because many verbs of declaration have this form. Indeed, there appear to be several different syntactical forms for explicit performatives of declaration. I believe the following three classes are the most important.

> 1. I find you guilty as charged.
>   I now pronounce you man and wife.
>   I appoint you chairman.
> 2. War is hereby declared.
>   I declare the meeting adjourned.
> 3. You're fired.
>   I resign.
>   I excommunicate you.

The deep syntactical structure of these three, respectively, is as follows:

> 1. I verb NP_1 + NP_1 be pred.

Thus in our examples we have

> I find you + you be guilty as charged.
> I pronounce you + you be man and wife.

    I appoint you + you be chairman.

    2. I declare + S.

Thus in our examples we have

    I/we (hereby) declare + a state of war exists.

    I declare + the meeting be adjourned.

This form is the purest form of the declaration: the speaker in authority brings about a state of affairs specified in the propositional content by saying in effect, I declare the state of affairs to exist. Semantically, all declarations are of this character, though in class 1 the focusing on the topic produces an alteration in the syntax which is exactly the same syntax as we saw in such representative verbs as "describe," "characterize," "call," and "diagnose," and in class 3 the syntax conceals the semantic structure even more.

    3. The syntax of these is the most misleading. It is simply

      I verb (NP)

as in our examples,

    I fire you.

    I resign.

    I excommunicate you.

The semantic structure of these, however, seems to me the same as class 2. "You're fired," if uttered as performance of the act of firing someone and not as a report means

    I declare + Your job is (hereby) terminated.

Similarly, "I hereby resign" means

    I hereby declare + My job is (hereby) terminated.

"I excommunicate you" means

    I declare + Your membership in the church is (hereby) terminated.

The explanation for the bemusingly simple syntactical structure of these sentences seems to me to be that we have some verbs which in their performative occurrence encapsulate both the declarative force and the propositional content.

John R. Searle

VI. Conclusions

We are now in a position to draw certain general conclusions.

1. Many of the verbs we call illocutionary verbs are not markers of illocutionary point but of some other feature of the illocutionary act. Consider "insist" and "suggest." I can insist that we go to the movies or I can suggest that we go to the movies; but I can also insist that the answer is found on page 16 or I can suggest that it is found on page 16. The first pair are directives, the second, representatives. Does this show that insisting and suggesting are different illocutionary acts altogether from representatives and directives, or perhaps that they are both representatives and directives? I think the answer to both questions is no. Both "insist" and "suggest" are used to mark the degree of intensity with which the illocutionary point is presented. They do not mark a separate illocutionary point at all. Similarly, "announce," "present," and "confide" do not mark separate illocutionary points but rather the style or manner of performance of an illocutionary act. Paradoxical as it may sound, such verbs are illocutionary verbs, but not names of kinds of illocutionary acts. It is for this reason, among others, that we must carefully distinguish a taxonomy of illocutionary acts from one of illocutionary verbs.

2. In section IV I tried to classify illocutionary acts and in section V I tried to explore some of the syntactical features of the verbs denoting member of each of the categories. But I have not attempted to classify illocutionary verbs. If one did so, I believe the following would emerge.

a. First, as just noted, some verbs do not mark illocutionary point at all, but some other feature, e.g., "insist," "suggest," "announce," "confide," "reply," "answer," "interject," "remark," "ejaculate," and "interpose."

b. Many verbs mark illocutionary point plus some other feature, e.g., "boast," "lament," "threaten," "criticize," "accuse," and "warn" all add the feature of goodness or badness to their primary illocutionary point.

c. Some few verbs mark more than one illocutionary point; for example, a protest involves both an expression of disapproval and a petition for change.

Promulgating a law has both a declarational status (the propositional content becomes law) and a directive status (the law is directive in in-

368

tent). The verbs of representative declaration fall into this class of verbs with two illocutionary points.

d. Some few verbs can take different illocutionary points in different utterances. Consider "warn" and "advise." Notice that both of these take either the directive syntax or the representative syntax. Thus

I warn you to stay away from my wife!	(directive)
I warn you that the bull is about to charge.	(representative)
I advise you to leave.	(directive)
Passengers are hereby advised that the train will be late.	(representative)

Correspondingly, it seems to me, that warning and advising may be either telling you *that* something is the case (with relevance to what is or is not in your interest) or telling you *to* do something about it (because it is or is not in your interest). They can also be, but need not be, both at once.

3. The most important conclusion to be drawn from this discussion is this. There are not, as Wittgenstein (on one possible interpretation) and many others have claimed, an infinite or indefinite number of language games or uses of language. Rather, the illusion of limitless uses of language is engendered by an enormous unclarity about what constitutes the criteria for delimiting one language game or use of language from another. If we adopt illocutionary point as the basic notion on which to classify uses of language, then there are a rather limited number of basic things we do with language: we tell people how things are, we try to get them to do things, we commit ourselves to doing things, we express our feelings and attitudes, and we bring about changes through our utterances. Often, we do more than one of these at once in the same utterance.

On What We Know

1

In spite of the repeated efforts of so many philosophers, since Plato's *Theaetetus*, to clarify the concept of knowledge, I am still dissatisfied with the result. There are a number of fairly obvious features of this concept which cannot be squared with the prevailing theories, and which are still being ignored by philosophers, who, as is their wont, are more interested in proving their theories by an arbitrary selection of facts than in the facts themselves which the theories are supposed to explain.

The most persistent, and still dominant, line of analysis tries to understand knowledge in terms of belief, true belief, true belief with adequate evidence, grounds, accessibility, or some other, often very elaborately and ingeniously stated condition. Whether such a claim is advanced as a reduction or just as a list of necessary conditions, I think it is still misleading and prejudices the issue. For it is taken for granted by the proponents of this view that knowledge (at least in the sense of *knowing that*) can have the same object as belief, that is, that it is possible to believe and to know exactly the same thing.

This, to me, is a highly questionable assumption. Granted, it is nonsense to say that one knows that p but does not believe it. It need not follow, however, that in this case one must believe that p. What is known may be something that cannot be believed or disbelieved at all. In other words, the incongruity of the sentence *I know that p but I do not believe it* may be due not to an implied inconsistency but to a category confusion similar to the one embedded in the sentence *I have a house but I do not covet it*. As one cannot be said to covet or fail to

AUTHOR'S NOTE: This paper was also read at a conference held at the University of North Carolina, Chapel Hill, in 1968, with Bruce Aune as the commentator. A modified version of it is included in my book *Res Cogitans* (Ithaca, N.Y.: Cornell University Press, 1972).

covet one's own property, it may be the case that one cannot believe or disbelieve what one knows. Again, from the fact that what one knows cannot be false, it does not follow that it must be true and hence that knowledge must entail true belief, if what is known is not a thing to which truth and falsity apply. A picture may be faithful or not faithful, not its object; yet it is the conformity with the object that makes a picture faithful. In a similar way conformity with things known may render beliefs true, without these things being true themselves. At this point I offer these considerations as mere possibilities; the task remains to justify the analogies I have suggested.

In recent years some other points have been added to our comprehension of the concept of knowledge, such as Ryle's distinction between knowing that and knowing how, and Austin's recognition of the performative aspect of the verb to know. These are valuable insights, but the features they single out do not account for the essence of the concept. Knowing that is distinct from knowing how, as it is distinct from knowing who, what, when, where, why, or whether, from knowing a story, a house, or one's friend. What is it, beyond the use of the same word, that is common to all these cases, or, at least, what are the interlocking similarities that would constitute a family resemblance? As to the performative aspect, its presence is not sufficient to make know a bona fide illocutionary verb; the intuition, moreover, that tells us that this verb, unlike, say, declare or promise, denotes a state and not a speech act is too strong to be ignored. In any case this aspect hardly applies beyond the domain of knowing that; consequently it too fails to account for the unity, no matter how loose, of this concept.

2

What do you know? Things of surprisingly many kinds. There are only a few verbs (among them another philosophers' darling, see) that display a similar versatility. In the previous section I mentioned in passing the main categories of the possible verb objects of know. The verb believe, which is supposed to help us in our task, is much more restricted. A comparison between these two verbs is indeed helpful to start with. In doing this, at the beginning, I shall operate on a rather unsophisticated level, restricting my observations to what some grammarians would call the surface structure of the noun phrases involved. As

371

we go on, the very nature of the investigation will force us to break through the crust and reveal more and more of the underlying structure.

There is a domain which appears to be shared by both verbs. This comprises the familiar *that*-clauses, i.e., nominals formed simply by prefixing *that* to an unaltered sentence. *I know that* and *I believe that* can be followed by any declarative sentence regardless of tense, modality, or structural variation. There is, on the other hand, a domain which is wholly owned by *know* to the exclusion of *believe*. One can be said to know, but not believe, birds and flowers, houses and cars, wines and detergents, cities and deserts. Practically any original noun will do with or without adjuncts such as the relative clause and its derivatives. Names and other phrases denoting people also qualify, of course, but at this point *believe* reenters the picture. After all, you can say that you believe Jane as well as that you know Jane. Needless to say, these two assertions have very little to do with one another. Believing a person may require knowing him (to some extent), but knowing him certainly does not entail believing him: the chief reason for not believing Jane may be the fact that you know her too well. At this point the reader will protest: "But, of course, *know* and *believe* operate in totally different ways in these cases!" In other words, the reader wants to peek below the surface. For the time being I shall thwart his desire.

There is another group of nouns which is appropriate to either *know* or *believe*, and which creates a similar situation. I think of *story*, *tale*, *explanation*, *theory*, and, perhaps, *opinion*, *suspicion*, *assumption*, and the like. All these things can be known or believed, but even if known, the question of belief remains open: it is perfectly normal to say, for instance: *I know the story but I do not believe it*. The reader might want to protest again and voice his intuition. I still resist, for we do not yet know enough to see the reasons for the intuition.

In connection with Ryle's *knowing how*, I have mentioned the other *wh*-forms, such as *what*, *when*, *why*, etc., that can introduce the verb object of *know*. This move, in general, fails with *believe*; whereas one can know where the treasure is hidden, one cannot believe where the treasure is hidden. There is one exception to this incompatibility, and that concerns *what*. I may believe what you said as I may know what you said. Clearly the relation of these two claims is similar to the one just encountered between knowing and believing stories or people. The knowledge of what one said does not imply belief, but the belief of what one said presupposes the knowledge of what one said. Since the word *what*, unlike the ones figuring in the previous examples, is a purely grammatical word, we can nourish the hope that in this case we shall be able to disambiguate the offending phrase on syntactical grounds alone, and then apply the result to the previous contexts relieving thereby the reader's pent-up frustration.

3

There are whats and whats. Consider the following three sentences:

(1)   Joe lost his watch.
(2)   I found what he lost.
(3)   I know what he lost.

Sentences (1) and (2) jointly entail that I found his watch. (1) and (3), however, do not entail that I know his watch. *What*, in (2), amounts to *that which* (or *the thing which*), i.e., a demonstrative pronoun (or a dummy noun) followed by the relative pronoun beginning a relative clause. As always, such a clause depends upon a noun-sharing between two ingredient sentences. The derivation of (2) can be sketched as follows:

(4)	I found (a watch)	(4a)	He lost (a watch)
(5)	I found (the watch)	(5a)	which he lost
(6)	I found that	(6a)	which he lost
(7)		I found what he lost	

Example (5a) is a relative clause obtained by replacing *a watch* by *which*. Since the clause is taken to be identifying, *watch* in (5) obtains *the*.[1] In (6) *that* replaces *the watch*. Finally, in (7), *that which* is contracted into *what*.

What, in (3), cannot be analysed into *that which*. It is not the watch he lost that I claim to know, but rather that it is a watch that he lost, although I put my claim in an indefinite form. *What he lost*, in this case, has nothing to do with a relative clause; it is a sentence nominalization on par with, say, *who lost the watch, when he lost it, how he lost it*, and the like. This nominalization operates by replacing a noun phrase or an adverbial phrase in the original sentence by *wh* plus the appro-

---

[1] Concerning relative clauses and the definite article, see my *Adjectives and Nominalisations* (The Hague, Paris: Mouton, 1968), chap. 1.

priate pro-morpheme. Since the same replacement is used in the corresponding question transformations, the resulting noun phrases are traditionally called "indirect questions." This name is misleading, however. Granted that after *wonder*, or some such verb, these nominals retain their interrogative flavor, but this is not true after *know, tell, learn,* or *realize.* This becomes clear as we contrast the sentences:

I wonder what he lost, namely, a watch or a ring or etc.

I know what he lost, namely, a watch.

The strings following *namely* are not interchangeable. It is easy to restore the underlying sentences from which these two are derived by the removal of redundancy. They are:

I wonder what he lost, namely (I wonder whether he lost) a watch or a ring or . . .

I know what he lost, namely (I know that he lost) a watch.

The *what*-clause in this last sentence does not incorporate a question but a claim, albeit an indefinite one, which, in turn, is respecified by the *namely* string. It appears, therefore, that the *what*-clause after *wonder* comes via an earlier step in the process of nominalization, i.e.,

whether he lost a watch or a ring or etc.

whereas the *what*-clause after *know* comes via a different intermediate step, to wit,

that he lost a watch.

The same ambiguity can be shown with respect to other *wh*-forms such as *who, when, why,* etc. Those coming through *whether* can indeed be called indirect questions, but the ones derived from the *that*-form should rather be called indefinite claims. Accordingly, the correct analysis for (3) will be the following:

(8)	I know . . .	(8a)	He lost (a watch)
(9)	I know . . .	(9a)	that he lost (a watch)
(10)	I know . . .	(10a)	what he lost
(11)		I know what he lost.		

Example (10a), to repeat, is not a relative clause but a *wh*-nominal formed out of (9a). The dots in (8) to (10) indicate the "noun gap," characteristic of container sentences, which is to be filled by an appropriate nominal.[2]

It is interesting to note that wh-clauses after the negation of know do not come through that but whether. The correct analysis of, say,

> I do not know what he ate

and of

> I do not know where he went

will show

> I do not know whether he ate . . . or . . . or . . .

and

> I do not know whether he went to . . . or to . . . or to . . .

rather than

> I do not know that he ate (fish)

and

> I do not know that he went to (Paris).

This is interesting linguistically: it shows that the negation precedes the nominalization in the generative process.

Wh-nominals are not confined to the object position; they can occur as subjects too, e.g.:

> Who killed her is uncertain.
>
> Why she went there is a mystery.

This possibility permits us to draw another interesting comparison between the two structures underlying what he lost.

(12)   What he lost is a watch.

(13)   What he lost is a mystery.

Sentence (12) is contracted from

> That which he lost is a watch

which is an extraction transform of

> He lost a watch.

Sentence (13), on the other hand, is certainly no transform of

> He lost a mystery.

---

[2] Ibid., chap. 2.

These facts enable us to explain the ambiguity of *what he said* in the sentences:

(14)   I believe what he said
(15)   I know what he said.

Believe cannot take *wh*-nominals, consequently, the analysis of (14) cannot follow the pattern of (3). (2) provides the correct analogy: the object of *believe* is a pronoun (or dummy noun) followed by a relative clause. In full:

(a) I believe (that p)   He said (that p)
    I believe that   which he said
              I believe what he said.

Roughly, the object of your saying and my believing is the same thing. Not so in (15). The object of my knowledge is not the object of his saying (*that p*), but, obviously, an indefinite version of *that he said that p*. Thus the derivation matches (8) to (11) above, i.e.:

(b) I know . . .   He said (that p)
    I know . . .   that he said (that p)
    I know . . .   what he said
              I know what he said.

The possibility of *believing what* (= *that which* or *the thing which*) is restricted to "things" that can be objects of belief. For this reason, such sentences as

    * I believe what he lost

are ruled out: the relevant co-occurrence sets of *believe* and *lose*, unlike those of *believe* and *say*, do not overlap. Roughly speaking, *believe* demands *that*-clauses, but *lose* requires object nouns. If so, the intelligent reader will ask, what saves (15) from being given the relative clause interpretation in addition to the other one, and from consequent ambiguity, since the object range of *say* and *know* widely overlaps in the domain of *that*-clauses? Twist as we might, (15) is not ambiguous. It seems, therefore, that the *that*-clauses following *say* are different from the *that*-clauses following *know*. And since the former are of the kind which is compatible with *believe*, it seems to follow that the *that*-clauses after *know* are different from the *that*-clauses after *believe*; i.e., that *know* and *believe* cannot have identical objects at all. As we go on, this suspicion will grow to certainty.

Just to rub in the point, consider another verb, *tell*. In this case the phrase *knowing what he told (somebody)* is indeed ambiguous. Although the sentence

I know what he told you

most likely will be interpreted in the sense of the *wh*-nominal (*I know that he told you that p*), I can elicit the other interpretation by saying, for instance,

I already knew what he just told you.

Here I claim to know that *p* (which he just told you). One cannot play the same trick with *say*. The sentence,

I already knew what he just said (to you)

is substandard, and the improved version,

I already knew what he would say . . .

once more selects the path of the *wh*-nominal. *Know* therefore is capable of absorbing the dummy for a *that*-clause, provided this latter is of the right kind. Now *tell*, but not *say*, seems to provide such. What, then, is the difference between *say* and *tell* in this respect? I shall take up this problem later on. For the time being, let us remember that as there are whats and whats, there are thats and thats.

4

It is time to return to the problem of knowing and believing people and stories. First I shall consider belief. It is clear that a sentence such as

(16)   I believe Jane

must be elliptical. For one thing, the breakdown of two common transformations (passive and extraction) shows that the sentence is "abnormal" for some reason or other:

    * Jane is believed by me
    ? It is Jane that I believe.

Intuition tells us that what (16) means is this:

I believe what Jane said (would or will say).

What, of course, is *that which*. This intuition mirrors a general deletion pattern that tends to substitute the subject of saying or doing something for the saying or doing itself. E.g.:

I refuted him (what he claimed)
I understood him (what he said)
I imitated him (what he did)
I heard him (his voice)

and so on. In a similar way, the sentence

I believe his story (explanation, etc.)

is but an ellipsis of

I believe what his story (contains, says, etc.)

The verbs in the parentheses are not to be taken too seriously. They are but idiomatic crutches to facilitate the move to the deep structure. What is essential is that this latter contains the elements,

I believe that p
His story (explanation, etc.) is that p

and that the two are fused into a relative clause construction by virtue of the identical noun phrase (*that p*).

It appears, therefore, that all occurrences of *believe* (I am not considering believing *in* somebody or something) can be reduced to *believing that*.

Nevertheless, this verb retains some latitude, inasmuch as it can take substitutes for the *that*-clauses (*it, thing, what*). This, interestingly enough, is not true of a cognate verb, *think*. This one takes *that*-clauses without discrimination, but refuses substitutes. One can answer, for instance, *I believe it* but not *I think it*, and owing to the exclusion of dummies, it is impossible to think a person, a story, or what one said. Later on I shall return to this difference between *think* and *believe*.

What, then, about knowing a story, an explanation, or other things of this sort? Knowing these things differs from believing them in exactly the same way as knowing what one said differs from believing what one said. Accordingly, say, knowing a story is but shorthand for knowing what the story is or how the story goes. And, as the parallel with *how* clearly indicates, this *what* is not *that which*. The same about tales, reasons, explanations, excuses, and theories offered by somebody or other. Incidentally, whereas these things, as we recall, can also be believed (but what a difference!), poems, jokes, names, foreign words, tongue twisters, and the like can be known but not believed. The reason is obvious. The possibility of believing, say, a story is founded upon the sentence in the deep structure:

The story is that p

which makes the relative clause inclusion possible. Poems, tongue twisters, etc., on the other hand, cannot be reproduced in *that*-clauses; consequently there is no way of connecting them with *believe*. Nothing prevents their being known, however. Knowing a poem, for instance, is knowing how it goes, or, if one is more ambitious, knowing how it is to be understood, interpreted, and what not.

The availability of the inexhaustible variety of *wh*-nominals the verb *know* can take makes it a relatively easy matter to explicate one element of the concept of knowing such things as a person, a house, a car, or a city. What, for instance, can you possibly mean when you say that you know Jane? There is a "minimal" sense of this claim, which is satisfied if you have ever met her in the flesh and, perhaps, talked to her. Yet in spite of such an encounter you may still insist that you do not know her ("I have met her, but I do not know her at all"). What is it that you disclaim in this case? What would be knowing her in this fuller sense? Well, it is an open-ended affair. It might be merely knowing what her full name is, where she comes from, what she does for a living, and other particulars of this sort. If you know her better, if you "really" know her, then you know what she thinks, how she feels about various matters, what she would do if . . . ; consequently you know how to treat her, and the like.

Knowing all these things *about* Jane will not, however, normally entitle one to claim that one knows her without the personal acquaintance previously mentioned. I do not know Mao Tse-tung, although I know many things about him. Yet some latitude remains in this respect. Churchill could have said truthfully, during World War II, "I know Hitler, he would destroy his country rather than surrender," even if he never met Hitler in the flesh. But, one might argue, they at least had some dealings with one another, which is not true of Mao and me. Again, the phrase *I used to know him many years ago* suggests an interruption of contact rather than of the flow of relevant information.[3]

[3] I am indebted to Paul Ziff for a clearer perception of these two aspects of knowing a person.

379

Mutatis muntandis, the same analysis works for knowing houses, cars, cities, and the like. Does the armchair geographer who knows a great deal about Lhasa know Lhasa? Does the little old lady from Dubuque who spent two days in Paris with a guided tour know Paris?

I suspect that the requirement of contact (acquaintance) is a hangover from the ancient sense of *know*, according to which knowing, say, pain and misery meant having these things, and which sense is also reflected in the phrase *carnal knowledge*. We are going to see, toward the end of this essay, that this element of contact with something actually "there" still pertains to the concept of knowledge throughout the entire domain of its application.[4]

The phrase I used above, *knowing how to treat her*, represents a new construction which I did not mention before among the possible objects of *know*. It is by no means restricted to *knowing how*. I know whom to blame, what to do, where to look, and when to stop in many situations. The transformational origin of these phrases is quite clear. The infinitive, *to* V+, is generally used in sentence nominalizations to code a noun-sharing between the subject of the nominalized sentence and either the subject or the direct object of the container sentence, provided there is a modal verb in the latter sentence. These features can be brought out with greater or lesser grammaticality in appropriate paraphrases; e.g.:

> I decided to go — I decided that I *should* go
> I persuaded him to go — I persuaded *him* that *he should* go
> I know where to go — I know where I *should* go
> I know how to solve the problem — I know how I *can* solve the problem,

and so forth. As we see, there is nothing special about *knowing how to*.

Reviewing our results, we find that the acquaintance sense aside, all verb objects of *know*, other than the *that*-clause, can be reduced to *wh*-nominals. Now these, themselves, are nothing but indefinite versions of *that*-clauses, formed, as we recall, by replacing a noun phrase or an adverbial phrase in the sentence following *that* by *wh* plus the appropriate pro-morpheme. Consequently whenever I claim that I know *wh* . . . , I guarantee that I could make another claim in which the *wh*-nominal is replaced by a corresponding *that*-clause. It makes perfect

---

[4] In many languages there are two separate words corresponding to *know*; one for noun-clause objects and the other for simple noun objects (*kennen–wissen*, *connaitre–savoir*, etc.).

sense to say that I know what he did, but I would not tell you; to say, on the other hand, that I know what he did, but I could not possibly tell you, is absurd.

It will be objected here that in some cases of knowing how it is impossible to tell, in words, what one knows. I know how to tie a necktie, but I could not tell you in words alone. I grant this, but point out that this situation is possible with nearly all the *knowing wh* forms. I know what coffee tastes like, what the color magenta looks like, where it itches on my back, when I should stop drinking, how the coast line of Angola runs, but I could not tell you in words alone. I must have, however, some other means to supplement words: pointing, offering a sample, a sketch, a demonstration, or saying "now." By these means I can tell you, or show you, what I know: I know that magenta looks like this (offering a sample), that it itches here (pointing), that I should stop drinking now. The need of supplementing words with nonlinguistic media affects knowing how, and knowing *wh* in general, exactly because it affects the corresponding knowing that.

I leave it to the imagination of the reader to account, along similar lines, for the meaning of phrases such as knowing geography, knowing Aristotle, or knowing Russian.

5

In the previous section we have concluded that the basic form of the verb object for both *believe* and *know* is the *that*-clause. Yet at the beginning I suggested that these verbs cannot have the same verb object. These two claims need not conflict, of course, if *that*-clauses can be ambiguous. And, indeed, we have already encountered one reason for thinking that they are. Roughly, the object of saying, a *that*-clause, can be the object of belief, but not the object of knowledge. In this section I shall gather the remainder of the evidence that points in the same direction.

The main argument I am going to advance involves a group of nouns that are normally joined to *that*-clauses by means of the copula, e.g.:

His suggestion is that p
That p is his prediction
That p is a fact
The cause of the phenomenon is that p.

Clearly there is a need to subdivide this class. Words like *suggestion*, *prediction*, *statement*, *confession*, *testimony*, *excuse* on the one hand, and *belief*, *opinion*, *assumption*, *view*, *theory*, and *suspicion* on the other, are derivatives either of illocutionary verbs or of verbs of propositional states. This is shown in the typical transformation exemplified by

He suggested that p — That p is his suggestion
He suspects that p — That p is his suspicion.

The *that*-clause, accordingly, is tied to a person and this tie is specified by the words I listed, and which I shall call "subjective" P-nouns. This class is to be distinguished from the class of "objective" P-nouns, which will comprise *fact*, *cause*, *result*, *outcome*, *upshot*, and a few others. Facts, causes, and the like do not belong to anybody, and the transformation just given has no parallel. Some P-nouns cross the line, e.g., *reason* and *explanation*: one can speak of Joe's reason or Joe's explanation versus *the* reason or *the* explanation.

These two groups behave quite differently with respect to *know* and *believe*. Subjective P-nouns can follow either verb, but the analyses of the resulting sentences follow quite different lines. If, for instance, someone's suggestion is that p, then believing his suggestion is believing that p; knowing that suggestion, however, never means knowing that p, but rather knowing what that suggestion is, i.e., knowing that the suggestion is that p. It appears, therefore, that *that*-clauses marked by subjective P-nouns are per se compatible with *believe*, but not with *know*. This latter verb cannot take on these *that*-clauses except in a roundabout way, via the *wh*-nominal.

In view of what we found before, it is easy to give the analyses of the two sentences involved:

(c) I believe (that p)   His suggestion is (that p)
    I believe that   which is his suggestion
         I believe what is his suggestion
         I believe his suggestion
(d) I know . . .   His suggestion is (that p)
    I know . . .   that his suggestion is (that p)
    I know . . .   what his suggestion is
         I know his suggestion.

The sentence with *believe* cannot follow the second pattern, since, as we recall, this verb cannot take *wh*-nominals. What is more interesting, and indeed decisive, is that the sentence with *know* does not conform to the first pattern. The clause *that p*, insofar as it is marked as a suggestion, that is, as something subjective, something produced by an illocutionary act, is not an appropriate object of *know*.

The situation is quite different with objective *P*-nouns. They naturally follow *know*, but only with great strain *believe*. You may know the facts about the crime, the cause of the explosion, the result of too much publicity, and the outcome of the trial. But what, possibly, could be meant by a sentence such as *I believe the cause of the explosion* or *I believe the outcome of the trial?* It might be objected that results are said to be believed or disbelieved: it makes sense to say *I do not believe the results of the autopsy.* We quickly realize, however, that in this case one speaks of the results submitted by some experts; they are "their" results. If no human authorship is involved, then believing or disbelieving results is impossible; it is nonsense to say, for instance, *I do not believe the results of the inflation.* In a similar way one may believe or not believe the "facts" submitted by the police, but not the facts about the crime.

The analysis of a sentence like *I know the cause of the explosion* is interesting. It goes via the *wh*-nominal: *I know what is the cause of the explosion*, i.e., *I know that the cause of the explosion is that p.* This, of course, entails that I know that *p*. If I know that the cause of the explosion was the overheating of the wire (that the wire got overheated), then I know that the wire got overheated. The words *cause, result, outcome*, and so forth are relative, inasmuch as they are followed by a genitive structure in nonelliptical sentences: causes, results, outcomes, and the like are causes, results, and outcomes of something or other. Hence knowing the cause of the explosion is not merely knowing that *p*, which is the cause, but knowing that that *p* is the cause, i.e., knowing what is the cause.

Facts are not relative in this respect. Consequently these two paths merge into one. Indeed, *I know that that p is a fact* or *I know, for a fact, that p* are but emphatic forms of *I know that p.* Of course one can relate facts to something or other and say, for example, *I know the facts about the crime*, which means *I know what are the facts about the crime*, i.e., *I know that the facts about the crime are that p and that q*, etc. This, naturally, entails that I know that *p*, that *q* etc.

At this point we should recall what we said about the relevant difference between *say* and *tell*. If you said that *p*, and I believe what you said, then I believe that *p*; knowing what you said, however, does not mean knowing that *p*, but knowing that you said that *p*. *Tell* works differently: if you told me that *p*, then knowing what you told me may mean, in a suitable context, that I know that *p*. This seems to suggest that *tell*, but not *say*, is an objective P-verb. It is not surprising to find, therefore, that one can tell, but not say, the facts about the crime, the cause of the explosion, the outcome of the trial, and so forth. *Tell* belongs to the family of *know*; *say* to that of *believe*.

The urge to find symmetries makes us ask the question, why is it that one cannot say suggestions, predictions, opinions, and the like. The answer is simple. *Say* like *think* can take *that*-clauses, but not their substitutes: you may believe that *p*, think that *p*, or say that *p*; but if I suggested that *p*, then you may believe my suggestions, but not think or say my suggestion.

But, you object, I can surely tell you my suggestion or opinion, and this move seems to cross the line between the subjective and the objective. No more, I reply, than knowing opinions or suggestions. The *wh*-nominal bridges over the gap. Telling your suggestion, or knowing your suggestion, is telling, or knowing, what that suggestion is (and *what*, here, is not *that which*). Telling your suggestion is not the same thing as suggesting or making a suggestion. A suggestion is a subjective entity. That one has made a suggestion, however, is something objective, it may be a fact.

Wh-nominals in general belong to the objective domain. We have seen that they are compatible with *know* but not with *believe*. Similarly they can follow *tell* but hardly *say*. I can tell you where I went yesterday and what I did there and why. Putting *say* for *tell* in such context, however, will yield ungrammaticality or at least substandard speech:

?He said where he went (. . . what he did, . . . why he did it).

But notice that negation changes the picture. The sentence

He did not say where he went (. . . what he did, . . . why he did it)

sounds distinctly better. We know, of course, that the presence of negation means that the subsequent *wh*-clause is to be derived not from *that p* but rather from *whether p* or *q* or . . . This may suggest that the objectivity of such clauses is somewhat problematic. A comparison with *P*-nouns will round out the picture. Sentences such as

> *His suspicion is why she did it
> *Who killed her is my opinion

are ungrammatical. On the other hand, it is easy to pair *wh*-clauses with objective *P*-nouns:

> What he did was the result of despair
> How he said it was the cause of the scandal.

It will be helpful to consider a very common verb, *state*, at this point. From our present point of view it is like *tell*, rather than *say*, since obviously one can state a fact, or the cause, the result or the explanation of something or other. What one states, therefore, is something objective. One's statement, on the other hand, is not. For whereas what one states may be a fact, one's statement, even if true, is never a fact. Accordingly, knowing your statement cannot be anything but knowing what that statement is; knowing what you stated, however, may amount to knowing that *p*, which you stated. Believing your statement, on the other hand, necessarily means believing that *p*, which you stated.

A little exercise for amusement's sake. What do the following sentences mean?

(17)   I believe what you believe.
(18)   I know what you believe.
(19)   *I believe what you know.
(20)   I know what you know.

Sentence (17) means that you and I share a belief. (18) means that I know what your belief is. (19) is deviant. (20) is ambiguous: it either means that I know what it is that you know, or (with a stress on *you*) that you and I share a piece of knowledge.

All these facts can readily be accounted for by the following simple hypothesis. There are two kinds of *that*-clauses, the subjective and the objective. They are distinct because, first, they have entirely different co-occurrence restrictions: one kind fits subjective *P*-nouns and subjective verbs such as *say* and *believe*, the other kind fits objective *P*-nouns and such objective verbs as *tell* and *know*; second, their transformational potential is different: objective *that*-clauses are open to the *wh*-nominalization, but subjective ones are not. Thus we see that the seemingly trivial and unexplainable "accident" of grammar that, for instance, it is possible for me to know what you ate, but not to believe what you ate, is, in fact, an important clue to the discovery of a fundamental distinction in linguistic structure and in our conceptual framework.

We have seen that *that*-clauses (or their pronoun substitutes) are incompatible with certain contexts as a result of their being embedded in a more immediate context. Since the incompatibility works both ways, we have to assume that the immediate context imprints a mark on the clause and this marker, subjective or objective, then decides its further co-occurrence restrictions and its transformational behavior with respect to the *wh*-nominalization. Take the phrase *believing what one knows*. Since *believe* rejects *wh*-nominals, *what* has to be *that which*. *That* is a pronoun substituting for a *that*-clause governed by the objective verb *know*, which clause, accordingly, bears an objective marker. This, however, precludes the context *believe*. The phrase, therefore, is ungrammatical; one cannot possibly believe what one knows.

6

Thus far, surprisingly enough for an essay on knowledge and belief, we have said very little about truth and falsity. The considerations just given concerning subjective and objective P-nouns and other propositional containers make it possible to fill this gap at this point. If one asks the question, what are the things that are paradigmatically true or false, the answer will contain the set of subjective P-nouns, i.e., *statement, assertion, testimony*, etc., on the one hand and *belief, assumption, suspicion*, etc., on the other. Turning to objective P-nouns, we find that although the adjective *true* can be ascribed to them, the resulting compounds require a rather special interpretation. Compare, for instance, *true statement* and *true result*. The analysis of the former phrase is simple and straightforward: a true statement is a statement which is true. The phrase *true result*, however, suggests a different interpretation: a true result is not a result which is true, but something which is truly ("really") the result of something or other. In this sense *true* (like *real*) contrasts not with *false* but with *alleged*. In the same way the true facts of the case will be contrasted with the alleged facts, true causes with alleged causes, and so forth. In all these contexts *true* can be replaced by *real* without any loss of meaning. *False*, moreover, hardly applies at all; what would be a false fact, false cause, or false result? Of course, we know from elsewhere that *true* in this adverbial sense is not the opposite of *false*: such phrases as *true fish* or *true North* do not have *false fish* or *false North* as opposites; nor do, for that matter, *false teeth* and *false hair* have *true teeth* and *true hair* for opposites. To argue, therefore, that since there are no false facts, false causes, or false results, all facts, causes, and results must be true, is to commit the same blunder as to conclude that since there are no false fish all fish must be true, or to insist that all hair must be false, since true hair does not exist. Insofar as *true* is opposed to *false* (i.e., used in an adjectival sense), facts and causes, results and outcomes, are neither true nor false.

We have found, however, that it is exactly these things that are the immediate objects of knowledge. It follows, then, that what is known is not something that can be true or false. What I say or what I believe is true or false, not what I know.

"But, surely — you object — what I know must be so, must be the case, or even must be the truth. Now do these phrases not mean the same thing as *must be true*?" No, they do not. Consider, once more, the verb *state*. We have shown that the two derivatives, *his statement* and *what he stated* are very different. The former belongs to the subjective domain: his statement may be true, but his statement cannot be a fact. What he stated, however, can be; people often state facts. And, I add now, what he stated may be so, may be the case, or may be the truth. His statement, on the other hand, is never the case or the truth. For the same reason, whereas it is possible to tell the truth or tell what is the case, it is not possible to say the truth or say what is the case. *Tell*, as we recall, is an objective verb, *say* a subjective one.

The phrases *is so*, *is the case*, *is the truth* are near synonyms with *is a fact*. The phrase *is true*, on the other hand, does not belong to this set. The thing which is true is not a fact, it only fits the facts, corresponds to what is the case, and, perhaps, agrees with the truth. Consequently, what I believe, or what I say, may fit the facts, in which case it is true; or may fail to fit the facts, in which case it is false. What I know, however, is the fact itself, not something that merely corresponds, or fails to correspond, to the facts.

"We make to ourselves pictures of facts."[5] We form, conceive, or

[5] L. Wittgenstein, *Tractatus Logico-Philosophicus*, 2.1.

adopt beliefs, opinions, and the like. And we issue such pictures for the benefit of others in making statements and suggestions, in giving testimonies, descriptions, and so forth. These are subjective things, human creations; they belong to people: we speak of Joe's beliefs or Jane's statements. Facts, results, causes, etc., are objective: they do not belong to anybody. They are "there" to be found, located, or discovered.

A picture is affected by the imperfections of the painter. Accordingly beliefs, statements, and other subjective things, even when true, bear the marks of human ignorance: they represent the facts from a certain point of view, in a given perspective. What I am driving at is the referential opaqueness of such contexts. Even if it is true that Joe believes that A. S. Onassis married Jacqueline Kennedy, it may be false that he believes that Onassis married the widow of the thirty-fifth president of the United States. Yet these two possible beliefs are true together, because they correspond to the same fact. Given this, we are faced with a serious difficulty. If the object of knowledge is not a subjective replica, but the objective fact itself, as we have claimed, then why is it that not only belief-contexts, but also knowledge-contexts are referentially opaque? For even if Joe knows that A. S. Onassis has married Jacqueline Kennedy, it is possible that he does not know that Onassis married the widow of the thirty-fifth president of the United States.

This is a very difficult problem, which requires the utmost care. Let us return to the picture analogy. Given two pictures of the same thing, say, of a rose, my claim that I see this or that picture of the rose does not entail that I see the rose itself. I do, of course, see the rose "in the picture," but seeing something "in the picture" is not the same thing as seeing something simpliciter. Therefore, seeing two pictures of the same thing does not entail seeing the same thing. Now consider seeing the rose itself from two points of view. Do I see the same thing? Yes and no. Yes, if I focus my attention on the object seen; no, if I focus my attention on the aspect (appearance) presented to me in the two glances.

I claim that the situation is similar with respect to belief versus knowledge. Two persons may hold different beliefs mirroring the same fact; they have, as it were, different pictures of it. If, however, two persons know the same fact, what they know, in one sense, will be the same thing, although what they know may appear to them in different perspectives. The person who knows that Onassis married Jacqueline Kennedy, and the person who knows that Onassis married the widow of the late president, without either of them knowing that Jacqueline is that widow, know the same fact, namely, both know *whom Onassis married*. A parallel move, as we recall, is impossible with *believe*. I cannot claim that two persons having beliefs that mirror the same fact, but differ in the referential apparatus, have the same belief, for example in our case, that they both believe *whom Onassis married*. The wh-nominal transcends referential opaqueness. It is not surprising, therefore, that its application is restricted to objective contexts. The possibility of wh-nominalization marks the objective domain of the language.

7

Now the intuitive pieces fall into a consistent pattern. The widely different ways in which we think of knowledge and belief, the difference in their conceptual environment, bears out the conclusion we established by more formal means. I shall select a few salient points for discussion, but the list could be continued and the details amplified to the extent of one's patience and curiosity.

I begin with the well-known difference in the form of the relevant questions: *Why do you believe . . . ?* and *How do you know . . . ?* *Why* demands reasons, but *how* asks for the manner of a successful achievement. Forming a belief, like painting a picture, or imagining something, is a human act, which can be recommended (*You ought to believe it*), praised (*. . . a reasonable belief*), or condemned (*. . . a foolish belief*), and for which one might and should have reasons. One does not say, on the other hand, that one has reasons for one's knowledge. Knowing something does not involve something like forming a picture; it is rather like seeing what is there. And you do not have reasons for seeing something either. The relevant question here is concerned with the source of the information: how did you find it out, "whence" do you know?[6]

Still, we can say *you ought to know* as much as *you ought to believe*. But what a difference! You ought to believe something if you have good reasons for doing so. *You ought to know* is ambiguous. One sense is this: you ought to learn it, or have learnt it (say, from a textbook). The other is more interesting: you had the opportunity (you were in a posi-

---

[6] "*Unde scis . . . ?*" asked the Romans and "*Woher wissen Sie . . . ?*" ask the Germans.

tion) to learn, realize, or find out. *You ought to believe* is a recommendation, *you ought to know*, in this interesting sense, is a reminder.

Then compare refusing to believe and (if it exists at all) refusing to know. The attitude of *I do not want to believe* is like a struggle against a compulsive image, but the attitude of *I do not want to know* is similar to closing one's eyes. Again, contrast the unbelievable with the unknowable. The unbelievable is something utterly unlikely, unexpected, or outrageous. The unknowable need not be any of these things; it may be quite simple. What makes the unknowable unknowable is not its internal nature, but the fact that we cannot have access to it. Think of the parallel contrast between what cannot be imagined (that defies the imagination) and what cannot be seen.

People have and share beliefs but not knowledges. This is so because the immediate object of believing is a belief, a picture of reality. The immediate object of knowing is not "a knowledge," a picture of reality, but reality itself.

Vendler on Knowledge and Belief

Philosophers commonly assume that it is possible to believe and to know exactly the same thing — for instance, to believe and to know that snow is white. Vendler tells us that this assumption is actually mistaken; in his view "one cannot possibly believe what one knows" because, among other things, the phrase 'believing what one knows' is ungrammatical. I shall explain why the arguments he offers are unconvincing.

Vendler churns up many interesting grammatical facts in the course of his essay, but his final conclusions regarding knowledge and belief are reached by a very obscure line of thought involving, apparently, two distinguishable stages. He tries to show, first, that certain 'that'-clauses introduced as grammatical objects of the verb 'believe' are not "per se compatible" with the verb 'know.' He then uses this conclusion, along with grammatical facts about other verbs and their objects, to support a general hypothesis about two kinds of 'that'-clauses, one "subjective" and the other "objective." The verb 'believes,' he says, takes subjective 'that'-clauses, and 'know' takes objective 'that'-clauses; consequently, there can be no "same thing" to serve as a common object of 'He believes' and 'He knows.'

As I see it, Vendler's first line of argument does not really support his conclusion regarding the compatibility of certain 'that'-clauses with both 'believe' and 'know,' and this latter conclusion, being unsupported, adds no weight to the general hypothesis he subsequently develops. His general hypothesis is, moreover, defective for reasons I shall mention toward the end of this paper.

The basic structure of his first line of argument appears to be the following. If someone's suggestion is that p, then believing his suggestion is believing that p. Yet knowing his suggestion is not knowing

AUTHOR'S NOTE: This paper was originally read at a symposium with Zeno Vendler held at the University of North Carolina in November 1968.

391

that p; it is knowing that he suggested that p. Given the patent differences between these implications, it appears that the 'that'-clauses marked by subjective P-nouns such as 'his suggestion' are per se compatible with 'believe' but not with 'know.'

As I have formulated it, Vendler's argument is an obvious non sequitur. If A's suggestion is that p, then knowing his suggestion — that is, knowing what his suggestion is — does not, admittedly, amount to knowing that p. But this hardly shows that the agent in question does *not* know that p (in the same sense of the clause 'that p') or that there is some kind of "incompatibility" between what A suggests (namely, that p) and the context 'He knows . . .'

What Vendler has shown is merely that whereas a certain inference involving 'belief' is valid, a superficially similar inference involving 'know' is not valid. The two inferences are only superficially similar because, as Vendler himself admits, 'S knows A's suggestion' means 'S knows what A's suggestion is' while 'S believes A's suggestion' means 'S believes that which A suggests.' Since the corresponding inferences are not logically analogous, the validity of one and the invalidity of the other is patently inadequate to demonstrate Vendler's conclusion.

Vendler is, of course, an astute philosopher, and one is inclined to think that there must be more to his argument than meets the eye, or at any rate has met my eye. I want therefore to look more closely at one of the entailments he emphasizes:

(1)   If A's suggestion is that p and S believes A's suggestion, then S believes that p.

Although this entailment is no doubt patently sound, exactly how, one might ask, can its truth be *proved*? A natural strategy is to argue that 'A's suggestion is that p' is an identity statement, of the form '$Q = R$,' and that (1) is consequently true by virtue of the principle of the substitutivity of identity. It is, admittedly, obvious that we cannot always make valid substitutions in accordance with this principle, but (1) is not, we may feel, a case in which the principle is inapplicable. In this regard (1) differs sharply from

(2)   If A's suggestion is that p and S knows A's suggestion, then A knows that p,

which is easily proved false by numerous counter-instances.

This approach to (1) and (2) may appear to support Vendler's conclusion. If we may validly substitute a particular 'that'-clause for 'A's suggestion' in one context but not in another, may we not express this fact by saying that the substituted clause is "compatible" with one context but not with the other?

The answer, as before, is "No": the truth of (1) but the falsity of (2) does not establish anything about an incompatibility. Although it is possible that the antecedent of (2) may be true while its consequent is false, it is equally possible that the consequent is true whenever the antecedent happens to be true. The following, in other words, is *not* a consequence of (2)'s failure:

(3)   If A's suggestion is that p and S knows A's suggestion, then it is not the case that A knows that p.

Although the strategy just considered does not support Vendler's conclusion, he may nevertheless believe that the truth of (1) hinges on the substitutivity of identity. For reasons that will shortly become clear, I want to discourage this opinion by a couple of brief remarks. First, the so-called referential opacity of contexts such as 'S believes or suggests that p' are bound to give us trouble if we attempt to demonstrate entailments like (1) by a substitution approach. An example of this is the failure of

(4)   If I say that p and A's suggestion is that p, then I say A's suggestion.[1]

Second, it is in any case implausible to construe such statements as 'A's suggestion is that p' as identities. For one thing, 'A's suggestion is that p' is equivalent (as Vendler notes) to 'A suggests that p'; and in this latter statement 'that p' seems adverbial to 'suggests', modifying the verb rather than denoting some object.[2] This appearance is supported by the fact that 'I believe S's belief' (which Vendler seems committed to accepting as grammatical, meaning 'I believe that which S believes') is adequately expressed by 'I believe as S believes.' Finally, related statements involving

---

[1] Note that if Vendler's argument discussed on pp. 391–92 of this paper were sound, an incompatibility between the objects of 'say' and 'believe' would be demonstrated by the fact that, even when A's suggestion is that p, saying that p is not saying A's suggestion. Yet Vendler holds that both 'believe' and 'say' take subjective 'that'-clauses.

[2] I have defended this view in "Statements and Propositions," Nous, 1(1967): 215–229.

indefinite *P*-nominals are plainly not identity statements: 'His belief is one that I cannot accept', '*S*'s belief concerns Mary', and so forth.

If, as I believe, it is a mistake to think that (1) is demonstrable in the way just suggested, the question immediately arises as to how that truth of (1) can be demonstrated. The answer, fortunately, is not difficult to give. If we employ what are called "substitution" quantifiers, we may construct the appropriate demonstrations very easily.[3] For anyone unfamiliar with these quantifiers, it is enough to say that they differ from the usual ones only semantically. Instead of ranging over a domain of objects, they are interpreted in accordance with the schema:

'$(\exists_s P)(\ldots P \ldots)$' is true just when there is an appropriate expression '*E*' which, when substituted for all free occurrences of '*P*' in '$(\ldots P \ldots)$', yields a true statement '$(\ldots E \ldots)$'.

Consider '*A* believes *C*'s suggestion,' which means '*A* believes that which *C* suggests.' A plausible way of interpreting this is as follows:[4]

(5) $(\exists_s Q)(C \text{ suggests } Q) \,\&\, (\forall_s R)(C \text{ suggests } R \supset A \text{ believes } R)$.

If we express the statement '*C*'s suggestion is that *p*' as '*C* suggests that *p*,' we may then immediately infer the desired conclusion,

(6)   *A* believes that *p*.

Consider now '*A* knows *C*'s suggestion,' which means '*A* knows what *C*'s suggestion is.' A plausible interpretation of this is as follows:

(7) $(\exists_s Q)(C \text{ suggests } Q) \,\&\, (\forall_s R)(C \text{ suggests } R \supset A \text{ knows that } C \text{ suggests } R)$.

Given the premise that *C* suggests that *p*, we may therefore infer that *A* knows that *C* suggests that *p*. We may not, of course, infer that *A* knows that *p* or even that *A* does not know that *p*. As already noted,

---

[3] See J. M. Dunn and Nuel Belnap, Jr., "The Substitution Interpretation of Quantifiers," *Nous*, 2(1968):177–185.

[4] Observe that, in line with Vendler's discussion, the context '*C* suggests' in formulas (5), (6), and (7) is to be understood as having a tacit rider such as 'at time *t*' or 'on the occasion *O*.' I have omitted this rider from my formulas merely for the sake of brevity.

My treatment of 'what (= that which) *C* suggests' amounts to treating it as a substitutional version of a so-called selective description. The context '*A* believes what *C* suggests' then has the form, '*A* believes $(\epsilon P)(C \text{ suggests } P)$,' which is equivalent to formula (5) in the text. The logic of selective description is discussed in R. M. Martin, *Truth and Denotation* (Chicago: University of Chicago Press, 1958).

these latter claims are logically independent of A's knowledge of what C suggests.

The importance of this approach to the entailments Vendler discusses is that we can demonstrate their respective truth and falsity without implying that the relevant 'that'-clauses differ from one another in some subtle or unsubtle way. In fact, even if we assume that the 'that'-clauses in 'He knows that snow is white' and 'He believes that snow is white' have exactly the same meaning and do not represent objects of any kind (whether "subjective," "objective," or neither), we may still consistently prove that (1) and (2) have precisely the status Vendler says they have, namely, true for (1) and false for (2).

If the facts thus far discussed do not really support Vendler's conclusion regarding the compatibility of 'know' with certain 'that'-clauses, what about the other grammatical peculiarities of 'know' and 'believe' that he mentions? Do not the following facts support his conclusion?

(8)   'Know' may take wh-nominals as grammatical objects, but 'believe' cannot.

(9)   'Believe' commonly takes the relative pronoun 'what' as a grammatical object, but 'know' does so only in special cases.

In my view these are very interesting facts about 'know' and 'believe,' but they certainly do not prove — or even, so far as I can tell, provide significant evidence for the conclusion — that a single, nonambiguous 'that'-clause (for example, 'that snow is white') may not serve as the verb object of both 'know' and 'believe.'

Take (8), for example. According to Vendler, a wh-nominal is a kind of indefinite substitute for a 'that'-clause. But the fact that 'know' but not 'believe' may take such a substitute for its grammatical object does not in any way imply that both verbs may not take as grammatical objects any particular 'that'-clause for which a wh-nominal may serve as an 'indefinite' proxy.

As for (9), this brings out an important difference of meaning between the verbs 'know' and 'believe,' but it scarcely establishes a difference of meaning or interpretation between the 'that'-clauses that may serve as grammatical objects of these verbs. We may indeed agree that while

(10)   If you said that p and I believe what you said, then I believe that p

395

is true,

(11)   If you said that p and I know what you said, then I know that
         p

is false. But this merely tells us something about the difference between
'believes what' and 'knows what' in otherwise similar sentence frames;
it tells us nothing about the interpretation of the ingredient 'that'-
clauses.

I now want to say something about Vendler's distinction between
objective and subjective P-nouns and verbs. As I see it, he has not
really succeeded in making this distinction clear. 'Tell,' he says, is an
objective P-verb, but 'say' is subjective. But why, exactly, is this so?
"Tell," in the sense of 'relate', can take wh-nominals as grammatical objects, as 'say' cannot; one can tell (relate) the facts but not say them,
and one can know what one relates but not what one says (or so Vendler believes). But is this really enough for objective status? If it is, then
these tests may conflict with the first test mentioned in connection with
subjective P-nouns.

Consider 'my suggestion,' 'my prediction,' and 'my doubt.' By the
first test, 'suggestion,' 'prediction,' and 'doubt' are subjective P-nouns;
yet the verbs 'suggest,' 'predict,' and 'doubt' all take objective wh-nominals:

> He predicted when, how, where, what would happen if . . .
> He suggested when, how
> He doubted whether

And what about 'wonder' and 'consider'? These seem as "subjective" as
'believe', but they both take wh-nominals.

Note, incidentally, that 'statement,' 'belief,' 'claim,' 'assertion,' 'opinion,' etc., form not only subjective P-nouns such as 'my statement' but
also objective-looking nouns such as 'the statement,' 'the belief,' and
'the assertion.' I believe, and indeed have argued in print,[5] that these
latter P-nouns are derivative from the former, but they nevertheless
seem as objective in Vendler's sense as 'the fact' or 'the result.' What
should be said about their status? Are they really subjective, or what?
If the statement that p is true, does it not follow that one's statement
that p is true as well?

[5] See my "Statements and Propositions," pp. 221–229.

I now want to comment on the alleged objects of knowledge and belief. There are at least two senses, obviously, in which one can speak of "an object of knowledge or belief." In one sense such an object is a linguistic expression: in a sentence such as 'He knows that snow is white' the clause 'that snow is white' may be the grammatical object of the verb 'knows.' In another sense an object of knowledge or belief is something apparently nonlinguistic. For Vendler, "the immediate object of believing is a belief, a picture of reality," while "the immediate object of knowing is . . . reality itself," or, as he also says, a fact.

The idea that there are objects of belief in the first sense is scarcely controversial. Although most philosophers would add that the objects (in this sense) of 'believe' and 'know' may be precisely the same, Vendler, as we know, would disagree.

The idea that there are objects of knowledge and belief in the second sense is, however, extremely controversial. Nominalists argue, for instance, that such objects simply do not exist. They will agree, of course, that one can know a person or a town, and that one can believe a friend when he says this or that. What they will deny is that one can know a fact or believe a picture of reality. Vendler evidently disagrees with the nominalists on this matter. The question is, "Does Vendler offer compelling reasons for his point of view?"

Given the grammatical arguments outlined in his essay, the question whether we must admit objects of knowing and believing (in this second sense) depends on whether 'that'-clauses must be understood as in some way representing such objects. This seems to follow from two theses of Vendler's essay. The first is that "all occurrences of *believe* [except for those concerning the belief *in* somebody or something] . . . can be reduced to believing that." The second is that "the acquaintance sense aside, all verb objects of *know*, other than the *that*-clause, can be reduced to *wh*-nominals . . . [which are in turn] nothing but indefinite versions of *that*-clauses . . ." Thus, although we do commonly say that so-and-so believes the story or knows the facts of the case, the import of such remarks may be expressed by saying, respectively, something like the following:

(12) $(\exists_s P)$ (the story "says" that P and so-and-so, as a result of reading or hearing the story, believes that P),

and

(13) $(\exists_s P)$ (the facts of the case are that P and so-and-so knows that P).

In objecting to the idea that entailment (1) is demonstrable by virtue of the substitutivity of identity, I said that it is implausible to construe statements such as 'A's suggestion is that p' as identity statements, as having the form '$Q = R$.' Alluding to a grammatical transformation mentioned by Vendler, I remarked that the statement is equivalent to 'A suggests that p' and that the 'that'-clause in this latter statement functions adverbially, modifying the verb 'suggests.' Although my remark, if sound, does not prove that the 'that'-clauses in such constructions do not somehow represent nonlinguistic objects, Vendler has plainly not given us good reason to accept this latter possibility. Until he does, however, he cannot expect us to assent to his unusual view that if a man believes that snow is white, his believing has something subjective as its object, and that if he knows that snow is white, his knowing has something objective as its object. For my part, there is little reason to assume that knowing and believing have extra-linguistic "objects" at all. To know the facts of the case or the results of an election is, as Vendler himself says, to know *that* something-or-other; but to know *that* something-or-other is not necessarily to know some extra-linguistic object.

It is with reluctance that I end these remarks, for there is a great deal more to be said about Vendler's stimulating essay. My conclusion, however, is this. Vendler has simply not shown (a) that the 'that'-clauses in such statements as 'He knows that snow is white' and 'He believes that snow is white' do not have exactly the same meaning, significance, or interpretation, and (b) that the objects of belief differ radically from the objects of knowledge in that while the former are subjective entities, the latter are objective entities. Since Vendler has not really established these points, he has not shown that the following statements (understood as involving a single, unequivocal use of the clause 'that snow is white') are not true:

(14)   If a man knows that snow is white, then he believes that snow is white.

(15)   If a man knows that snow is white, then it is true that snow is white.

(16)   If a man knows that snow is white, then he is justified in thinking that snow is white.

If these statements are indeed true (as I believe they are), then the philosopher's standard approach to the analysis of knowledge is at least not off the track from the very beginning.

Reply to Professor Aune

It must be obvious to the reader that Professor Aune's approach is so different from mine that we hardly can be expected to agree — or even sensibly disagree — at the end. What he seems to be after is the representation, in some system of notation or other, of certain inferences that arise out of our talk about knowledge and belief. My aim, on the other hand, is to account, in terms of a coherent linguistic theory, for the grammaticality, meaning, and implications of the relevant sentences and phrases. And I do not think that his remarks bear upon this problem at all.

At the beginning of his comments he accuses me of failing to show that knowing one's suggestion (that p) entails not knowing that p. Nobody could show this, of course, since knowing that p is perfectly compatible with knowing somebody else's suggestion (that p). If Jim has suggested that Joe is the culprit, my knowledge that Joe is the culprit does not prevent me from knowing Jim's suggestion. But then I know two things: first, that Joe is the culprit, and second, what Jim's suggestion is. And these two are pretty independent: I may know the one without the other. Not so with believing Jim's suggestion. I cannot achieve this without believing that Joe is the culprit. Aune's logic may exhibit *how* the respective inferences go, but he does not explain *why* the inference patterns are distinct.

Another point. Aune (and he is not alone) claims that the verb object of *believe* (*suggest, predict,* and so forth) functions adverbially. As far as I can see, the only evidence for this view is the fact that not only the pronoun *it,* but also the pro-adverb *so* can follow *believe: I believe so* is as grammatical as *I believe it.* This prop, however, is too slender to support any conclusion. First I counter it by pointing out that in order to elicit the appropriate verb-object we ask *what,* and not *how,* somebody believes, suggests, or predicts. *How* will be answered by things like *firmly* and not by *that Joe is guilty*. Again, the relative pronoun, *which*, applies to such a verb-object without fail: *He believes that p, which is false.* On the other hand *so* is not necessarily a pro-adverb. Consider *Joe kicked the tire and so did Jim.* Here, again, *so* is interchangeable with *it*: *Joe kicked the tire and Jim did it too.* Now, surely, *kicking the tire* is not an adverb, but a noun phrase. And so are the objects of *believe*, etc.

Incidentally, *believe so* and *believe it* do not quite mean the same thing. If you ask me *Will it rain tomorrow?*, I may reply *I believe so.* If you say, however, *Joe said that it will rain tomorrow*, my concurring reply will be *I believe it.* The reason is that in this second case the appropriate verb object of *believe*, *that it will rain tomorrow*, precedes the occurrence of *believe* in the discourse, so it can be replaced by the pronoun (*it*). In the first case there is only a question before, and that is not an appropriate object of *believe*. So the pronoun would be out of order, and faute de mieux we fall back on the less demanding *so*.

Concerning *say* versus *suggest*, I agree with Aune that his sentence (4) fails. It does so, however, not because these verbs are incompatible, but simply because *say*, like *think* and unlike *suggest* or *believe*, does not take nominalizations other than the *that*-clause. Whereas I can suggest a solution or believe an explanation, I cannot say or think a solution or an explanation. They take, however, pronouns, hence *Say what you think* or *I believe what he said* pass all right. And notice *what* here is *that which*, unlike in *I know what he said*. Thus *say* and *believe* are compatible, *say* and *know* are not.

The most serious objection Aune levels against me concerns the possibility of putting *wh*-nominals after "subjective" verbs like *predict*, *suggest*, and *doubt*. Notice, however, that my subjective words were *prediction*, *suggestion*, etcetera. And remember that I myself have pointed out the analogous ambivalence of *state*. I repeat: what one states is something objective (may be a fact); one's statement, however, even if true, is not a fact. It is something subjective. Mutatis mutandis the same thing holds of *predict* and *suggest*. What one predicts or suggests is objective (hence the possibility of the *wh*'s); one's prediction or suggestion is not. Interestingly enough *think* and *believe* are not objective even in their straight verb form. Hence we do not have *I think (believe) where (when, how . . .)* . . . This is an interesting difference worth further studies.

Finally, I am not worried about incurring the displeasure of nominalists because of my burgeoning ontology. To me, my beliefs, other people's ideas, and the facts I know, are as "real" (if not more) as any cat or dog, utterance, or inscription in the world. There remains a world of difference between Aune and me.

Brain Writing and Mind Reading

What are we to make of the popular notion that our brains are some-how libraries of our thoughts and beliefs? Is it *in principle* possible that brain scientists might one day know enough about the workings of our brains to be able to "crack the cerebral code" and read our minds? Phi-losophers have often rather uncritically conceded that it *is* possible in principle, usually in the context of making some point about privacy or subjectivity.[1] I read Anscombe to deny the possibility. In *Intention*[2] she seems to be arguing that the *only* information about a person that can be brought to bear in a determination of his beliefs or intentions is information about his past and future actions and experiences; a per-son's beliefs and intentions are whatever they must be to render his behavioral biography coherent, and neurological data could not pos-sibly shed light on this. This is often plausible. Suppose Jack Ruby had tried to defend himself in court by claiming he didn't know (or be-lieve) the gun was loaded. Given even the little we know about his biography, could we even make sense of a neurologist who claimed that he had scientific evidence to confirm Ruby's disclaimer? But in other cases the view is implausible. Sometimes one's biography seems com-pletely compatible with two different ascriptions of belief, so that the Anscombean test of biographical coherence yields no answer. Sam the reputable art critic extols, buys, and promotes mediocre paintings by his son. Two different hypotheses are advanced: (a) Sam does not believe the paintings are any good, but out of loyalty and love he does this to help his son, or (b) Sam's love for his son has blinded him to the faults of the paintings, and he actually believes they are good. Presum-

AUTHOR'S NOTE: Earlier drafts of this paper were read at the University of Maine, Tufts University, and the University of Cincinnati Colloquium on Brain and Mind, November 1971.

[1] See, in another context, A. I. Melden's use of the notion in *Free Action* (New York: Humanities, 1961), pp. 211–215.

[2] G. E. M. Anscombe, *Intention* (2nd ed.; Oxford: Blackwell, 1963).

ably if (a) were true Sam would deny it to his grave, so his future biography will look the same in either case, and his past history of bigheartedness, we can suppose, fits both hypotheses equally well. I think many of our intuitions support the view that Sam really and objectively had one belief and not the other, and it goes against the grain to accept the Anscombean position that in the absence of telltale behavioral biography there is simply nowhere else to look. Couldn't the brain scientist (in principle) work out the details of Sam's belief mechanisms, discover the *system* the brain uses to store beliefs, and then, using correlations between brain states and Sam's *manifest* beliefs as his Rosetta Stone, extrapolate to Sam's covert beliefs? Having deciphered the brain writing, he could read Sam's mind. (Of course, if we could establish this practice for Sam the art critic, we would have to reopen the case of Jack Ruby, but perhaps, just perhaps, we could then devise a scenario in which neurologists *were* able to confirm that Ruby was the victim of a series of unlikely but explainable beliefs — as revealed by his "cerebroscope.")

I admit to finding the brain-writing hypothesis tempting,[3] but suspect that it is not coherent at all. I have been so far unable to concoct a proof that it is incoherent, but will raise instead a series of difficulties that seem insuperable to me. First, though, it would be useful to ask just why the view is plausible at all. Why, for instance, is the brain-writing hypothesis more tempting than the hypothesis that on the lining of one's stomach there is a decipherable record of all the meals one has ever eaten? Gilbert Harman offers the first few steps of an answer:

We know that people have beliefs and desires, that beliefs and desires influence action, that interaction with the environment can give rise to new beliefs, and that needs and drives can give rise to desires. Adequate psychological theories must reflect this knowledge and add to it. So adequate models must have states that correspond to beliefs, desires and thoughts such that these states function in the model as psychological states function in the person modeled, and such that they are representational in the way psychological states are representational. Where there is such representation, there is a system of representation; and that system may be identified with the inner language in which a person thinks.

This reduces the claim that there is an inner language, which one

---

[3] I claimed it was a distinct possibility with regard to intentions in "Features of Intentional Actions," *Philosophy and Phenomenological Research*, 29(1968):232–244.

thinks in, to the trivial assertion that psychological states have a representational character.[4]

The first point, then, is that human behavior has proven to be of such a nature that the only satisfactory theories will be those in which inner *representations* play a role (though not necessarily a role that is not eliminable at another level of theory). Diehard peripheralist behaviorists may still wish to deny this, but that is of concern to historians of science, not us. It is Harman's next point that strikes me as controversial: where there is representation there is system, and this system *may be identified with* a person's inner language. Are all representations bound up in systems? Is any system of representations like a language? Enough like a language to make this identification more useful than misleading? Or is Harman's claim rather that whatever sorts of representations there may be, the sorts we need for human psychology must be organized in a system, and this system must be more like the system of a language than not? Assuming Harman's claim survives these questions, we still would not have an argument for the full-fledged brain-writing hypothesis; two more steps are needed. First, we need the claim that these psychological models with their language-style representations must be realized in brainware, not ectoplasm or other ghostly stuff.[5] This ought to be uncontroversial; though psychologists may ignore the details of realization while elaborating and even testing their models, the model-making is ultimately bound by the restriction that any function proposed in a model must be physiologically or mechanically realizable one way or another. Second, it must be claimed that it will be possible to determine the details of such realizations from an empirical examination of the brainware and its causal role in behavior. This second point raises some interesting questions. Could the functional organization of the brain be so inscrutable from the point of view of the neurophysiologist or other physical scientist that no fixing of the representational role of any part were possible? Could the brain use a system that no outsider could detect? In such a case what would it mean to say the brain used a system? I am not sure how one would go

---

[4] "Language Learning," Nous, 4(1970):35. See also his "Three Levels of Meaning," Journal of Philosophy, 65(1968):590–602, esp. p. 598.

[5] Cf. Wilfrid Sellars, "Notes on Intentionality," Journal of Philosophy, 61(1964): 663, where he discusses mental acts as tokens expressing propositions, and claims that all tokens must be sorts of tokens and must have a determinate factual character, and proposes identifying them with neurophysiological episodes.

about giving direct answers to these questions, but light can be shed on them, I think, by setting up a crude brain-writing theory and refining it as best we can to meet objections.

Again Harman gives us the first step:

> In a simple model, there might be two places in which representations are stored. Representations of things believed would be stored in one place; representations of things desired in the other. Interaction with the environment would produce new representations that would be stored as beliefs. Needs for food, love, etc., would produce representations to be stored as desires. Inferences could produce changes in both the set of beliefs and the set of desires. (*Ibid.*, p. 34)

No doubt we would also want to distinguish more or less permanent storage (belief and desire) from the more fleeting or occurrent *display* of representations (in perception, during problem solving, sudden thoughts, etc.). In any case we already have enough to set some conditions on the brain-writing hypothesis. Some formulations of it are forbidden us on pain of triviality. For instance, claiming that there is brain writing, but that each representation is written in a different language, is just an oblique way of asserting that there is no brain writing. I think the following six conditions will serve to distinguish genuine brain-writing hypotheses from masqueraders.

(1) The system of representations must have a generative grammar. That is, the system must be such that if you understand the system and know the finite vocabulary you can generate the representations — the sentences of brain writing — you haven't yet examined. Otherwise the language will be unlearnable.[6] Only if there were a generative grammar could the investigator get himself into a position to extrapolate from manifest beliefs and desires to covert beliefs and desires. There need not be a single generative grammar covering all representations, however. Just so long as there is a finite number of different "languages" and "multilingual" functional elements to serve as interpreters, the learnability condition will be met.

(2) Syntactical differences and similarities of the language must be reflected in physical differences and similarities in the brain. That is, the tokens of a syntactical type must be physically distinguishable by finite test from the tokens of other syntactical types. That does not

---

[6] See Donald Davidson, "Theories of Meaning and Learnable Languages," in Y. Bar-Hillel, ed., *Logic, Methodology, and Philosophy of Science* (Amsterdam: North-Holland, 1965), pp. 383–394.

mean that all tokens of a type must be physically similar. What physical feature is peculiar to spoken and written tokens of the word "cat"? There must simply be a finite number of physical sorts of token of each type. Tokens and "strings" of tokens may of course align themselves in physical dimensions other than those of natural language. For instance, lexical items might be individuated not by shape but by spatial location, and ordering in the strings might be accomplished not by a sequence in space or time but by degree of electric potential.

(3) Tokens must be physically salient. This is a "practical" point. Tokens might bear physical similarities, but similarities so complex, so diffuse and multidimensional that no general *detection* mechanism could be devised; no frequency filters, stereo-locators, litmus papers, or simple combination of these could be built into a token detector. If tokens turned out not to be physically salient — and this is rather plausible in the light of current research — the brain-writing hypothesis would fail for the relatively humdrum reason that brain writing was illegible. It is worth mentioning only to distinguish it from more important obstacles to the hypothesis.

(4) The representation store must meet Anscombe's condition of biographical coherence. The sentences yielded by our neurocryptographer's radical translation must match well with the subject's manifest beliefs and desires, and with common knowledge. If too many unlikely beliefs or obvious untruths appear in the belief store, we will decide that we have hit upon something strange and marvelous — like finding the Lord's Prayer written in freckles on a man's back — but not his belief store. To give a more plausible example, we might discover that certain features of brain activity could be interpreted as a code yielding detailed and accurate information about the relative tensions of the eye muscles, the orientation of the eyeball, the convexity of the lens, etc., and this might give us great insight into the way the brain controlled the perceptual process, but since a man does not ordinarily have any beliefs about these internal matters, this would not be, except indirectly, a key to his belief store.[7]

---

[7] Discovering such a code is not establishing that the information the code carries for the scientist is also carried for the person or even for his brain. D. H. Perkel and T. H. Bullock, in "Neural Coding" (in F. Schmitt, T. Melnechuk, et al., eds., *Neurosciences Research Symposium Summaries*, vol. 3 [Cambridge, Mass.: M.I.T. Press, 1969]), discuss the discovery of a code "carrying" phasic information about wing position in the locust; it is accurately coded, but the "insect apparently makes

(5) There must be a reader or playback mechanism. It must be demonstrated that the physical system in which the brain writing is accomplished is functionally connected in the appropriate ways to the causes of bodily action, and so forth. Of course, if we were to find the cortex written all over with sentences expressing the subject's manifest beliefs, we would be convinced this was no coincidence, but until the operation of the mechanisms that utilized the writing was discovered we would not have a theory. (A person who discovered such a marvel would be roughly in the same evidential position as a clairvoyant, who (we can imagine) might be able to predict with uncanny accuracy what a person would say, etc., and yet could not be supposed to have any authority — in a court of law, for instance — about a person's beliefs.)

(6) The belief store must be — in the main — consistent. If our translation manual yields sentences like "My brother is an only child" and pairs of sentences like "All dogs are vicious" and "My dog is sweet-tempered" one of several things must be wrong. If the subject declines to assert or assent to these anomalous sentences we will discredit the translation manual (cf. Quine on radical translation); if the man does issue forth with these sentences we will conclude that we have discovered a pathological condition, and our brain-writing system will be viewed as a sort of assent-inducing tumor.[8]

A more graphic way of looking at this point is to ask whether the neurocryptographer could do a bit of tinkering and thereby *insert* a belief in his subject. That is, if he can *read* brain writing he ought to be able to *write* brain writing. Let us suppose we are going to insert in Tom the false belief: "I have an older brother living in Cleveland." Now can the neurocryptographer translate this into brain writing and do a bit of re-wiring? Let us suppose he can do any rewiring, as much and as delicate as you wish. This rewiring will either impair Tom's basic rationality or not. Consider the two outcomes. Tom is sitting in a bar and a friend asks "Do you have any brothers or sisters?" Tom says, "Yes, I have an older brother living in Cleveland." "What's his name?" Now what is

no use of this information." (Blocking this input and substituting random input produces no loss of flying rhythm, ability, etc.)

[8] This condition of rationality has some slack in it. We do permit some small level of inconsistency, but large-scale illogicality must be indicative of either a defect in the subject so serious as to disqualify him as a believer at all, or a defect in our translation hypotheses. See my "Intentional Systems," *Journal of Philosophy*, 68(1971):87–106.

going to happen? Tom may say "Name? Whose name? Oh, my gosh, what was I saying? I don't have an older brother!" Or he may say, "I don't know his name," and when pressed he will deny all knowledge of this brother, assert things like "I am an only child and have an older brother living in Cleveland." In neither case has our neurocryptographer succeeded in wiring in a new belief. This does not show that wiring in beliefs is impossible, or that brain writing is impossible, but just that one could only wire in one belief by wiring in (indefinitely?) many other cohering beliefs so that neither biographical nor logical coherence is lost.[9]

Now suppose we have a brain-writing theory that meets all of these conditions: we have a storage facility functionally tied to behavior that is somehow administered to preserve logical and biographical coherence, and the mode of storage involves elements having physically salient syntactical parts for which we have a generative grammar. This system is going to take up some room. How much room do we need? Marvin Minsky has an optimistic answer: "One can't find a hundred things that he knows a thousand things about. . . . I therefore feel that a machine will quite critically need to acquire the order of a hundred thousand elements of knowledge in order to behave with reasonable sensibility in ordinary situations. A million, if properly organized, should be enough for a very great intelligence."[10] If Minsky's estimate were realistic, the brain, with its ten billion neurons or trillions of molecules would be up to the task, no doubt. But surely his figure is much too low. For in addition to all the relatively difficult facts I have mastered, such as that New York is larger than Boston and salt is sodium chloride, there are all the easy ones we tend to overlook, like New York is not on the moon, or in Venezuela, salt is not sugar, or green, or oily, salt is good on potatoes, on eggs, tweed coats are not made of salt, a grain of salt is smaller than an elephant. . . . Surely I can think of more than a thousand things I know or believe about salt, and salt is not one of a hundred, but one

---

[9] I examine this case more fully in "Mechanism and Responsibility," in Ted Honderich, ed., *Essays on Freedom of Action* (London: Routledge and Kegan Paul, 1973), pp. 157–184. Joan Straumanis has pointed out to me that there is some experimental evidence that suggests that another outcome of the rewiring experiment could be that Tom spontaneously and unconsciously fabricates a web of cohering beliefs to "protect" the inserted belief and his others from each other (a sort of pearl-in-the-oyster effect).

[10] M. L. Minsky, ed., *Semantic Information Processing* (Cambridge, Mass.: M.I.T. Press, 1968), p. 26.

of thousands upon thousands of things I can do this with. Then there is my knowledge of arithmetic; two plus two is four, twenty plus twenty is forty. . . . My beliefs are apparently infinite, which means their storage, however miniaturized, will take up more room than there is in the brain. The objection, of course, seems to point to its own solution: it must be that I *potentially* believe indefinitely many things, but I *generate* all but, say, Minsky's hundred thousand by the activity of an extrapolator-deducer mechanism attached to the core library. So let us attach such a mechanism to our model and see what it looks like.

It has the capacity to extract axioms from the core when the situation demands it and deduce further consequences. If it is to do this, it will need to have an information store of its own, containing information about what items it would be appropriate at any time to retrieve from the core, and, for instance, the metalinguistic information it needs to analyze the contradiction in "all cats are black" and "my cat is brown." Now perhaps it does this by storing the information that what is black is not brown, or maybe *that* information is in the core storage, and the metalinguistic information stored in the extrapolator-deducer mechanism is to the effect that the core element, "what is black is not brown," is relevant to an analysis of the contradiction. Now how will the extrapolator-deducer mechanism store its information? In its own core library of brain-writing sentences? If it has a core library, it will also need an extrapolator-deducer mechanism to act as librarian, and what of *its* information store? Recalling Lewis Carroll's argument in "What the Tortoise Said to Achilles," [11] we can see that the extrapolator-deducer will be hamstrung by a vicious regress if it must always rely on linguistically stored beliefs which it must retrieve and analyze about what can be deduced from what. This a priori point has been "empirically discovered" in the field by more than one frustrated model-builder. As one team sums it up: ". . . a memory that merely stores *propositions* leads to technological, or organic, monstrosities and frustrates, rather than facilitates inductive operations." [12]

The conclusion is that writing — for instance, brain writing — is a

[11] *Mind* (1895), reprinted in I. M. Copi and J. A. Gould, eds., *Readings on Logic* (New York: Macmillan, 1964). Harman offers a similar argument in "Psychological Aspects of the Theory of Syntax," *Journal of Philosophy*, 64(1967):75–87.
[12] H. von Foerster, A. Inselberg, and P. Weston, "Memory and Inductive Inference," in H. L. Oestreicher and D. R. Moore, eds., *Cybernetic Problems in Bionics* (New York: Gordon and Breach, 1968).

dependent form of information storage. The brain must store at least some of its information in a manner not capturable by a brain-writing model. Could it do without brain writing altogether? I think we can get closer to an answer to this by further refining our basic model of belief.

Representations apparently play roles at many different levels in the operation of the brain. I have already mentioned the possibility of codes representing information about the tension of eye muscles and so forth, and these representations do not fall into the class of our beliefs. At another level there is the information "we use" to accomplish depth perception. Psychophysicists ascribe to us such activities as *analyzing depth cues* and *arriving at conclusions* about distance based on *information we have* about texture gradients, binocular interaction, and so forth. Yet it is nothing conscious that I do in order to perceive depth, and if you ask me what beliefs I have about texture gradients I draw a blank. Closer to home, a child can demonstrate his understanding of addition by reeling off sums without being able to formulate or understand propositions about the commutativity of addition. His performance indicates that he has caught on to commutativity, but should we say that among his beliefs is the belief that addition is commutative? To give one more case, while driving down a familiar road I suddenly am struck by the thought that its aspect has changed — somebody has painted his shutters or a tree has blown down, or something. Do I have a *belief* about how it used to be that grounds my current judgment that it has changed? If so, it is a belief to which I can give no expression and about which I am quite in the dark. Somehow, though, the information is there to be used.

Suppose we partition our information store into the part that is verbally retrievable and the part that is not. I would not want to claim that this separates our beliefs from everything else. Far from it. Our pre-analytical notion of belief would permit young children and dumb animals to have beliefs, which must be verbally irretrievable. Perhaps, though, a strong case can be made out that at least our verbally retrievable beliefs are stored in brain writing. The picture that emerges is not, I think, implausible: there are on the one hand those representations that are available for our conscious, personal use and apprehension, and on the other hand those that operate behind the scenes to keep us together. If any representations are stored in brain writing, the former will be, for they are in intimate relation to our natural languages. Included in this group will be the bits of factual knowledge we pick up by asking questions and reading books, especially the facts that only language-users could apprehend, such as the fact that Thanksgiving is always on Thursday. With regard to this group of representations Minsky's figure of a hundred thousand looks more realistic, provided we have an extrapolator-deducer mechanism.[13]

If ever it seems that we are *storing sentences*, it is when we are picking up facts in this verbal manner, but are these things we pick up our beliefs? Sometimes we salt away a sentence because we like the sound of it, or because we will later be rewarded for producing it on demand, or just because it has a sort of staying power in our imagination. In Chekhov's *Three Sisters*, Tchebutykin, reading a journal, mutters: "Balzac was married in Berditchev," and then repeats it, saying he must make a note of it. Then Irina dreamily repeats it: "Balzac was married in Berditchev." Did they acquire a belief on that occasion? Whether *they* did or not, the sentence has stuck in *my* mind, and yet I wouldn't say it was one of my beliefs. I deliberately have not looked it up in the encyclopedia; probably it's true — why would Chekhov insert a distracting falsehood, for mischief? No doubt if someone offered me a thousand dollars if I could tell him where Balzac was married, I'd say Berditchev (wherever that is), but it would be wrong for him to conclude that this was a belief of mine.

If brain writing served only for such storage of words and sentences that we pick up for various reasons, at least we could all breathe a lot easier about the prospects of evil scientists reading our every seditious thought and turning us over to the authorities. Imagine the Loyalty Commissar asking the neurocryptographer if the man in the cerebroscope is a true patriot. "Let's see," says the scientist, "Here is the sentence we've been looking for: 'I pledge allegiance to the flag . . .'" Would finding the sentence, "America's the greatest land of all," satisfy the Commissar? I think not.

The matter of verbally retrievable beliefs is in any case more complicated than the picture we've just been examining. Whereas if I am

---

[13] Minsky's anthology, *Semantic Information Processing*, is a collection of several brilliant attempts to provide working models for just such question-answering, extrapolating systems. It is a gold mine of philosophically tantalizing suggestions and problems.

asked who won the Super Bowl in 1969 it does seem a bit as if I am searching for a ready-made sentence to utter in response, in other cases this sort of account does not ring true at all. Suppose I am watching the shell game, intent on which shell the little pea is under. At any moment it seems to be true that I have a belief about where the pea is, and can tell you if you ask, but it does not seem plausible that this is accomplished by a rapid writing and erasing of successive sentences: "now it's left, now it's center, now right" and the flashing on and off of the negation sign in front of "it's under the center shell." For one thing, if asked to give you my perceptual beliefs of a moment I may have to work a bit to formulate them, yet the perceptual representation was what it was before I was asked. The representationality — or intentionality — of something (e.g., a belief or perception) is compatible with its being vague or indeterminate in some respects.[14] The effort of retrieval is often an effort to formulate a sentence that is an approximation of a belief, and we are often distressed at the hard edge of determinacy our verbal output substitutes for the fuzziness of our convictions.

The answer we formulate, the judgment we find an expression for when asked for our belief, is determinate and individuated, because it consists of a specific string of words in our natural language, whether we then speak it aloud or not. These representations, not the beliefs to which we have verbal access but the occurrent, datable judgments themselves, have the syntactic parts we have been looking for, and about these the brain-writing hypothesis looks much more workable. Not only are judgments determinate; they are, as Harman has pointed out, lexically and syntactically unambiguous.[15] If it occurs to me that our mothers bore us, I know for sure whether I am thinking of birth or ennui. So it is proper to view a judgment not as a sentence simpliciter but as a deep structure or sentence under an analysis. Judgments, unlike beliefs, occur one at a time; we have at any moment indefinitely many beliefs, but can be thinking just one thought. We saw that the brain-writing hypothesis with regard to storage of beliefs did not really effect any economies of design, because however systematic and efficient one's grammar is, one still needs infinite space to store infinitely many tokens, but with

---

[14] See G. E. M. Anscombe, "The Intentionality of Sensation: A Grammatical Feature," in R. J. Butler, ed., *Analytical Philosophy, Second Series* (Oxford: Blackwell, 1965).

[15] "Language Learning."

regard to representation of judgments the situation is different. A finite mechanism incorporating a generative grammar would be an efficient means of representing, one at a time, any of an infinite set of propositions.

The interesting thing about judgments is that although each of us is authoritative about the content of his judgments and although each of us is authoritative about his sincerity or lack of sincerity in giving outward verbal expression of a judgment, we are not in a privileged position when it comes to the question of whether our judgments are reliable indicators of our beliefs.[16] Normally there is harmony between our judgments and our behavior, and hence between our judgments and our beliefs, but when we are afflicted by Sartre's *mauvaise foi*, our sincerest judgments can be lies about our beliefs. I may judge to myself that a man is innocent, while believing him guilty.

This suggests that even if we were to discover a brain-writing system that represented our judgments, the mind reading that could be accomplished by exploiting the discovery would not uncover our beliefs. To return to the case of Sam the art critic, if our neurocryptographer were able to determine that Sam's last *judgment* on his deathbed was "My consolation is that I fathered a great artist," we could still hold that the issue between the warring hypotheses was undecided, for this judgment may be a self-deception. But at this point I think we are entitled to question the intuition that inspired the search for brain writing in the first place. If discovering a man's judgments still left the matter of belief ascription undecided, and if in fact either ascription of belief will account for, explain, predict Sam's behavior as well as the other, are we so sure that Sam determinately had one belief or the other? Are we sure there is a difference between his really and truly believing his son is a good artist, and his deceiving himself out of love while knowing the truth in his heart of hearts? If there were brain writing, of course, there would have to be a physical difference between these two cases, but now, what reasons do we have for supposing there is brain writing?

We are thrown back on our conviction that the brain must be an organ that *represents*, but I hope it is no longer obvious that the brain must represent in sentences. In fact we know that at least some of the

---

[16] See John Vickers, "Judgment and Belief," in K. Lambert, ed., *The Logical Way of Doing Things* (New Haven, Conn.: Yale University Press, 1969), and A. W. Collins, "Unconscious Belief," *Journal of Philosophy*, 66(1969):667–680.

representation must be accomplished in some more fundamental way. Must there be a *system* for such representation? I cannot yet see that there must. In particular, I cannot yet see that there must be a *learnable* system, in Davidson's sense, for it is not clear to me that the brain must — or can — learn (the way a child learns a language) its own ways of representing. Certainly information can be transmitted by means of unlearnable languages. Consider a string of nine light bulbs in a row; there are 512 different patterns of lit and unlit bulbs possible in this array, and so we can use the array to transmit the natural numbers from 0 to 511. There are all sorts of systems one might use to assign patterns to numbers. An obvious and efficient one would be binary notations: 000000001 is 1, 000000010 is 2, and so forth. Once a person knows the system he can generate the numbers he hasn't yet observed, but suppose instead of using this or any other system, the patterns are assigned numbers randomly; a random assignment will carry the information just as well; the outsider will simply not be in a position to predict the balance of the assignments, having learned some of them. Can the brain "use" information carried by unlearnable systems? At some levels of "coding" it obviously does — where the "codes" carry very specific and limited information. In general can the brain get along without learnable representation systems? Until we can say a lot more about what it is for a system to use representations, I for one cannot see how to answer this question. If the answer is no, then there must be brain writing, but how it overcomes the difficulties I have raised here is beyond me. If the answer is yes, then the only way translation of representation can be accomplished is "sentence by sentence," assigning meaning to representations by determining their functional role in the behavior of the whole system. But where competing translations are behaviorally indistinguishable, the content of the representation will be indeterminate.

INDEXES

Name Index

Achinstein, Peter, 156n
Albritton, Rogers, 161, 162
Alston, William, 356
Anscombe, G. E. M., 249, 346, 403n, 404, 407, 413
Aristotle, 262–263, 264, 265
Aune, Bruce, 370n, 400–402
Austin, J. L., 344, 349–352, 356, 359, 363, 365, 371
Ayer, A. J., 174, 284

Babel, 7
Bar-Hillel, Y., 284
Bayle, François, 301
Belnap, Nuel, 394n
Benacerraf, Paul, 185
Bennett, Jonathan, 11, 12
Bloch, B., 52
Bloomfield, Leonard, 42–45, 52, 84n, 283
Bohr, Niels, 132
Bolinger, Dwight, 295
Boyd, Julian, 356
Boyd, R., 155n
Bullock, T. H., 407n
Butler, R. J., 413n

Carnap, Rudolf, 16n, 18, 87, 88, 123, 124, 126, 134, 135, 138, 156n, 175, 183–185, 269, 284
Carroll, Lewis, 194
Chisholm, R., 245n
Chomsky, Noam, 19n, 46, 109n, 124n, 131n, 172, 270, 273–280, 282, 283, 290, 321, 323, 326, 333–336, 338, 341
Church, Alonzo, 76, 79n
Collins, A. W., 414n
Copi, I. M., 410n

Danto, A., 156n
Davidson, Donald, 63–76, 88, 101, 121, 127, 132, 179–181, 183, 210–211, 270, 284–286, 288, 406n, 415
Descartes, René, 136, 258
Deutscher, Max, 245n
Dewey, John, 154
Donnellan, Keith, 94–97, 103, 196–197, 211n, 233–234, 236–237, 263
Duhem, Pierre, 36
Dunn, J. M., 394n

Feigl, Herbert, 163n
Fodor, J. A., 50, 73n, 74n, 112, 113, 188–189, 216n, 270, 274, 281, 286, 288–290, 294, 295, 321
Frege, G., 64–65, 72, 74, 76, 88, 117–119, 121, 127, 134, 138, 354

Goodman, Nelson, 122–123, 125, 195n, 243n, 275, 311, 312, 314
Gould, J. A., 410n
Grice, H. P., 4n, 177, 246, 248n, 283, 289, 337–339

Halle, Morris, 279
Hampshire, Stuart, 284
Harman, Gilbert, 19n, 63n, 274, 293, 296, 305, 306, 404–406, 410n, 413
Harman, Lucy, 297
Harnish, R. M., 66n, 111–113
Hintikka, J., 121
Honderich, Ted, 409n
Hook, Sidney, 275
Humboldt, Wilhelm von, 273, 274
Hume, David, 87

Inselberg, A., 410n

Jeffrey, Richard, 16n

419

Subject Index

421